D1063846

Archaeological
Chemistry

CHEMICAL ANALYSIS

A SERIES OF MONOGRAPHS ON
ANALYTICAL CHEMISTRY AND ITS APPLICATIONS

VOLUME 55

A WILEY-INTERSCIENCE PUBLICATION

JOHN WILEY & SONS
New York / Chichester / Brisbane / Toronto

Archaeological Chemistry

A Sourcebook on the Applications of Chemistry to Archaeology

ZVI GOFFER

Soreq Nuclear Research Centre

A WILEY-INTERSCIENCE PUBLICATION

JOHN WILEY & SONS
New York • Chichester • Brisbane • Toronto

Library of Congress Cataloging in Publication Data:

Goffer, Zvi.
  Archaeological chemistry.

  (Chemical analysis; v. 55)
  ''A Wiley-Interscience publication.''
  Bibliography: p.
  Includes index.
  1. Archaeological chemistry. I.   Title. II.   Series.

CC79.C5G63   930'.1'028   79-1425
ISBN 0-471-05156-X

Printed in the United States of America

10 9 8 7 6 5 4 3 2 1

296/85

FOREWORD

Archaeology is a natural bridge between the humanities and science. Although the goal of the archaeologist is the reconstruction of ancient history, he bases his deductions largely on *observations* of mute remains rather than on the written word which has been handed down. The scientist also deals with observations: either of natural processes or of those set in motion by him in the laboratory (experiments). Yet there is an obvious and profound difference in the methods used.

The archaeologist, using his classical methods, is confined to what can be seen with the eyes. Out of such detailed observations has come an intricate mosaic of inferences as to who made the artifacts that have been found, when they were made, and often why they were made. Trade connections between peoples can be established, evidence for conquests discerned, and many cultural aspects of daily life revealed. No one realizes better than the thoughtful archaeologist that these methods have limitations in deducing vital details and, in fact, that the clues are not always free of ambiguity.

The essence of science rests in its ability to discern and measure properties of matter that are completely closed to visual observation. It has therefore been left to the scientist to see what information of use to the archaeologist lies buried within an artifact. This field of study has diverse facets and has in recent years taken on sufficient stature to warrant a distinguishing title: *archaeometry*. Some materials lend themselves to being dated. Others contain information as to where they were made. Still others, such as metals, can reveal the processes of ancient technology, and this list can be greatly expanded.

Out of these studies is coming much new insight of great value to archaeologists, and the literature has assumed impressive proportions. At the same time these advances have created a severe problem for the practicing archaeologist and the student.

The data obtained by the scientist do not necessarily give a "yes or no" answer to an existing archaeological problem, because often new "facts" are brought to light that the archaeologist could not have thought about. For the archaeological interpretation of the findings, the archaeologist must first be able to understand exactly what the data mean in order to assess their pertinence and reliability. Unfortunately, this understanding is severely hindered by terminology and concepts for which he has no background. The absence of independent judgment leads to unfortunate consequences which are not conducive to good scholarship.

This dilemma will not dissolve quickly, since it involves the education of archaeologists, explaining pertinent scientific concepts in a language that is understandable. This underlines the importance of textbooks and other writings addressed specifically to the archaeology student. This volume on chemical techniques and applications is an important and welcome entry in this educational process.

PROFESSOR I. PERLMAN

Institute of Archaeology
The Hebrew University of
Jerusalem, Israel

PREFACE

The application of the natural sciences to archaeology is a relatively recent development which is having a most profound and dramatic effect on the practice of archaeology. A few years ago, I was asked to give a series of lectures on this subject at the Institute of Archaeology of the University of Tel Aviv. On setting out to compile information for the series, I very soon realized that there was no single published work dealing comprehensively with the application of chemistry to archaeology. Moreover, the texts that sought to relate the physical sciences in general to the practice of archaeology devoted very little space to chemical applications.

The present work is an attempt to make up this deficiency in some degree. It is addressed primarily to archaeologists and to students of archaeology. Its purpose is to provide them with a guide to the help they can obtain from chemistry in the pursuit of their calling. Natural scientists (and laymen) who are interested in archaeology may find the book useful if they want to know what has been—and is being—done in the field.

Throughout the book I have aimed, as far as possible, for a unified treatment, and in addition to descriptions of methods and techniques, I have included many case histories to illustrate possible applications.

A major difficulty that usually arises in interdisciplinary programs is that of communication. Each discipline has its own jargon. Here, the chemist must learn the jargon of the archaeologist, and vice versa. In addition, the archaeologist must find out to what sort of questions, in his subject, the chemist can provide answers. Once a collaborative atmosphere has been established, the two disciplines can conjoin to form a new and distinct one—*Archaeological chemistry*.

I make no apology for sacrificing some scientific rigor in the interests of clarity and simplicity. I have tried to keep technical terminology and scientific notation to a minimum. Anyone with a high-school graduate's knowledge of chemistry should be able to follow the text with little difficulty. This approach reflects my belief that a general understanding of techniques and their applications is more important than a familiarity with the minutiae of work. Whoever wishes to delve further into the subject should have no difficulty in doing so after reading the book.

The bibliographical lists at the ends of chapters do not pretend to be exhaustive. Their purpose is sometimes to mention background reading, occasionally to give credit for original work, but most often to indicate directions for further study.

In a book intended to be a work of reference, it should be easy to locate the definition of an idea or the statement of a problem. Therefore, the subject index is as complete as possible.

With regard to the book's contents, it has continually been a problem to decide what to put in and what to omit; I have tried to avoid making the book too unwieldy, while fulfilling as far as possible the aims set out above. I am not wholly satisfied with the result; some readers may be even less so. If I have left out material that to others seems vital, it must be for later writers in this field to rectify such omissions. I hope this book will provide them with a stimulus.

The research for and writing of this book were greatly facilitated by my home institution, the Soreq Nuclear Research Centre, Yavne, Israel. It is during a sabbatical term at Bell Laboratories, Murray Hill, New Jersey, that the book is being completed and readied for publication. I am grateful to these two institutions; in both I have met with sympathy and understanding towards my interest in this subject.

So many friends and colleagues gave me encouragement and help that I cannot name them all here. A special word of appreciation is due to the library staff at the Soreq Nuclear Research Centre for locating hard to find books and journals. Bill Biolsi gave valuable counsel in the preparation of the index. Antonio Tubal assisted with the artwork. My wife, Chava, worked devotedly with me throughout, and especially in preparing the manuscript and correcting the proofs. My friend Lionel Holland is to be thanked for numerous improvements and corrections to the text.

Borrowed illustrations are acknowledged in the customary way. I thank the following for permission to print their photographs: R. H. Brill, Figures 9.4 and 9.8; The British Museum, Figures 9.5, 9.6, and 22.2; I. Carmi, Figure 17.2; F. de Korosy, Figure 8.1; R. L. Fleisher, Figures 19.2 and 19.4; I. Friedman, Figure 6.2; Israel Museum, Figure 8.7; Metropolitan Museum of Art, Figure 22.3; I. Perlman, Figure 6.9; M. Prausnitz, Figure 6.1.

For whatever shortcomings remain in the book—I am wholly responsible.

ZVI GOFFER

Murray Hill, New Jersey
March 1979

CONTENTS

PART I. CHEMISTRY

PART II.   MATERIALS AND TECHNOLOGIES

CHAPTER 10.   COLOR: PIGMENTS, DYES, AND INKS   167

PART III.   DECAY AND RESTORATION OF ARCHAEOLOGICAL MATERIALS

PART IV. DATING

PART V.   ANALECTA

Archaeological Chemistry

PART

1

CHEMISTRY

Chemistry is a subject of immense potential value to archaeologists. In order to be able to use it for the elucidation of archaeological problems, a familiarity with some basic principles is necessary. Some of these are set out in the following section.

The reader who has long been a stranger to the principles of chemistry may be — understandably — a little afraid of this section at first. However, he will find little in it to daunt him: and his eventual rewards — in the form of new insights into his own discipline — will be well worth the effort.

ARCHAEOLOGICAL CHEMISTRY

The physical sciences in general, and chemistry in particular, play an important role in modern archaeology, which is the study of past civilizations through the examination of their material remains. Whereas other branches of archaeology concern themselves with characteristics such as the function, shape, size, and aesthetic qualities of objects studied, archaeological chemistry deals with the actual nature of the materials used for making them — specifically, their composition, structure, and properties.

As a systematic science, chemistry is young in years, but archaeology is even younger. Ancient materials and artifacts recovered from excavations intrigued chemists long before archaeology took on the organized, systematic nature it has today. The primary motive of the early investigators in archaeological chemistry seems, in fact, to have been personal curiosity.

The first significant results of chemical analysis of ancient objects were published in 1796 by M.H. Klaproth, a pioneer in the chemical investigation of antiquities, as well as in the field of chemical analysis as a whole. He determined the composition of Greek and Roman coins, other ancient metallic objects, and specimens of Roman glass (Klaproth 1796).Some distinguished chemists such as Sir H. Davy, J.J. Berzelius, and M. Berthelot and many others now almost forgotten were concerned with the analysis of ancient materials (Davy 1815, 1817; Berzelius 1836; Berthelot 1906). One of these, C. C. T. F. Gobel, was (according to E.R. Caley) the first to suggest that the results of chemical analysis of archaeological materials could be of service to archaeology (Gobel 1842; Caley 1967).

It was in 1853, however, that the possibility of cooperation between archaeology and chemistry was first demonstrated. In that year was published the first account of archaeological discoveries which had an appendix describing the chemical examination of excavated artifacts (Layard 1853). This publication marks the beginning of collaboration between archaeology and chemistry. It is the earliest concrete evidence of the appreciation by archaeologists of the value of chemical information on materials and objects recovered from excavations.

The first attempt to trace a correlation between chemical composition and the possible provenance of ancient materials was suggested by J. E. Wocel (Wocel 1853, 1856). He was also the first to suggest that approximate or relative dates of ancient metal objects might be estimated from their composition.

3

Archaeological excavations were being conducted in a systematic way by the last quarter of the nineteenth century. This led to wider appreciation of the services of chemistry. The results of chemical investigation began to appear in the body of archaeological reports; some archaeologists also published the results of chemical analysis, and chemists' reports began to be published as separate articles in archaeological journals (Schliemann 1878).

By the end of the nineteenth century A. Carnot had investigated the fluorine content of ancient bones and suggested that the results might be used as a means of dating such remains (Carnot 1892). However, the validity of this method of archaeological dating, which is widely accepted today, was not tested until more than 40 years later.

The systematic application of chemical principles and techniques to the restoration and preservation of antiquities only started at the end of the nineteenth century. The pioneer in this field was F. Rathgen, who set up a laboratory at the Staatlichen Museum in Berlin and later published the first book dealing with practical procedures for the conservation of antiquities (Rathgen 1905).

In the first three decades of the present century the usefulness of chemistry in archaeology continued to grow. After World War I archaeological chemistry laboratories began to increase in number, and through painstaking research, sober facts gradually replaced conjecture regarding the technical achievements on ancient civilizations. However, few new principles were introduced, and contributions consisted mainly of refinements and extensions of known techniques. One of the few innovations was the now widely used electrolytic method for the restoration of ancient bronzes (Fink and Eldridge 1925).

In the last 50 years the rapid development of scientific instrumentation and the application of radioactivity and nuclear techniques have brought about a revolution in archaeological methods and resulted in the establishment of archaeological chemistry as a separate discipline employing specialized techniques. The contribution of chemistry to archaeology was highlighted by W. F. Libby's development of the technique of radiocarbon dating (Libby 1952). But this was by no means the only important contribution, and at present chemistry is playing an increasingly important role in archaeological research. Today's archaeologist is unlikely to fully understand an archaeological report unless he has a general scientific understanding and some knowledge of chemical principles and techniques (Riederer 1976).

Archaeological chemistry, which is the chemical study of materials used in antiquity, can be divided into two main branches: *chemistry of materials*, also called *descriptive chemistry*, which is the study and tabulation of the characteristic qualities of materials important in archaeology, and *analytical chemistry*, which studies the quantitative composition of these materials. The discovery that radioactive materials can provide "clocks" for dating events has added a new dimension to the chemical — as well as the physical — approach to the solution of archaeological problems.

The chemical investigation of ancient materials may occasionally contribute to chemical knowledge itself. More frequently, and to a greater degree, the contribution is to archaeological knowledge. But the acquisition of chemical and technological information concerning materials and objects from antiquity is, in in itself, of value.

REFERENCES

Berthelot, M., *Mem. Acad. Sci. Paris*, **49**, 4 (1906).

Berzelius, J., *Ann. Word. Oldkundighed*, **104**, (1836-1837).

Caley E. R.,*J. Chem. Ed.*, **26**, 242 (1949).

Caley E. R.,*J. Chem. Ed.*, **28**, 64 (1951).

Caley E. R.,*J. Chem. Ed.*, **44**, 120 (1967).

Carnot, A., *Comptes Rendus,* **115**, 243 (1892).

Davey, H., *Phil. Transact.*, **105**, 97 (1815).

Davey, H., *Archaeologia*, **18**, 222 (1817).

Fink, C. G., and Eldridge, C. H., *The Restoration of Ancient Bronzes and other Alloys*, The Metropolitan Museum of Art, New York, 1925.

Gobel, F., *Uber den Einfluss der Chemie auf die Ermittelung der Volker der Vorzeit oder Resultate der Chemischen Untersuchung Metallischer Alterthumer Insbesondere der in den Ostseegouvernements Verkommenden Behufs der Ermittelung der Volker von Welchen Sie Abstammen*, Erlangen, 1842.

Klaproth, M. H., *Phil. Transact. Roy. Soc.*, 395 (1796).

Layard, B., *Discoveries in the Ruins of Nineveh and Babylon*, 1853.

Libby, W. F., *Radiocarbon Dating*, Chicago Press U.P., 1952.

Rathgen, F., *Die Konservierung von Alterthumsfunden*, Berlin, 1905.

Riederer, J., *Berliner Beitrage zur Archaeometrie*, **1**, 33(1976).

Schliemann, H., *Mycenae*, John Murray, London, 1878.

Wocel, J. E., *Sitz. der Kaiser. Akad. Wiss, Wien, Phil. Hist. Classe*, **11**, 716 (1853).

Wocel, J. E., *Sitz. der Kaiser. Akad. Wiss, Wien, Phil. Hist. Classe*, **16**, 169 (1855).

CHEMISTRY—FUNDAMENTAL CONCEPTS

Chemistry belongs to the group of studies called the *physical sciences*. Together with physics and geology, chemistry is concerned with the study of matter. It deals with those properties of matter that are independent of shape (e.g., composition and structure), with the ways in which these properties change, and with the principles governing such changes. Why are ceramics durable? What makes metals pliable? How does copper corrode? Can composition "fingerprint" glass? Does chemical alteration make it possible to date obsidian? Problems such as these—some easily solved, others still waiting solution—are part of the discipline called *chemistry* .

MATTER AND MATERIALS

Matter is anything (whether solid, liquid, or gas) that occupies space. The noun *material* is used with reference to any particular kind of matter: wood, oxygen, iron, water, clay, are all materials. Some materials are naturally occuring (i.e., they occur in nature in the same form as that in which they are used by man), and others are *manufactured* (i.e., their properties are altered in some way during or after extraction from their natural surroundings, and before use). The first materials used by primitive men were naturally occurring—wood, hides, and stones—picked up from the ground. In time, however, men began to manufacture materials appropriate to their needs and thus the foundations of technology were laid (see Table 2-1).

Materials can be either homogeneous or heterogeneous. *Homogenous* materials are those having the same properties throughout, such as water, silica, and the metals. *Heterogeneous* materials are nonuniform in composition (see Fig. 2.1). They are usually *mixtures*, mechanical combinations of two or more homogeneous materials in which each of the latter retains its characteristic properties. Pottery— an aggregation of fired clay, mineral fillers and organic materials—is an obvious example of an heterogeneous material. Others are illustrated in Fig. 2.2.

Homogeneous materials are of two kinds: solutions and substances. *Solutions* are homogeneous mixtures at the molecular level. They do not have a definite chemical composition since the proportions of their components can be infinitely varied. The term "solution" is generally applied to liquids, but in practice solution

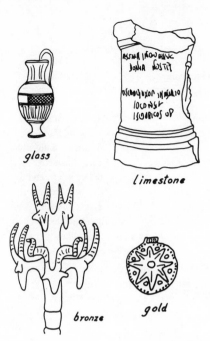

glass

limestone

bronze

gold

Fig. 2.1. Homogeneous materials.

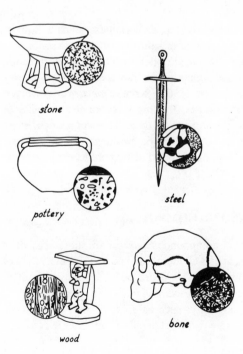

stone

pottery

steel

wood

bone

Fig. 2.2. Heterogeneous materials.

7

TABLE 2.1   Some Important Archaeological Materials

Material	Naturally Occurring	Manufactured
Inorganic		
Metals	Gold, silver, iron (the less active metals)	All metals and alloys excluding those occurring naturally
Building materials	Limestone, basalt, marble, alabaster, etc.	Lime, plaster of Paris, baked brick, and tile
Ceramics	Clay, obsidian, flint	Clay body ceramics, glass, faience
Pigments	Ocher, chalk, malachite	Egyptian blue, lead white
Organic		
Vegetable	Wood, resin, pollen	Gum, cosmetics, dyes, lakes, fibers
Animal	Skin, bone, horn, ivory, feathers, hair, blood, fat, wax	Glue, leather, fibers
Other	Coprolites	

may exist of gases in gases, gases in liquids, liquids in liquids, solids in liquids, and solids in solids. Metal alloys are examples of solutions of solids in solids.

Substances are homogeneous materials having definite chemical composition and physical properties. Different specimens of the same substance, whatever their origin, always have the same properties. *Chemical properties* are involved when substances undergo chemical changes. Rusting is a chemical property of iron, and fading of color is a chemical property of dyes. *Physical properties* are used to identify different substances; these include hardness, density, luster, crystal structure, ductility, malleability, melting and boiling points, and electrical and thermal conductivity. See pages 00 and 00.

ELEMENTS AND COMPOUNDS

Substances may be either elements or compounds. *Elements* are substances all of whose atoms are identical. At present 103 elements are known (Table 2.9), 92 of which are found on earth in the natural state. These are the basic constituents of all matter. The other 11 are unstable and have been synthesized. Some of the 92 natural elements are familiar to everybody, such as gold, copper, and carbon. Others, called *trace elements,* are so rare that most people are unlikely ever to hear of them, let alone encounter them. They are important, however, in that the relative quantities in which they occur in nature vary from one location to another, and this variation can be used as a means of identification of archaeological objects. The relative abundance of the different elements in the earth's crust is shown in Table 2.9. Only a few elements, most of them metals, have been used uncombined; the first were those found in nature in the native state. The best known examples are gold and copper.

Most elements combine chemically to give rise to compounds, the second great group of substances. *Compounds* are substances that may be decomposed into two or more simpler compounds or elements. Compounds may be of natural or synthetic origin. *Minerals,* from which many elements are extracted, are chemical compounds of natural origin. Minerals have well-defined composition and properties.

ATOMS AND MOLECULES

The smallest part of an element that still retains the specific properties of that element is an *atom*. Atoms are made up of even smaller, *subatomic* particles, of which the best known are the electron, proton, and neutron.

Electrons are negatively charged particles; each carries a single unit of negative electric charge. *Protons* also carry an electric charge, equal in quantity but opposite in sign to the electron. Protons are 1840 times heavier than electrons. *Neutrons* are also heavy particles, weighing the same as protons. They have no electrical charge at all; hence their name.

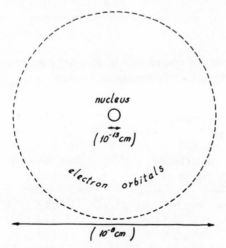

Fig. 2.3. The atom. The very small nucleus consists of protons (and usually also neutrons). Electrons occupy the space around the nucleus.

PHYSICAL PROPERTIES OF ARCHAEO—

DENSITY AND SPECIFIC GRAVITY

The *density* of a material is its mass per unit volume. Sometimes *specific gravity*, defined as the ratio of the density of a substance to the density of water, is used instead of density.

TABLE 2.2   Density

Material	Density (g/cm^3)	Material	Density (g/cm^3)
Alabaster		Ivory	1.9
Egyptian (calcium carbonate)	2.7	Lead	11.3
Modern (calcium sulfate)	2.3	Leather	0.8-1.0
Amber	1.1	Limestome	2.7
Bone	1.7-2.0	Marble	2.6-2.8
Brass	8.2-8.7	Silver	10.5
Bronze	8.7-8.8	Slags	2.4
Clay	1.8-2.6	Speculum metal	8.6
Copper	8.9	Steel	7.6-7.8
Flint	2.6	Tin	7.1
Glass	2.4-2.8	Wood	
Gold	19.0	Pine	0.43
Gypsum	2.3	Oak	0.79
Iron	7.8	Zinc	7.0

The proportions of the constituents of an alloy may be assessed with some accuracy if its density or specific gravity are known.

TABLE 2.3   Specific Gravity of Gold Alloys

Specific Gravity (g/cm^3)	Gold Content (percent)
19.0	98 ± 0.5
18.0	93 ± 1.0
17.0	86 ± 2.3
16.0	79 ± 3.5
15.0	71 ± 5.0
14.0	62 ± 6.5

TABLE 2.4   Specific Gravity of Silver Alloys

Specific Gravity (g/cm^3)	Silver Content (percent)
10.5	100
10.4	95
10.3	89
10.2	84
10.1	78
10.0	72

—LOGICALLY IMPORTANT MATERIALS

THERMAL CONDUCTIVITY

Thermal conductivity is the ability of a material to transmit heat energy. Different materials can be rated in terms of their relative thermal conductivities.

TABLE 2.5   Relative Thermal Conductivity (Silver Taken as Standard of 100)

Good Thermal Conductors	Insulators	Good Thermal Conductors	Insulators
Silver 100	Brick 0.01	Brass 28	Wood 0.05
Copper 93	Glass 0.15	Iron 19	Wool 0.01

DUCTILITY AND MALLEABILITY

The *workability* of a material denotes its ability to undergo deformation without failure. It embodies the properties of *ductility,* the ability of a material to be drawn into wires, and *malleability,* the property that allows a material to be hammered and rolled out into thin sheets.

TABLE 2.6   Metals Arranged in Order of Decreasing Workability

Ductility	Malleablity	Ductility	Malleablity
Gold	Gold	Copper	Lead
Silver	Silver	Tin	Tin
Iron	Copper	Lead	Iron

HARDNESS

Hardness is the ability of a material to resist abrasion, scratching, or penetration. It is assessed using Mohs's scale, where 10 minerals are arranged in order of ascending hardness. The hardness of a material can be rated on this scale, where a mineral high in the scale scratches one lower than itself.

TABLE 2.7   Mohs's Scale of Hardness

Hardness	Material
1	Talc
2	Gypsum
3	Calcite
4	Fluorite
5	Apatite
6	Orthoclase/feldspar
7	Quartz
8	Topaz
9	Corundum
10	Diamond

TABLE 2.8   Hardness of Materials

Material	Hardness	Material	Hardness
Agate	6-7	Gold alloys	2.5-3
Alabaster	1.7	Gypsum	1.5-2
Amber	2.3	Iron	4-5
Brass	3.5	Lead	1.5
Copper	2.5-3	Marble	4-3
Flint	6.8	Silver alloys	2.5-7
Glass	4.5-6	Steel	5-8

It is sometimes convenient to visualize an atom as somewhat resembling a solar system; in the center of the atom is a very small and heavy *nucleus* consisting of protons (and usually some neutrons), around which electrons move in orbits, something like planetary orbits. The number of electrons in the atom is the same as the number of protons in its nucleus; hence the net electric charge of the atom is zero (Figs. 2-3, and 2-4).

TABLE 2.9   The Chemical Elements

Name	Symbol	Atomic number	Atomic weight	Abundance in Earth's Crust (ppm, by weight)
Actinium	Ac	89	(227)	3×10^{-10}
Aluminum	Al	13	26.9	81,300
Americium	Am	95	(243)	—
Antimony	Sb	51	121.8	1
Argon	Ar	18	39.9	0.04
Arsenic	As	33	74.9	5
Astatine	At	85	(210)	—
Barium	Ba	56	137.3	250
Berkelium	Bk	97	(249)	—
Beryllium	Be	4	9.0	6
Bismuth	Bi	83	208.9	0.2
Boron	B	5	10.8	3
Bromine	Br	35	79.9	1.6
Cadmium	Cd	48	112.4	0.1
Calcium	Ca	20	40.1	36,300
Californium	Cf	98	(251)	—
Carbon	C	6	12.0	320
Cerium	Ce	58	140.0	46.1
Cesium	Cs	55	132.9	7
Chlorine	Cl	17	35.4	314
Chromium	Cr	24	51.9	200
Cobalt	Co	27	58.9	23
Copper	Cu	29	63.5	70
Curium	Cm	96	(247)	—
Dysprosium	Dy	66	162.5	4.5
Einsteinium	Es	99	(254)	—
Erbium	Er	68	167.2	2.5
Europium	Eu	63	151.9	1.1
Fermium	Fm	100	(253)	—
Fluorine	F	9	19.0	900
Francium	Fr	87	(223)	—
Gadolinium	Gd	64	57.2	6.4

TABLE 2.9 The Chemical Elements

Name	Symbol	Atomic number	Atomic weight	Abundance in Earth's Crust (ppm, by weight)
Gallium	Ga	31	69.7	15
Germanium	Ge	32	72.6	7
Gold	Au	79	169.9	0.005
Hafnium	Hf	72	178.5	4.5
Helium	He	2	4.0	0.003
Holmium	Ho	67	164.9	1.1
Hydrogen	H	1	1.0	1,300
Indium	In	49	114.8	0.1
Iodine	I	53	126.9	0.3
Iridium	Ir	77	192.2	0.01
Iron	Fe	26	55.8	50,000
Krypton	Kr	36	83.8	9.8×10^{-6}
Lanthanum	La	57	138.9	18.3
Lawrencium	Lr	103	(257)	—
Lead	Pb	82	207.2	16
Lithium	Li	3	6.9	65
Lutetium	Lu	71	174.9	0.7
Magnesium	Mg	12	24.3	20,900
Manganese	Mn	25	54.9	1,000
Mendelevium	Md	101	(258)	—
Mercury	Hg	80	200.6	0.5
Molybdenum	Mo	42	95.9	15
Neodymium	Nd	60	144.2	23.9
Neon	Ne	10	20.2	7×10^{-5}
Neptunium	Np	93	(237)	—
Nickel	Ni	28	58.7	80
Niobium	Nb	41	92.9	24
Nitrogen	N	7	14.0	46.3
Nobelium	No	102	(253)	—
Osmium	Os	76	190.2	0.005
Oxygen	O	8	15.9	466,000
Palladium	Pd	46	106.4	0.01
Phosphorus	P	15	30.9	1,180
Platinum	Pt	78	185.1	0.005
Plutonium	Pu	94	(242)	—
Polonium	Po	84	(210)	3×10^{-10}
Potassium	K	19	39.1	25,900
Praseodymium	Pr	59	140.9	5.5
Promethium	Pm	61	(145)	—
Protactinum	Pa	91	(231)	8×10^{-7}
Radium	Ra	88	(226)	13×10^{-6}
Radon	Rn	86	(222)	—

TABLE 2.9 The Chemical Elements

Name	Symbol	Atomic number	Atomic weight	Abundance in Earth's Crust (ppm, by weight)
Rhenium	Re	75	186.2	0.001
Rhodium	Rh	45	102.9	0.001
Rubidium	Rb	37	85.5	310
Ruthenium	Ru	44	101.1	0.004
Samarium	Sm	62	150.3	6.5
Scandium	Sc	21	44.9	5
Selenium	Se	34	78.9	0.09
Silicon	Si	14	28.1	277,200
Silver	Ag	47	107.9	0.1
Sodium	Na	11	22.9	28.300
Strontium	Sr	38	87.6	300
Sulfur	S	16	32.0	520
Tantalum	Ta	73	180.9	2.1
Technetium	Tc	43	99	—
Tellurium	Te	52	127.6	0.002
Terbium	Tb	65	158.9	0.9
Thallium	Tl	81	204.3	3
Thorium	Th	90	232.0	11.5
Thulium	Tm	69	168.9	0.2
Tin	Sn	50	118.7	40
Titanium	Ti	22	47.9	4,400
Tungsten	W	74	183.8	69
Uranium	U	92	238.0	4
Vanadium	V	23	50.9	150
Xenon	Xe	54	131.3	1×10^{-6}
Ytterbium	Yb	70	173.0	2.6
Yttrium	Y	39	88.9	28.1
Zinc	Zn	30	65.4	132
Zirconium	Zr	40	91.2	220

The physical and chemical properties of atoms are determined by the number of protons in their nuclei. All the atoms of any particular element are identical. The lightest and simplest atoms is that of hydrogen; its nucleus contains one proton. Heavier elements have more complex atoms, with two, three, four, and so on, protons in their nuclei. The number of protons in the nucleus of the atoms of an element is called its *atomic number*. The elements can be arranged in an ascending ordinal series (one, two, three, and so on up to 103) according to their atomic numbers. Another commonly used method of tabulating the elements is the

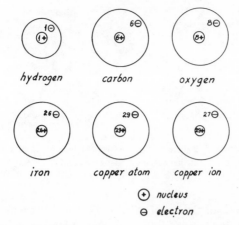

Fig. 2.4.   Simplified two-dimensional representation of the electron configuration of atoms.

periodic table, where the elements are in sequence according to their atomic numbers, in seven horizontal "periods." When arranged in this way, vertical columns of elements having similar chemical properties are produced.

As mentioned previously, the nucleus of an atom is surrounded by negatively charged electrons moving in successive concentric layers or orbits. The chemical properties of an atom are determined by the behavior of the electrons in the outermost layers. In metals the outer electrons wander freely between the atoms; this is what makes metals, as a rule, good conductors of electricity.

Atoms, both like and unlike, combine to form *molecules.* When the combining atoms are of different elements, a chemical *compound* is formed. Whereas atoms are the smallest particles of elements, molecules are the smallest particles of compounds. The molecules of a given chemical compound always contain the same atoms in the same proportion and in the same spatial arrangement. In *crystalline* solids, the overall spatial arrangement of all the atoms and molecules comprising a single crystal is fixed and regular and specific for a given compound. Noncrystalline solids, called *amorphous* materials, do not have any regular spatial arrangement of the constituent atoms and molecules (Fig. 2-5).

SYMBOLS, FORMULAS, AND CHEMICAL NOMENCLATURE

Scientists have agreed on a convenient, internationally accepted system of symbols, formulas, and related nomenclature to describe the elements, the compounds they form, and the reactions between them (see Table 2-9 for the symbols of the elements). Few of the symbols for the common elements are likely to cause

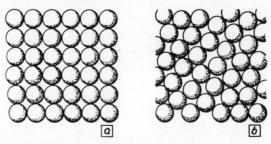

Fig. 2.5. Crystalline and non-crystalline solids: (*a*) in crystalline solids the component particles (whether atoms, ions or molecules) are in a regular repeated pattern. (*b*)   in non-crystalline solids the component particles are not in any orderly arrangement.

difficulty, except those of the metals (Table 2-10). The metals known to the ancients were given symbols derived from their Latin names. The name of compounds of two elements usually end in *-ide* (Table 2-11). The commonest compounds of three elements contain oxygen. Their names usually end in *-ate* (Table 2-11).

TABLE 2.10   Chemical Symbols

Metal	Symbol	Latin Name
Gold	Au	Aurum
Silver	Ag	Argentum
Mercury	Hg	Hydrargyrium
Copper	Cu	Cuprum
Lead	Pb	Plumbum
Tin	Sn	Stannum
Iron	Fe	Ferrum

TABLE 2.11   Chemical Names

Common Name	Symbol	Chemical Name
Water	H_2O	Hydrogen oxide
Common salt	NaCl	Sodium chloride
Pyrolusite	MnO_2	Manganese dioxide
Flint, silica	SiO_2	Silicon dioxide
Chalk, limestone	$CaCO_3$	Calcium carbonate
White lead	$PbCO_3$	Lead carbonate
Copperas, green vitriol	$FeSO_4$	Iron sulfate

All the atoms of any given element have the same weight. Since individual atoms are too small to weigh directly, a relative standard of *atomic weights* is used whereby the weight of the atom of each element is compared to the weight of a chosen standard, carbon-12(see Table 2.9 for the atomic weights of the elements).

The composition of chemical compounds is described by *chemical formulas* in which the atoms of the elements are represented by their symbols, and the number of atoms or molecules present is shown by numbers respectively suffixed or prefixed to the symbols. Chemical reactions are described by equations resembling algebraic equations.

ORGANIC AND INORGANIC SUBSTANCES

Chemical compounds are divided into *organic* and *inorganic*. Originally the name "organic" was applied to substances derived from living organisms. This arose from the misconception that some "vital force," identified with life itself, was necessary for the formation of these substances. This belief is no longer held. Organic compounds owe their existence to the peculiar properties of carbon, which is able to combine not only with other elements but with itself to form molecules in the form of chains and rings. Millions of different compounds of carbon exist, including the complex molecules that make up living tissue. Nowadays the term *organic chemistry* is used to designate that branch of chemistry which deals with the compounds of carbon; *inorganic chemistry* deals with the elements other than carbon, and with their compounds.

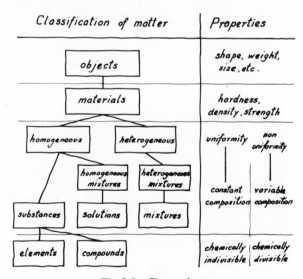

Fig. 2.6.   Classes of matter.

Complex organic molecules can form *isomers*, that is different molecules in which the constituent elements and their relative proportions are alike, but their spatial arrangement in relation to one another inside the molecule is different. The isomerism of certain compounds is of archaeological importance in connection with the dating of bones (see Chapter 16).

The classification of matter discussed in the present section is summarized in Fig. 2.6.

CHEMICAL COMPOUNDS OF ARCHAEOLOGICAL IMPORTANCE

Inorganic compounds of archaeological importance vary though a wide range of both naturally occurring and manufactured materials, including stones, pigments, glass, pottery, and metals.

Organic compounds are more scarce in archaeological research, since they are more readily subject to external attack and decay. Wood and fibers, leather, dyes, an oils become more and more scarce as the archaeologist's interest moves further back in time. It is safe to state that the archaeological study of ancient times is mostly concerned with inorganic materials, not because ancient men did not use organic materials, but because the surroundings have slowly and steadily contributed to their disappearance.

ANALYTICAL CHEMISTRY

One branch of chemistry is concerned with obtaining information about the composition and identity of materials in terms of the elements or compounds that they contain. This branch is called *analytical chemistry*.

The results provided by the chemical analysis of ancient materials can be used by archaeologists in many practical ways. Positive identification by chemical analysis makes collation of archaeological materials more complete and precise; analytical results can be used to gain information on manufacturing techniques, dating, and provenance. Chemical analysis may also provide clues as to the authenticity of allegedly ancient objects, so important to the museum expert, whereas restoration of ancient artifacts often calls for knowledge of their chemical composition (Musty 1976).

The sort of information that the analytical chemist may be asked to provide may vary greatly from one case to another. When examining a metal artifact, for instance, he may be concerned with its elemental constituents. When dealing with a lump of ancient dye, his aim will most probably be to identify the substance or substances, that are responsible for its color.

ASSEMBLAGES OF ANTIQUITIES

The results obtained from the analysis of a large number of well-chosen samples may have statistical significance and hence yield information on a number of problems of archaeological interest. For example:

1. Information on the technological process of a culture can be obtained from a study of how materials were produced or artifacts manufactured.
2. The study of trade routes can be facilitated by elemental analysis of artifacts or by the measurement of the physical properties of the materials used in their manufacture.
3. Historical conclusions can be drawn from the results of chemical analysis of ancient coins. Sudden compositional variations may be associated with corresponding economic changes; mints can be identified and counterfeits exposed.

The *systematic analysis* of materials or artifacts can be undertaken by analyzing either the matrix or the trace elements present (Table 3.1). The *matrix* consists of the major consitituents of a sample. Matrices may be of inorganic materials such as minerals, clay, glass or metals or organic materials like wood and amber. *Trace elements* are minor impurities present within the basic matrix structure. Ordinarily, concentrations of trace elements are measured in parts per million, parts per billion, or even smaller units. No prediction can be made about the trace elements to be found in a matrix, and only by experimentation can their presence be ascertained.

TABLE 3.1   Matrix and Minor Elements (Relative Concentration)

Matrix or Major Elements	Minor Elements	Trace Elements
over 1 percent	1-0.01 percent	less than 0.01
	Units of Concentration	
percent	percent	ppm
		ppb

PROVENANCE STUDIES

The study of the provenance of archaeological artifacts, based on chemical similarity, offers a typical example of systematic analysis. In any given location the presence of trace elements in minerals may have arisen haphazardly rather than as the result of the geochemical process of their formation. Often the concentration of the trace elements is characteristic of a particular site (Zeller and Gordon 1970; Hertogen and Gijbels 1971). When studying obsidian artifacts, for example, the results of trace-element analysis constitute a kind of ''fingerprinting'' pattern that can be used for identification and classification.

The chemical composition of clay from different sources may differ because of differences between the mineral progenitors of clay or because of different geochemical conditions involved in the process of clay formation. Since pottery is made from fired clay, the analysis of potsherds is very similar to that of mineral samples. Thus by analyzing the trace-element composition of a sufficient number of potsherds, information can be obtained on their provenance.

The determination of the provenance of glass through trace-elements analysis is more difficult. Trace elements are contributed to the glass by each of the three main raw materials: sand, modifiers, and colorants. Since each of these may come from a different geographical source, it is not easy to interpret the information yielded by chemical analysis.

An even more complicated undertaking is determining the provenance of metallic objects by chemical analysis. Metals are usually smelted and extracted from ores; during smelting and refining the relative concentrations of trace elements may be completely altered. Confusion may be compounded by the possibility of alloying (the components of the alloys may come from different sources) and of remelting of previously used metals. Thus the results of trace-element analysis of metal objects are likely to be archaeologically inconclusive. As a rule, the analysis of the major components of metals is much more likely than trace element analysis to yield useful information about provenance.

METHODOLOGY OF SYSTEMATIC ANALYSIS

How many elements to analyze in each sample, how many samples to test, and what accuracy to aim at, are questions that always recur. The objective in systematic analysis should be the determination of the largest possible number of elements in the largest number of samples, with the greatest possible accuracy.

Since systematic analysis involves the statistical evaluation of results, the examination of a small number of samples may lead to erroneous interpretation. Many materials used in antiquity are not homogeneous in composition; thus there are often compositional variations in artifacts originating from the same source. In such a case the research would lay special emphasis on the examination of the maximum number of specimens possible, even at the cost of some sacrifice in accuracy. The course of action adopted in systematic analysis is thus, of necessity, influenced by the particular circumstances under consideration. Economy in time and equipment may be factors limiting the usefulness of a technique.

The range and scope of problems which may be investigated using systematic analysis is almost unlimited. The development of modern methods of physical and chemical analysis requiring small samples, and the emphasis laid in recent years on nondestructive methods have provided the investigator with a very large selection of useful analytical techniques.

Two broad categories of chemical analysis are recognized: *qualitative analysis,* the identification of the constituents of a sample, and *quantitative analysis,* the determination of the relative amounts of the constituents. The range of problems within each of these two categories is very large. For the archaeological chemist it extends from the determination of infinitesimal amounts of impurities (e.g., when establishing provenance by means of trace elements) to the identification of metals present in an alloy and to the dating of skins and rocks.

CHEMICAL ANALYSIS

The analytical process involves four steps: (1) sampling, (2) sample preparation, (3) determination of constituents and measurement of their relative amounts, and (4) calculation, evaluation, and interpretation of results.

Most of the analytical work is done by the chemist in the laboratory. The involvement of the archaeologist in the analytical process will mostly be confined to setting the problem and interpreting the results obtained. There is one important step, however, where the archaeologist may play a very decisive part. This is in sampling. The way in which a material is sampled may critically affect, not only the results of the analysis, but also their interpretation and historical implications. It is for this reason that the process of sampling will be discussed here at some length.

SAMPLING

Whenever a chemical analysis is to be carried out, a sample of the material must first be obtained. *Sampling* consists of extracting, from a relatively large quantity of material, a small portion which is truly representative of the composition of the whole material. If the sample is not representative, the subsequent analytical operations will be of little meaning, since the properties of a material may vary from place to place. Slags from ancient smelting sites, for example, contain particles of different sizes, which may differ in composition in a manner related to their size. Special precautions, conditioned by the need for representativeness of the whole, must be observed when sampling such materials for analysis.

Homogeneous materials such as metals and glass generally offer no sampling problems; the sampling merely consists of separating a portion of the material. Metals, for example, can be sampled by cutting, sawing, or drilling, preferably at more than one point, both at the surface and through the bulk. Ancient metallic objects, however, are usually corroded to a greater or lesser extent, their homogeneity thus being destroyed. Hence to assure representative samples, ancient metals should be considered as heterogeneous materials and sampled accordingly.

Sampling of heterogeneous materials such as slags, pottery and bones is a complex operation, sometimes the most difficult part of the whole analysis. The underlying principle lies in the selection of a large number of portions, in a systematic manner, from different parts of the bulk, which are then combined. The following rules should be observed when sampling heterogeneous materials:

1. The weight of the sample should be as large as possible.
2. If the material consists of particles of different sizes, the large and the small particles should be sampled separately, and in the respective proportions in which they are present in the bulk.

3. Individual samples should contain as many particles as possible, and subsequent reduction in size of the particles—by crushing and grinding—should be left to the chemical laboratory staff.

The preservation of samples in an unchanged condition during transport and storage requires special attention. The container should be as chemically inert as possible with respect to the sample, and the atmosphere in which it is kept should be conducive to preservation. Dug-out metals—especially copper and bronze—and organic materials, tend to decompose rapidly after exposure to the moisture and oxygen of the atmosphere. In containers holding such materials the atmosphere around the sample should be rendered inert by replacing air by a nonreactive gas such as nitrogen.

Sampling Archaeological Materials

Sampling from archaeological materials raises some very special problems. In many instances no visible damage to the object being studied is permitted. In others only very small amounts of samples are available, and sometimes the size of the specimen may have considerable bearing on the sampling technique.

A very simple method of sampling homogeneous materials, without causing visible damage, consists of rubbing across the specimen to be analyzed a microscope slide with a roughened section at its center. A few micrograms of the specimen often suffice to produce a streak on the rough glass surface. This streak can then be dissolved and analyzed. Other sampling techniques whose effect on the specimen can hardly be seen may be mentioned: piercing the sample with an hypodermic needle and withdrawing a small core of material or, again, rubbing a streak of material off, onto a piece of etched quartz and analyzing *in situ*. Before choosing a particular sampling technique it is important to consider whether the analysis is intended to involve only the surface, body, or the whole of the object or materials and to sample accordingly.

SAMPLE PREPARATION AND DETERMINATION
OF CONSTITUENTS

A sample submitted for chemical examination is seldom in a form suitable for analysis. Thus operations are required to prepare the sample for analysis.

Essentially, *sample preparation* means getting the sample into such a condition that those properties or constituents in which the analyst is interested can be most readily measured. Materials that are irrelevant or might actively interfere with analyisis must be removed where possible. Sometimes the sample has to be dissolved or converted from a solid to a liquid or gas. Two main factors determine the mode of sample preparation: (1) the properties of the sample to be analyzed and (2) the analytical method to be used.

ANALYTICAL DETERMINATION AND MEASUREMENT

The procedures by which a constituent of interest is qualitatively identified, quantitatively measured, or both constitute the heart of the analytical process.

Chemical analysis is, essentially, the determination of the magnitude of some property of a system. Almost every element has some physical properties or will give rise to chemical reactions that, under certain conditions, are specifically characteristic of that element. These make it possible to distinguish the element sought from all others. The property measured may be weight, volume, wavelength, intensity of radiation, or any one of many others. The important thing is that it is related, in a clearly defined way, to the constituents under study.

Up to about 30 years ago chemical analysis depended almost entirely on the painstaking separation of one element from another by means of their characteristic chemical reactions. The invention of the spectroscope, the development of electrical and electronic measuring devices, and the discovery of radioactivity led to an entire new range of analytical methods. Today, almost any physical property of an element or of a compound can be made the basis of a physical method of chemical analysis.

For historical reasons, analytical techniques based on chemical reactions are known as *classical methods of chemical analysis*. Accurate quantitative analysis was first made possible through the development of the precise balance, accurate enough for *gravimetric* determinations. At about the same time the development of accurately calibrated glassware enabled *volumetric* analysis.

The more modern *physical methods* have the great advantage of being based on the direct measurement of physical properties, obviating the necessity for detailed separation of the desired components. Nevertheless, classical methods of analysis are still widely used in analytical work, especially in the determination of the major components of a sample.

In any given analytical problem the choice of method is governed by a number of considerations, including the accuracy necessary, the speed with which the results are needed, and the number of tests required. Priorities may have to be established between conflicting interests (e.g., speed vs. accuracy), and the analyst must have a wide knowledge of the advantages and limitations of different techniques if he is to choose properly. The methods of which today's analyst must have a working knowledge cover an enormous range, form classical gravimetry and volumetry to a long list of instrumental methods that were unknown less than a decade ago. The main categories into which analytical techniques are classified are shown in Table 3.2.

GRAVIMETRIC AND VOLUMETRIC ANALYSIS

Gravimetry and volumetry are the two classical procedures of chemical analysis. In *gravimetric analysis* the physical property measured is mass; the desired constituent of a given sample is isolated and determined by measuring its weight or the weight of some chemical derivative to which it is related in a precisely known

TABLE 3.2 Analytical Chemistry Techniques

Method	Property Measured
Classical	
Gravimetry	Mass
Volumetry	Volume
Physical	
Spectroscopy	Absorption or emission of radiation
Nuclear	Radioactivity
Mass spectrometry	Charge: mass ratio
Thermal	Heat content
Electrical	Electric charge transfer

manner. Hence the desired constituent must be one that can be isolated in a weighable form. To achieve this, the sample is first brought into solution. By the addition of suitable reagents to the sample solution, removal of the desired component can by precipitation can be induced, either as itself or—more frequently—as a known derivative that can be weighed after drying, or sometimes after ignition.

Volumetry is the general term used for methods of analysis where the physical magnitude measured is volume. A reagent, in the form of a solution of known concentration, is added incrementally to the sample solution until the point is ascertained (by visual or instrumental indication) at which all of the desired component has been transformed into a well-known derivative. From the amount of reagent used during the addition—usually called *titration*—the amount of the desired component in the sample can be calculated.

For the archaeological chemist, gravimetry and volumetry still provide the best means by which the proportions of major constituents can be determined with high accuracy in metal, slags, glass, building materials, and so on. They are also the only means by which nonmetallic elements—such as carbon, sulfur, and nitrogen—can be accurately determined. These methods have the advantage of not requiring expensive equipment, whereas their chief disadvantages are the relatively large samples and the time usually required, disadvantages that become especially apparent when a large number of analyses must be made.

PHYSICAL METHODS OF CHEMICAL ANALYSES

Physical methods of analysis are concerned largely with energy, in contrast to gravimetry and volumetry, which are concerned with matter. For almost every manifestation of energy there is a corresponding analytical method.

In archaeological research the most important physical methods of analysis are those based on the emission or absorption of radiation or of energetic particles. These phenomena are discussed in further detail in the following two chapters.

The use of physical methods has led to the development of some modern techniques of especial importance to archaeology, namely: (1) microanalysis and ultramicroanalysis, (2) trace analysis, and (3) nondestructive testing.

MICROANALYSIS

Microanalysis and ultramicroanalysis are techniques that provide a solution to the problem of analyzing very small samples of any nature or composition. They involve highly specialized working procedures and apparatus, necessary for carrying out experiments with small quantities of substances and small volumes of fluids

Confusion frequently arises as to the distinction between "micro" and "ultramicro" analysis. Table 3.3 indicates the respective sample sizes and the minimum amounts detectable with each method.

TABLE 3.3   Scales of Analytical Methods

Method	Sample Size (mg)	Minimum Detectable Amount of Substance (mg)
Macro	>100	10
Semimicro	~ 10	1
Micro	~ 1	0.01
Semiultramicro	0.01-0.1	10^{-3}
Ultramicro	10^{-4}-10^{-3}	10^{-5}
Submicro	10^{-6}-10^{-5}	10^{-7}

Micro and ultramicro techniques are invaluable when extremely small samples are to be examined or when damage is not permitted in larger samples. Using ultramicro analysis, for example, the pigments used to decorate Chinese oracle bones were successfully analyzed, even though subsequent microscopic examination failed to show that a sample had been taken for analysis.

TRACE ANALYSIS

Trace analysis is the determination of those chemical constituents present in a sample in very small quantities or traces. A *trace* is any constituent present in a concentration of less than about 100 ppm of weight (0.01%). The term *ultratrace* has been coined to designate those materials whose concentration is less than 1 ppm.

Optical, radioactive, or electro magnetic methods are used for trace analysis. Figure 3.1, summarizes in graphical form the detection capabilities of the various trace techniques.

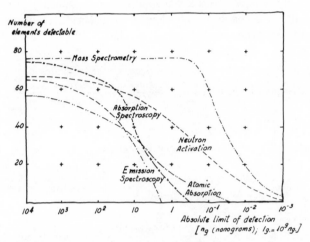

Fig. 3.1. Detection capabilities of trace techniques of analysis. Absolute limits of detection are plotted against the number of elements for which data have been published. It can be seen that all the techniques permit the determination of most elements in quantities as small as 100 nanograms.

The use of trace analyses of artifacts made from metals, obsidian, pottery, or glass has helped provide valuable information regarding authenticity, provenance, trade routes, economic trends, and other features or archaeological interest. (De Bruin, Korthoven, et al. 1976; Lahanier 1975). Specific cases are discussed in the relevant chapters of this book.

NONDESTRUCTIVE TESTING

Nondestructive testing may be defined as the examination of materials without damaging them or imparing their desirable properties. The age-old practice of "ringing" pots against one another to detect the presence of cracks may well be the earliest known example of nondestructive testing. Little further progress was made in this field until recently, when major advances were made in the application of optics, penetrating radiation, and atomic energy. Today, nondestructive test methods such as X-ray fluorescence, X-ray microprobe, beta-ray backscatter, and neutron-activation analysis are available that leave absolutely no marked after investigation. (Barrandon, Callu and Brenot 1977; Fullbright 1976; Reimers, Lutz and Segebade 1977).

CALCULATION, EVALUATION, AND INTERPRETATION OF ANALYTICAL RESULTS

The purpose of quantitative analysis is to determine the quantity of a particular constituent in a sample. Thus the amount of the constituent under test must be calculated after a test has been carried out and expressed in a suitable manner.

Analytical results are usually reported as relative quantities, as expressed by the formula:

$$\text{Relative quantity of constituent} = \frac{\text{Quantity of desired constituent}}{\text{Quantity of sample}} \times \text{coefficient}$$

Coefficients of 100, 1000, 1,000,000 and so on can be used, thus giving the most common ways of expressing relative quantities:

$$\text{As a percentage} = \frac{\text{Weight of constituent}}{\text{Weight of sample}} \times 100 = \% \text{ constituent (by weight)}$$

$$\text{As parts per million} = \frac{\text{Weight of constituent}}{\text{Weight of sample}} \times 1,000,000 = \text{ppm constituent}$$

$$\text{As parts per billion} = \frac{\text{Weight of constituent}}{\text{Weight of sample}} \times 1,000,000,000 = \text{ppb constituent}$$

Parts per thousand ($\%_o$) are also sometimes used.

ACCURACY OF RESULTS

The recipient of analytical results is interested in their accuracy, that is, the degree of precision with which they describe the material under analysis. Any measurement, however exact, has some degree of inherent imprecision. Hence, to define the accuracy of any given measurement, it is necessary to introduce the concept of error. *Error* is the difference between an observed result and a true one:

$$\text{Error} = \text{observed result} - \text{true result}$$

The term *accuracy* is used to describe the degree of error present in a measurement. A high degree of accuracy implies a low error, and a low accuracy, a high error. The degree of accuracy required from a set of results should be no more than that sufficient for the purpose in hand. This may range all the way from 1 part in 10 (as in the determination of the gross composition of mortars) to 1 part or less in 100 (as required for age determination or provenance studies).

The amount of work required to obtain results of high accuracy is generally many times greater than that required to obtain less accurate ones. Thus if low accuracy is sufficient, special efforts to achieve high accuracy may be wasted. It should be remembered, however, that accurate data on the composition of ancient objects and materials may become important later, in connection with future investigations, and that the opportunity to analyze a given object may occur only once and never be repeated. Thus it is probably advisable to select the most accurate available method practicable under a given set of circumstances.

PRESENTATION OF ANALYTICAL DATA

Many of the reports of chemical analysis of ancient objects that find their way into print are of little or no value at all to other investigators. Their failure lies in the inadequacy of the description of the exact provenance or source of the objects, their condition, the method of sampling, or even the method of analysis employed. As a consequence, analytical data of different investigators cannot be compared. These oversights hamper other investigators who want to use the data, but more serious is the possibility that they lead archaeologists into pitfalls. In an attempt to correct this lamentable situation, a statement listing the information essential to a report was issued by the International Committee of Museums (ICOM) and is reproduced here (Organ 1971).

Information Required in a Published Report of Analysis for the Report to Possess Value as Evidence for Archaeological Purposes

1. Description of object, provenance or attribution, and location at time of report. [The analysis of an unidentified object is worthless. It must be made possible to find the object analyzed, uniquely identified by its accession number or by its photograph, at some subsequent date.]
2. Location, on the surface of the object, of source of sample, or site of non-destructive examination, described relative to some uniquely identifiable feature or, better, identified in a scaled photograph. [A sample may have been taken from a position that later studies of the type of object reveal to have been unrepresentative.]
3. Method of sampling, e.g., filing or scraping (state area abraded and weight removed), drilling (state diameter and material of drill), coring or trepanning (state diameters). Describe sample, for example: fine powder, long spiral drilling. [This information may have bearing on the trace elements reported by the analyst and on the possibility that corrosion products were present.]

4. Estimated depth from which portion of sample actually used for analysis was withdrawn. [From this the reader will be able to infer to what degree the small sample may represent the composition of the whole.]

5. Nature of preparation of sample, if any, e.g., removal, by means of a magnet or steel abraded from drill, discard of first drillings, or other. [This part of the procdure may have considerable influence on the precision reported in section 8.]

6. Details of method of analysis, including size of sample used and number of replicates, sufficient to enable the work to be repeated by another investigator.

7. Number of separate analyses made.

8. Result with estmated precision and accuracy, stated as standard deviation of the method, with a note on how this was estimated.

Archaeologists and archaeological chemists reporting their work are advised to adhere as closely as possible to the recommendations of this body. Much waste of valuable information may be avoided thereby.

REFERENCES

Barradon, J. N., Callu, J. P., and Brenot, C., *Archaeometry, ***19**, 173 (1977).

DeBruin, M., Korthoven, P. J. M., Van der Steen, A. J., and Houtman, J. P. W., *Archaeometry, ***18**, 75 (1976).

Fullbright, H. J., Report LA-UR-76-622, 1976.

Hertogen, J., and Gijbels, R., *Analytica Chim. Acta, ***56**, 61 (1971).

Morrison, G. H., and Skogerboe, R. K., in G. H. Morrison, ed., *Trace Analysis, *Interscience, New York, 1965.

Lahanier, C., ICOM Comm. for Cons., 4th Trienial Meeting, Venice, 75/4/6-1, 1975.

Musty, J., *Proc. Anal. Div. Chem. Soc., *96 (1970).

Organ, P. M., *Archaeometry, ***13**, 27 (1971).

Reimers, P., Lutz G. J., and Brenot, C., *Archaeometry, ***19**, 167 (1977).

Zoller, W. H., and Gordon, G. E., *Anal. Chem., ***42**, 257 (1970).

SPECTROSCOPY

The composition of almost anything that emits, absorbs, transmits, or reflects light can be studied by *spectroscopy,* the complex of analytical techniques based on the interaction of energy with matter. In some cases a tiny sample, invisible to the naked eye, is sufficient for spectroscopic analysis; to analyze an object, it often is not even necessary to touch it. For these reasons—and many others that are explained in this chapter—spectroscopy is widely used in the analysis of materials and objects of archaeologicl interest.

Light, heat, and X-rays are essentially identical in nature in that they are all forms of *radiant energy.* Light is, however, the only form of energy visible to the naked eye and hence many of its properties are familiar to everybody. Before discussing the interaction of radiant energy with matter it seems appropriate, therefore, to look at some of the properties of visible light.

LIGHT

If a beam of ordinary sunlight is passed through a glass prism, it is *refracted,* or bent, and different colors are produced, as illustrated in Fig. 4.1. Bands of red, orange, yellow, green, blue, and violet blend into each other, without sharply defined boundaries between them, forming a *continuous spectrum.* To the naked eye, the spectrum appears to end at red and violet, respectively. However, beyond both red and violet there is also radiation, even though it cannot be seen; a sensitive thermometer is heated by *infrared* radiation, which occurs beyond the red portion of the spectrum; invisible light beyond the violet, hence called *ultraviolet light* will darken an ordinary photographic plate.

ELECTROMAGNETIC RADIATON

There are two different ways in which radiant energy can be described in precise mathematical terms. Some radiation phenomena can be explained only by defining radiation as a flow of discrete particles traveling through space at the speed of light. Other phenomena can only be explained by regarding radiation as a *wave motion,* called *electromagnetic waves;* and although particles and waves seem to

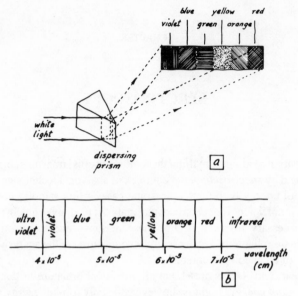

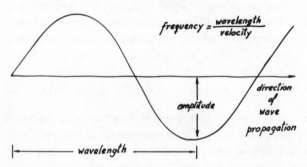

Fig. 4.1. (*a*) Dispersion of white light into a continuous spectrum. (*b*) Spectrum of visible light.

be incompatible, the *particle-wave duality* must be invoked to explain the behavior of radiant energy.

Electromagnetic waves (see Fig. 4.2) are distinguished from each other by two parameters: (1) The *wavelength,* the distance from the crest of one wave to that of the next, and (2) the *frequency,* the number of crests passing through a given point in unit time.

These two parameters are related by a mathematical expression:

$$\text{Velocity of light} = \text{Wavelength} \times \text{frequency}$$

Fig. 4.2. Electromagnetic waves; profile of a traveling energy wave at an instant of time.

which can be written using mathematical symbols:

$$c = \lambda\nu$$

where c is velocity of light, 3×10^{10} cm/second; λ is wavelength measured in centimeters; and ν is frequency, expressed in cycles per second.

Various units of length are used to define wavelengths, depending on the position of the waves in the energy spectrum. Table 4.1 lists these units and their equivalents in the metric system.

TABLE 4.1   Units of Length used in Spectroscopy

Unit	Symbol	Size (m)
Angstrom	Å	10^{-10}
Micron	μ	10^{-6}

THE ELECTROMAGNETIC SPECTRUM

The entire range of radiation, with wavelengths from infinitesimal to kilometers, constitutes the electromagnetic spectrum (Fig. 4.3).

The full electromagnetic spectrum can be conveniently broken down into several regions. Wavelengths of 4000-7000 Angstrom units (0.4-0.7 microns) are the only ones to which the human eye is sensitive. They make up the spectrum of *visible light*. Next to the visible spectrum, infrared rays have wavelengths longer than light, of 7250-10.000.000 Angstrom units (0.725-1000 microns); they are perceived by the human body as heat rather than color. Still longer wavelengths above 100,000,000 Angstrom units are characteristic of *radio waves,* used in telecommunications. Radiation of shorter wavelength than that of visible light, in the range 10-4000 Angstrom units, is called *ultraviolet light*. Beyond this there are X-rays, with wavelengths between 10^{-1} and 10 Angstrom units.

CONTINUOUS AND LINE SPECTRA

It is well known that hot substances radiate heat. It is equally well known that if their temperature is sufficiently high, they also emit light. When the heated substance is a solid or a liquid, the emitted radiation has a *continuous spectrum;* this means, simply, that radiation of all wavelengths is emitted.

The intensity and the wavelength of the radiation emitted by an incandescent body depend on the temperature of the body, and not on its nature. As the

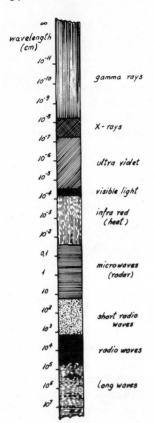

wavelength
(cm)

∞

10^{-11}

10^{-10} gamma rays

10^{-9}

10^{-8}

10^{-7} X-rays

10^{-6} ultra violet

10^{-5}

10^{-4} visible light

10^{-3} infra red
                (heat)

10^{-2}

0.1

1   microwaves
               (radar)

10

10^{2} short radio
               waves
10^{3}

10^{4} radio waves

10^{5}

10^{6} long waves

10^{7}

Fig. 4.3. The electromagnetic spectrum. The entire known spectrum is shown. It is probable, however, that still longer and/or shorter electromagnetic waves will be discovered in future.

temperature rises, the intensity of radiation increases, whereas the prevailing wavelengths tend, at the same time, to grow shorter (Fig. 4.4). Thus the color of an incandescent body will change from dull red, through bright red, to blue and violet, as its temperature increases.

When the source of electromagnetic radiation is a gas, however, the spectrum of the emitted radiation is of an entirely different character; it consists of bright lines of clearly defined wavelengths on a dark background and is called a *line spectrum* (see Fig. 4.5). The position of the spectral lines is specific to a given element. The line spectrum of a simple atom like hydrogen is relatively simple, whereas that of iron, a complex atom, is very complicated, containing thousands of lines. Any solid or liquid substance can be converted into a vapor or gas if it is heated to a sufficiently high temperature. Thus its line spectrum can be used to identify the elements that it contains.

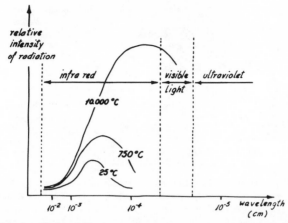

Fig. 4.4. Black body radiation. Distribution of intensity of energy radiated by a heated black body. At low temperatures the intensity of radiation is greatest in the infra-red region of the spectrum. As temperature increases, the peak intensity of radiation shifts to shorter wavelengths (i.e., to the visible light region).

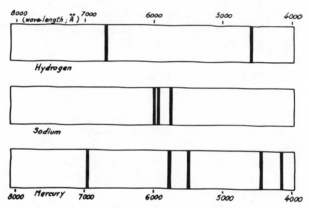

Fig. 4.5. Line spectra of some elements. For most elements the spectrum consists of bright lines (each of characteristic wavelength) on a dark background. Only the most often used lines are shown in the diagram.

ORIGIN OF ELECTROMAGNETIC RADIATION

To understand the reason for the occurrence of line spectra, it is advisable to consider the atoms of elements as resembling miniature solar systems in which electrons revolve in fixed orbits about a central nucleus without absorbing or emitting energy. If energy is introduced into the atom, (e.g., by heating), some of the electrons become *excited:* they leave their normal orbits and jump into others further away from the nucleus (Fig. 4.6). This situation is inherently unstable, and the excited electrons tend to rapidly fall back into their original orbits. As they do

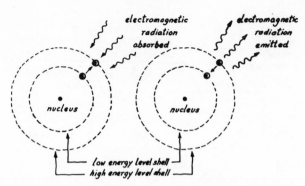

Fig. 4.6. When an atom absorbs radiant energy from an outside source, its electrons are excited ("jump") to higher energy levels. Subsequently, as the electron jumps back from the higher to the lower energy level, the atom *radiates* energy.

so, they lose their energy of excitation, which is given off as radiation. This may take the form of visible light, UV, X-radiation, depending on the *electron transitions,* that is, the way the electrons have been moving between orbits. Some of the most important electron transitions are listed in Table 4.2. The amount of energy needed to move an electron from one given orbit to another determines the energy, hence the wavelength, of the radiation given off.

TABLE 4.2　Atomic and Molecular
Transitions Pertinent to Spectral Regions

Transition	Spectral Region
Inner-orbit electrons	X-rays
Outer-orbit electrons	Ultra violet and visible
Molecular vibrations	Infra red

It a great deal of energy is introduced into an atom (e.g., by very strong heating), many electrons in it will be excited to varying degrees. Each of these electrons, at each different level of excitation, will give off radiation of a particular wavelength; that is why the emission spectrum of a relatively large atom, like iron, may contain several thousand lines, with each line produced by light of a different wavelength.

ABSORPTION SPECTRA

Under certain circumstances atoms are also capable of absorbing radiation. Atoms subjected to radiation absorb energy at the same wavelengths as they emit when excited. Thus if a beam of white light is passed through a sample of a particular material and subsequently separated into its spectral components, a series of dark lines appears in the spectrum at those wavelengths where the light has been absorbed. The position of these absorption lines for a particular material will naturally be the same as the position of the bright lines in its emission spectrum. A spectrum obtained in this way is called an *absorption spectrum*.

Whereas the atoms of elements can be studied by means of their emission spectra, the molecules of complex chemical compounds can be more readily studied by their absorption spectra. Molecules do not give rise to sharp spectral lines but to *bands,* each band extending over a considerable range of wavelengths and frequently superimposed with a fine structure of separate lines.

X-RAYS

Heating a material is not the only way of exciting its electrons. They may also be excited by bombardment with electrons from an outside source, especially if these have been *accelerated,* that is, given extra power (e.g., by the influence of a magnetic field). When accelerated electrons strike a target, electrons in the atoms of the target material become excited and give off X-rays. These are emitted as a continuous band across the entire range of X-ray wavelengths, with superimposed peaks (i.e., points of high emission) at certain wavelengths (see Fig. 4.7). The

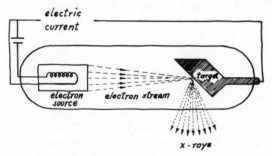

Fig. 4.7. Diagram of an X-ray tube. In operation, electrons stream from the heated electron source (cathode) toward the target (anode). The speed of impact is dependent on the potential difference which for archaeological purposes may range 10,000-100,000 volts or higher. The atoms of the target, excited by the electron stream, give off X-rays.

position of these peaks, or emission lines, in the X-ray spectrum is characteristic for each chemical element and can be used to identify the latter in the same way as the lines in the visible emission spectrum.

X-Irradiation can be used for analytical purposes in a number of ways:

1. Absorption of X-rays (*radiography* and *fluoroscopy*) gives information on the absorbing material.

2. Measurement of the wavelength and intensity of emitted X-rays can be used to identify and quantitatively determine elements in a sample. *X-Ray fluorescence* and *X-ray microprobe* are emission techniques (see later).

3. *Diffraction of X-rays* (see Fig. 4.8) is used for the identification and analysis of crystalline substances.

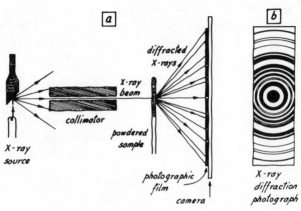

Fig. 4.8. X-Ray diffraction: (*a*) Diagram of an X-ray-diffraction camera: crystals of a finely powdered sample bend (or ''diffract'') X-rays into conical envelopes at various angles, which fall upon a photographic film. (*b*) X-ray-diffraction photograph. The developed photographic film reveals a concentric pattern of circular lines. Measurements of the thickness of the lines and the distances between them yield information about the crystalline structure of the sample. Parameters characteristic of the crystal structure of the sample can be calculated.

RADIOGRAPHY

Radiation of different wavelengths varies in its ability to penetrate matter. The shorter the wavelength of the radiation, the greater its penetrating power. X-Rays have much greater power than visible light to penetrate and pass through solid bodies. Radioactive elements give off another form of radiation called *gamma rays*; these have even shorter wavelength (0.001-0.1 Angstrom units) and greater penetration power through considerable thicknesses of solids such as metal and stone. The penetrating power of X- and gamma-rays makes it possible to discover and locate hidden details in objects and even in complex structures. The best-known techniques for doing this are *fluoroscopy* and *radiography*.

Fluoroscopy produces a temporary image on a screen coated with a fluorescent substance, that is, one capable of converting the energy of the penetrating radiation into visible light (see Fig. 4.9). Often the object under examination can be rotated and moved within the path of the beam of X-rays, so that every part can be scrutinized.

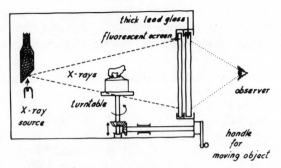

Fig. 4.9. Fluoroscopy. Diagram of apparatus used for examining objects by fluoroscopy. The X-ray image of the object is seen on the screen, through the protective lead-glass window.

Radiography uses photographic materials. A radiograph is a permanent photographic record, actually an X-ray or gamma-ray "photograph" (see Fig. 4.10).

The extent to which radiation impinging on an object penetrates it or is absorbed depends on four factors: (1) wavelength of the radiation, (2) Composition of the material, (3) density of the material, and (4) thickness of the material.

Rays of long wavelength, produced at low voltages, have less penetrating power than do those of short wavelength that are produced at high voltages. The extent of absorption of X- and gamma-rays by any material depends on the atomic number of its component elements. Elements of low atomic number absorb less radiation than do those of higher atomic number. Thus a layer of carbon-black paint absorbs less radiation than does an equivalent layer of lead-white paint. A dense material

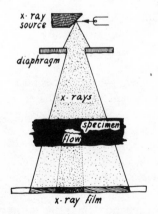

Fig. 4.10. Radiography. Schmematic diagram of the fundamentals of radiographic exposure. The dark region of the film represents the more penetrable part of the object and the light regions, the more opaque.

absorbs more radiation than does a less dense material. And a thick layer of any material absorbs more radiation than does a thin layer of the same material. Thus the energy of radiation used varies with the substance radiographed (Table 4.3).

TABLE 4.3   Energy of X-Ray Radiation

Energy of Radiation (Kilo Volts)	Application
1-50	Radiography of thin foils, leather, paper
50-250	Radiography of up to 10 cm metal thickness; also medical radiography
200-500	Radiography of thicker metal sections

Radiography and fluoroscopy are used in archaeology to provide information that could be otherwise obtained only by damaging the object. The range of subject matter is endless—from papyri to metal castings, from fossils to mummies. An Egyptologist studying a mummy in its yards of cloth wrappings can radiograph it without removing the wrappings and thus obtain information about burial posture, skeletal anatomy and pathology and, sometimes, enclosed objects. Radiography of architectural structures with portable X-ray equipment was used as an aid to dating and structural analysis (Hart 1977). An anotated bibliography on the uses of radiography (and other X-ray techniques) in archaeological studies is available (Mansell 1977).

Three-demensional vision—stereoscopic radiography—is sometimes helpful in that it permits location of detail of gross structure below the surface. Using this technique, two radiographs are obtained that, when viewed simultaneously, blend to produce a three-dimensional picture (Loose 1960).

SPECTROSCOPY

The interaction of radiant energy with matter can be used to obtain information about substances. Measurement of the controlled emission or absorption of radiant energy by a sample of matter can give information on the nature and concentration of the elements or compounds present in it. The branch of science that deals with the measurement of radiant energy and its interaction with matter is spectroscopy. *Emission spectroscopy* is concerned with the measurement of the relative amounts of energy emitted by atoms in the excited state at each wavelength. High-voltage electric currents, high-temperature flames, or high-energy radiation are used to bring about the necessary excitation. Inorganic substances usually are studied by emission techniques. Organic compounds, on the other hand, tend to decompose when subjected to the high energies required for emission spectroscopy. They are, therefore, usually studied by *absorption spectroscopy* in which, as explained in a preceding section, a beam of radiation is directed at a sample and the wavelength and intensity of the absorbed radiation are measured.

Since the wavelength of radiation, either emitted or absorbed, is characteristic for each element and substance, it is possible, using spectroscopic techniques, to identify materials by finding out what elements or compounds they contain and in what proportions.

Spectroscopic techniques are not confined to visible light; other kinds of radiation—in infra red, X-ray, gamma—can also be used. Spectroscopy is, consequently, an extremely powerful analytical tool that can help in the solution of many problems.

SPECTROSCOPY IN ARCHAEOLOGICAL RESEARCH

Spectroscopic analysis may be used in many ways in archaeology. The principal advantages of using spectroscopic techniques lie in the rapidity with which results can be obtained and in the fact that the samples needed are usually very small. For many years archaeologists were unable to explain the origin of the purple color of gold sequins found on mummies in Egyptian tombs. Spectroscopic analysis revealed that the color was due to certain impurities present naturally in some samples of gold. Identifying these impurities made it possible to determine where the gold was mined and to duplicate the purple color. Since only a minute amount of material was available for analysis, the spectrograph proved to be particularly useful; not a single sequin had to be destroyed to give the answer.

Qualitative and quantitative analyses made by emission-spectroscopic techniques are used extensively for determining the presence of trace elements in metals, ceramics, and glass. Absorption spectrophotometry is useful in testing organic materials. The pages that follow contain information intended to be helpful in: (1) understanding the reasons for selecting a given method of analysis for a given problem, (2) understanding the principles involved, and (3) interpreting the results obtained. From considerations of space, it is possible to discuss here only those techniques that have already found direct application in archaeological studies, with perhaps a mention of others that seem to offer promise in this field.

EMISSION SPECTROSCOPY

The emission of radiation by chemical elements under suitable excitation is utilized for the identification and quantitative determination of metallic and many nonmetallic elements. The wavelength of the radiation emitted allows the qualitative identification of the elements present; quantitative determinations are based on an empirical relationship between the intensity of the emitted radiation at some particular wavelength and the quantity of the corresponding element in the sample.

The high energies required to excite the atoms sufficiently to cause them to radiate are produced either by electrical discharges—*emission spectrography*—or by the combustion of gases—*flame photometry*.

EMISSION SPECTROGRAPHY

Emission spectrography is the measurement of radiation emitted by atoms excited by electric discharges. In qualitative emission spectrography the wavelength of the radiation emitted is measured and serves to characterize the elements in question. When the total energy, or the intensity, of the radiation emitted is measured, the basis is obtained for quantitative measurements.

Emission spectrography is particularly suitable for the study of materials of archaeological interest, since it combines versatility with economy. Over a score of elements can be identified both qualitatively and quantitatively over a wide range of concentrations from a relatively small sample. The method has been used in studies of ancient metallurgy, coins, ceramics, glass, and other substances.

The equipment required for spectrographic analysis is illustrated in Fig. 4.11. It consists basically of a stand for holding the electrodes, where the sample is heated

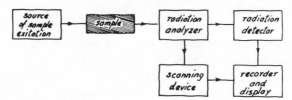

Fig. 4.11.   Emission spectroscopy. Block diagram of the instruments required for emission spectroscopy measurements.

to high temperature, an optical system for dispersing the radiation emitted by the sample, and a detector, usually a photographic plate or film, sometimes a light-sensitive cell. Peripheral equipment includes a power pack and transducers, indicators, and recorders for transforming the output of the instrument into a readable signal.

To produce emission spectra, a powerful electrical discharge is passed between two portions of a sample or, most frequently, between the sample and a counter-electrode that does not contain the elements being tested for. This latter is usually made of very pure graphite.

Given the proper equipment, all that is necessary to identify an element is to ascertain the wavelengths of the photographically recorded spectral lines obtained from the sample. This step can be facilitated if the spectrum obtained from the

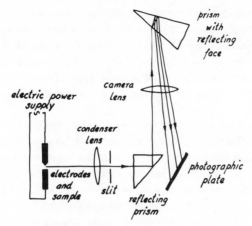

Fig. 4.12. Emission spectrograph (schematic).

sample is compared with a reference spectrum made in the same way from a material of precisely known composition and placed alongside it. In addition, compilations of the known lines of all the chemical elements are available and can be consulted. (Fig. 4.12).

The degree of blackening of the photographic plate at any given wavelength depends on the concentration of whatever element is emitting that particular wavelength. In quantitative determinations it is necessary, therefore, to ascertain the blackening of the different lines in the photographic plate. For semiquantitative work the visual estimation of the degree of blackening is usually sufficient, but in accurate quantitative work the numerical evaluation of blackening is carried out with a special instrument, the microphotometer.

Emission spectrography provides a rapid method of analysis for which only a very small quantity of sample—about 20 milligrams—is needed for an extensive analysis. When the conditions of the analysis are properly chosen, the sensitivity can be very high, and many elements can be detected when no more than millionth of a gram is present in the sample. Hence the method is very useful for finding elements that occur as minor or trace constituents. At the upper limit of concentration, on the other hand, it is inconvenient to determine concentrations in excess of 10 percent without having to recourse to special techniques.

LASER MICROPROBE ANALYSIS

The introduction of the laser as a source of radiation has made possible the analysis of very small areas of conductive and nonconductive solids, including those of low atomic number. A laser beam of high intensity can be focused on a minute area of a sample, as small as 50 microns in diameter and a quantity of material as small as 1 microgram (0.000001 gram) can thus be vaporized and analyzed. The energy of the laser beam is so high that the vaporized material rises from the sample as a small luminous plume that has a temperature approximating to 15,000°C. The bright light beam is relayed into the optical system where it is resolved into the different wavelengths present.

A 4000-year-old Hittite bronze figurine of a bull from Anatolia had several stripes on its back that were notable in having resisted corrosion during millenia of burial. Using a laser microprobe, the stripes were shown to be relatively rich in arsenic (Brech and Young 1965). The pits made by the laser are so minute as to not interfere with the display of the object in a museum or with subsequent scientific examination (see Fig. 4.13). The damage done by the laser microprobe is so small that to all intents and purposes the method can be considered nondestructive.

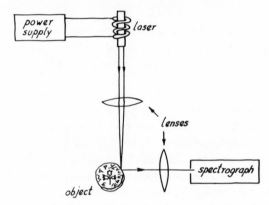

Fig. 4.13.   Laser microprobe (schematic). The laser beam can be focused on minute areas of a sample.

FLAME PHOTOMETRY

Flame photometry (see Fig. 4.14) is a useful technique for determining minute quantities of elements by measuring directly the intensity of their flame-produced radiation. The sample to be analyzed must first be brought into solution. The solution is then introduced into a flame in the form of a fine spray. In the flame the solvent evaporates, followed by the solids, whose constituent atoms become excited and emit visible radiation that can be measured. If the conditions remain constant, the intensity of the radiation emitted is proportional to the amount of the element sought. The maximum temperatures attained in flame photometry (~3,500°C) are relatively low in comparison with those achieved in emission

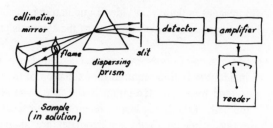

Fig. 4.14.   Flame photometer (schematic).

spectrography (~9,500°C). Consequently, flame spectra are easily produced with relatively simple equipment.

Flame photometry is an almost indispensable technique for determining certain elements, particularly the alkali metals (sodium, potassium, etc.). Most of the potassium determinations used in potassium-argon dating are carried out by flame photometry.

X-RAY FLUORESCENCE SPECTROSCOPY

When a sample is irradiated with a beam of "primary" X-rays—that is, rays of shorter wavelength than that of the characteristic X-radiation of the component elements—the atoms of the target gain energy. The energy gained does not remain in them but is lost by several mechanisms; one mechanism is reradiation of X-rays of characteristic wavelength. The atoms in the sample are said to fluoresce; they reemit X-radiation of the same wavelength they would emit if made the target of an X-ray tube. The phenomenon is called *X-ray fluorescence* and is similar, in essence, to optical emission. However, whereas in optical emission the wavelength range of the radiation is 2000-5000 Angstrom units, in X-ray fluorescence the wavelength of the radiation measured the range is 0.1-10 Angstrom units. The

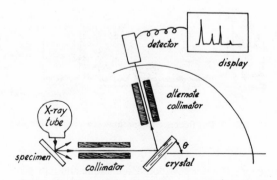

Fig. 4.15.   X-Ray fluorescence spectrometer (schematic).

wavelength of the fluorescent radiation provides a qualitative indication, and the relative intensity gives a quantitative measurement of the elements present in the sample.

The instrument used for X-ray fluorescence analysis is called an *X-ray spectrograph*. The primary incident radiation required for excitation of the atoms of the sample is usually provided by a tungsten-target X-ray tube. The sample under analysis is placed in the path of the X-ray beam and emits fluorescent radiation in all directions. Those rays directed to the spectrometer are singled out by a *collimator*—a collection of thin, parallel metals plates—and emerge from it as a beam of directed rays with a mixture of wavelengths.

A crystal is used to separate the various wavelengths: for every angle of incidence of the X-radiation (Figure 4.15), the only wavelength that is reflected is the one that conforms with the mathematical condition known as Bragg's formula:

$$n\lambda = 2d \sin \theta$$

where n is a whole number: $1, 2, 3, \ldots$; λ is the wavelength of the X-ray radiation used; d is a constant, characteristic of every crystalline substance; and θ is the angle on incidence of the X-radiation on the sample. By rotating the crystal relative to the direction of the incident beam it is possile to select any desired wavelength. A detector measures the intensity of radiation of any chosen wavelength. The variations of intensity of radiation at different values of Bragg's angle θ can be presented in the form of a graph, as shown Fig. 4.16. From the graph, both qualitative and quantitative information can be obtained simultaneously.

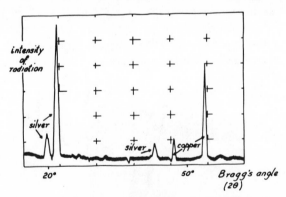

Fig. 4.16. X-Ray fluorescence spectrum of a silver-copper alloy.

X-Ray fluorescence is a nondestructive technique and thus provides a convenient means for the analysis of a large variety of archaeological materials. The manipulations are simple and the results are rapidly obtained. The disadvantages of the method are: the limited number of elements that can be determined, and the shallow penetration of the soft X-radiation used, into matter. Elements of atomic number lower than 22, such as sodium, aluminum and potassium, cannot be determined by X-ray fluorescence, since the radiation they emit is wholly or partly absorbed by air before it can reach the detector. X-Ray fluorescent radiation is also strongly absorbed by solid matter; in consequence, the technique is most useful for the examination of the surfaces of materials. Thus it is ideally suited for the study of enamels, glass and metals. X-Ray fluorescence can be used for the determination of major and minor constituents—in the range 1-100 percent—and sometimes, under favourable conditions, of trace impurities down to a few parts per million.

Nondispersive X-Ray Fluorescence Spectroscopy

Nondispersive X-ray fluorescence spectroscopy is based on *energy discrimination,* a system in which a detector—in this case a semiconductor diode—receives simultaneously all the wavelengths emitted by the excited sample. The detector converts the energy carried by the X-rays into electrical signals. These are fed into electronic integrating circuits so that the corresponding outputs will increase steadily as the exposure to radiation is continued over a period of time.

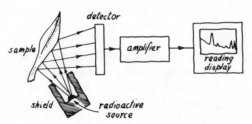

Fig. 4.17.   Non-dispersive X-ray fluorescence spectrometer (schematic).

The system illustrated in Fig. 4.17, is highly sensitive, much more so than spectrometers using optical systems. A powerful X-ray tube is no longer necessary; the amount of excitation energy required is so small that it can be provided by a small radioactive source such as americium-241, cadmium-109, or cobalt-47.

The extremely compact system comprises of only a radioactive source (see Fig. 4.18), sample holder, semiconductor detector, and complementary electronic equipment for amplification and measurement of the electrical outputs (Bowman, Hyde, et al. 1966). Portable instruments for nondispersive X-ray fluorescence have great possibilities for field archaeology since the components are

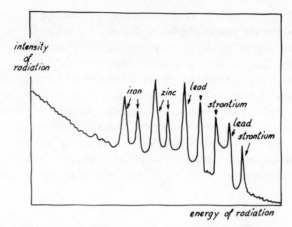

Fig. 4.18.   X-Ray fluorescence spectrum induced by a radioactive source.

independent of an external power source. In at least one archaeological expedition portable equipment was tested at the excavation site. The results obtained were only semiquantitative because of differences in texture, shape, and composition of the objects encountered. Nevertheless, it was possible to analyze objects of almost any geometry without special preparation and without damage (Frierman, Bowman, et al., 1969). In glasses, glazed ceramics, and metals, elements present in concentrations of 0.1-30 percent can be detected with an accuracy of 10-20 percent.

Energy discriminating X-ray fluorescence analysis has become an increasingly useful tool for the analysis of archaeological objects. Its application to the analysis of metals, ceramics and other materials is well documented (Metcalf 1970; Hall 1973; McKerrel 1976; Carlson 1977).

ELECTRON-PROBE X-RAY MICROANALYSIS

Electron-probe X-ray microanalysis uses X-ray fluorescence to examine small solid areas of 0.1-3 μ in diameter. The instrument used has been variously called *electron-probe microanalyzer, electron microprobe, scanning electron microprobe,* or simply *electron probe.* It combines the technique of X-ray fluorescence spectroscopy with that of optical microscopy.

The instrument, illustrated in Fig. 4.19, consists of four basic component systems: (1) electron optical system, (2) X-ray optical system, (3) viewing system, and (4) translation mechanism.

The electron optical system focuses a narrow beam of electrons onto the specimen, causing excitation of the atoms and the consequent emission of X-rays.

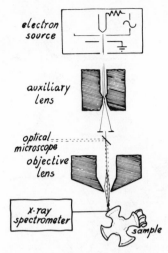

Fig. 4.19. The basic components of an electron-probe microanalyzer (schematic). An optical microscope is incorporated so that the operator can locate precisely the spot to be examined.

The emitted X-radiation is analyzed by the X-ray optical system according to wavelength and intensity, thus allowing point-to-point chemical analysis—both qualitative and quantitative—of the sample's surface. The viewing optical system is used by the analyst in the selection and direct observation of the area under analysis. The specimen may be moved by a translation mechanism so that any desired area may be brought into position for analysis.

The technique is subject to the same limitations as X-ray fluorescence analysis; elements of low atomic number, (below number 22) for example, cannot be analyzed. However, not all specimens suitable for fluorescence analysis can be successfully adapted to electron-probe examination. The surface analyzed must be smooth; otherwise variation in X-ray output will lead to spurious results. It must also be electrically conducting, either naturally, or by deposition of a thin metallic layer prior to analysis.

The electron microprobe has been proven of value in special applications such as the identification of small inclusions in ceramics and metals or variations in the composition of surfaces; it is ideal for the study of surface enrichment of metals. Superimposed thin layers of paints can be separately examined by the electron microprobe (Patersen and Ogilvie 1960): a "core" of pigment is taken from a painting by passing a hypodermic needle through it; part of the needle is then cut away, after which a section of the core can be mounted for electron probe examination, as shown in Fig. 4.20.

The X-ray fluorescence techniques described so far make use of optical systems to disperse the different wavelengths of radiation emitted by the sample and are called *dispersive* X-ray techniques. The optical system, however, can be

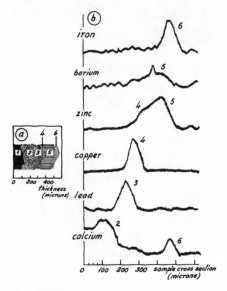

Fig. 4.20. Electron-microprobe analysis of pigments in layers of paint: *(a)* cross section of a layered paint sample. The thickness of the layers is measured by means of a microscope (the layers are numbered 1-6, from ground layer to surface, respectively); *(b)* microprobe scan of a vertical section through the layers (numbers above the peaks refer to the layers; i.e., the number 3 above the lead peak indicates that lead is present in layer 3).

Results

Layer No.	Microscopic Examination	Elements Detected by Microprobe	Identification
1	Wood base	—	Wood
2	White ground	Calcium	Calcium carbonate
3	White layer	Lead	White lead
4	Bronze-colored layer	Copper and zinc	Imitation gold leaf
5	White layer	Zinc and barium	Lithophone (zinc sulfide (30 percent) plus barium sulfate (70 percent)].
6	Dark yellow layer	Iron	Yellow ocher

dispensed with; instead of scanning sequentially the dispensed radiation, the mixture of wavelengths can be simultaneously analyzed in a *nondispersive* X-ray spectrometer.

ABSORPTION SPECTROSCOPY

Atomic Absorption Spectroscopy

Analysis by *atomic absorption spectroscopy* involves the breaking down of molecules of their consitutent atoms and the study of the radiant energy absorbed by these. The sample is first brought into solution and then vaporized in such a manner that, the elements sought exist as neutral atoms in the vapor produced. In this state each separate element in the vapor absorbs radiation of characteristic wavelength. Under controlled conditions the absorption of light of a particular wavelength can be used to determine the presence and concentration of elements sought (see Fig. 4.21). Atomic absorption is a highly sensitive and precise method of analysis, with which many elements can be determined in all concentrations down to a few parts per million or even less. The equipment is simple to operate and requires very little operator training.

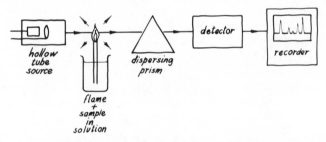

Fig. 4.21.   Atomic absorption spectrophotomer (schematic).

A typical atomic absorption spectrometer consists of a radiant energy source, a sample cell, a monochromator, and a detector. Light from the radiant source—a special kind of lamp called a *hollow cathode lamp*—traverses an atomized spray of the sample solution and falls on the monochromator, where light of appropriate wavelength—characteristic of the element sought—is separated from other radiation emitted by the source and finally falls on the detector. Electronic equipment amplifies and records the output from the detector. For quantitative measurements, the difference between the intensity of unabsorbed radiation and that of radiation after absorption is related to the concentration of the element in question.

The main disadvantages of the method are its limited appliation and the impossibility, at present, of simultaneous multiple analysis. About 50 elements, all metals, can presently be detected; equipment available today cannot detect nonmetals or metalloids. Although equipment for multiple analysis exists, it is not much used and the separate successive determination of each element makes the method laborious and time consuming. Thus atomic absorption has, up to the present, not been much used in the study of archaeological materials; its promising qualities, however, warrant the inclusion of the method in any list of techniques available to the archaeological chemist.

Infrared Absorption Spectroscopy

The spectroscopic methods discussed so far deal either with the emission or the absorption of radiation by atoms. Single atoms neither emit nor absorb radiation in the IR region of the spectrum. On the molecular level, however, the absorption of radiation in the 2.5-16-μ wavelength range can be of great importance to the analyst. Nearly all molecules show some degree of selective absorption of energy in the IR region of the spectrum. The molecular absorption spectrum is usually a complicated pattern of numerous narrow bands, such as shown by the active component of the ancient dye indigo (Fig. 4.22).

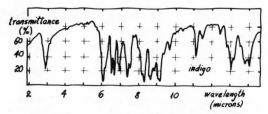

Fig. 4.22.   Infrared absorption spectrum of indigo.

Infrared absorption is a powerful technique for the identification of complex molecules and can be used with liquid as well as solid samples. To obtain an infra-red spectrum, radiation from an infra-red source is divided in a spectrometer into two equivalent beams; one passes through the sample cell and the other, through a reference cell (Fig. 4.23). The two beams are then recombined and

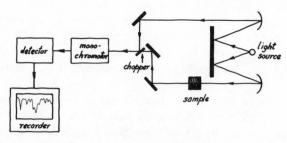

Fig. 4.23.   Automatic "double beam" infra-red absorption spectrophotometer (schematic).

focused on a monochromator slit, where the different wavelengths are separated, and finally to fall on a detector, causing it to emit electrical signals that can be amplified and recorded. The difference in the absorption between sample and reference gives rise to the absorption spectrum.

Quantitative determinations with this technique are not practical. The importance of infra-red spectroscopy lies mainly in the qualitative identification of substances, either pure or in mixtures. The method is extremely versatile for the identification of organic samples, and less so for inorganic ones. Infrared spectroscopy has found extensive application in the identification of dyes, pigments, oils, resins, and a variety of other archaeological materials.

REFERENCES

Bowman, H. R., Hyde, E. K., Thompson, S. G., and Jared, R. C., *Science,* **151**, 562 (1966).

Brech, F. and Young, W. J., in W. J. Young, ed., *Application of Science in Examination of Works of Art,* Museum of Fine Arts, Boston, Mass., 1965, p. 233.

Carlson, J. H., *Archaeometry,* **19**, 2 (1977).

Frierman, J. D., Bowman, H. R., Perlman, I., and York, C. M., *Science,* **164**, 588 (1969).

Hall, E. T., Schweizer, F., and Toller, P. A., *Archaeometry,* **15**, 53 (1973).

Hart, D. M., *Technology and Conservation Art, Architecture and Antiquities,* **77**(2), 10 (1977).

Loose, L., *Studies in Conservation,* **5**, 85 (1960).

McKerrel, H., in *Applications of Nuclear Methods in the Field of Works of Art,* Rome, 381, 1976.

Mansell, H., *ICCM Bulletin,* **3**(2), 17 (1977).

Margoshes, M., and Scibner, B. F., *Anal. Chem.,* **38**, 297R (1966).

Metecalf, D. M., and Schweizer, F., *Archaeometry,* **12**, 173 (1970).

Patersen, N. L., and Ogilvie, R. E., *Transact. AIME,* **218**, 439 (1960).

RADIOACTIVITY AND NUCLEAR ENERGY

A surprisingly large number of the applications of chemistry to archaeology depend on radioactivity and the use of nuclear techniques. Natural radioactive isotopes provide inherently reliable clocks used for "nuclear timing" of archaelogical events. The use of stable isotopes has made possible the differentiation and classification of ancient materials according to their place of origin. The application of nuclear techniques of chemical analysis to ancient materials has provided information on their geographical source, on the techniques of making artifacts and tools and on their dissemination to near and remote countries.

The use of radioactivity and nuclear techniques marks, in a way, the beginning of the serious and methodical application of science to archaeology; hence it seems appropriate to review the principles of radioactivity, properties of radioactive substances, and uses of nuclear techniques important to archaeological research.

Radioactivity is a natural phenomenon occurring as a result of changes that take place in the nuclei of atoms. *Nuclear science* deals with atomic nuclei and their changes.

THE NUCLEI OF ATOMS

The central core of an atom is its *nucleus*. This core is unbelievably small, 10,000-50,000 times smaller than the atom itself. However, almost the entire weight of the atom is concentrated in the nucleus. The latter is also electrically uncharged, with a positive charge equal in value—but opposite in sign—to the number of electrons surrounding it in the atom.

The simplest atomic nucleus is the *proton*. This is the nucleus of the atom of hydrogen, the lightest element. There are several hundred known kinds of atomic nuclei, and the detailed structures of some of these are as yet unknown. It is generally accepted, however, that all known atomic nuclei can be considered as being built up of protons and neutrons.

The *neutron* is a nuclear particle that has neither a negative nor a positive electric charge. It is electrically neutral; hence its name. It weighs the same as a proton,

1840 times as much as an electron. In the atomic weight scale, the weight of a neutron or proton is one unit of mass (Table 5.1).

TABLE 5.1   Elementary Particles in the Atom

			Weight		
Particle	Symbol	Charge	Atomic Weight Scale	Relative (atomic units)	Actual weight (grams)
Electron	e	−	0	1	9.1×10^{-28}
Proton	p, H$^+$	+	1	1	1.7×10^{-24}
Neutron	n	0	1	1	1.7×10^{-24}

ISOTOPES

All atoms of the same element have the same number of protons in their nuclei and of electron surrounding them. But they do not necessarily have the same number of neturons; extra neutrons do not change the charge of the nucleus, which is what determines the atomic number of the atom and its chemical properties. Therefore, depending on the number of neutrons in their nuclei, atoms of one element—that is, atoms having the same atomic number—may weigh more, or less, than others. Atoms that differ only in the number of neutrons in their nuclei, but not in their chemical properties, are called *isotopes* (see Fig. 5.1).

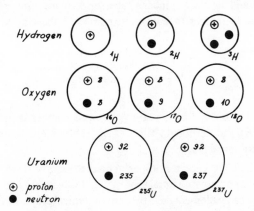

Fig. 5.1.   Isotopes. Symbolic representation of the structure of the isotopes of various elements; nuclei only are illustrated.

In the same way as the atomic number of an element defines the number of protons in a nucleus, the *mass number* defines the total number of particles, protons and neutrons, in it. The difference between the mass number and the atomic number gives the number of neutrons in an isotope's nucleus.

Mass number = Number of neutrons + number of protons
Atomic number = Number of protons
_____
Difference = Number of neutrons

The atomic number of oxygen, for example, is 8. The mass numbers of its isotopes are 16, 17, and 18. The number of neutrons in each isotope can be calculated from the above given difference; thus:

	Oxygen-16	Oxygen-17	Oxygen-18
Mass number	16	17	18
Atomic number	8	8	8
Number of neutrons	8	9	10

The symbols used to identify the chemical elements do not identify particular isotopes. Since it is, sometimes, desirable to distinguish between isotopes of the same element, a set of symbols, using atomic numbers and mass numbers, has been devised for this purpose. In a typical isotope symbol, the atomic number is a subscript on the left and the mass number a superscript on the right. For the isotope of carbon that has mass number 14 and atomic number 6, the symbol is:

$$_{8}C^{12}$$

12 ← Mass number
C ←— Symbol of element
8 ←——— Atomic number

Sometimes, when discussing isotopes, only the mass number is indicated, as in oxygen-18 (or 0-18) and carbon-14 (or C-14). Table 5.2 lists some isotopes important in archaeological research.

There are more than 350 known isotopes in nature, most of them stable. Some elements have only one known natural isotope, and others have as many as 10. Since isotopes of the same element have the same chemical properties, they cannot be separated by chemical means. The mass spectrometer and the mass spectrograph (Fig. 5.2) have been designed to separate isotopes and to determine their relative abundance in a given sample of an element.

TABLE 5.2   Isotopes Important in Archaeological Research

Isotope	Symbol	Half-life	Use
Argon-40	A^{40}	Stable	Potassium-argon dating
Carbon-13	C^{13}	Stable	Radiocarbon dating
Carbon-14	C^{14}	5730 years	Radiocarbon dating
Cesium-137	Cs^{137}	30 years	Gamma radiography
Lead-204	Pb^{204}	Stable	Determination of provenance of lead
Lead-206	Pb^{206}	Stable	and lead glass
Lead-207	Pb^{207}	Stable	
Oxygen-16	O^{16}	Stable	Study of paleotemperatures
Oxygen-18	O^{18}	Stable	Determination of provenance of marbles
Potassium-40	K^{40}	1.31×10^9 years	Potassium-argon dating
Thulium-170	Tm^{170}	127 days	Gamma radiography
Uranium-235	U^{235}	7.1×10^8 years	Source of energy in atomic reactors
Uranium-238	U^{238}	4.5×10^8 years	Fission-track dating, α-recoil dating

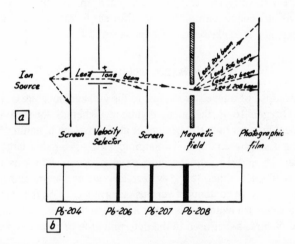

Fig. 5.2. Mass spectrometry: (*a*) mass spectrometer (schematic): a narrow beam of ions, all having the same velocity and the same electrical charge, is deflected by a magnetic field; atoms of greater mass are deflected less than lighter ones. (*b*) mass spectrum of lead.

STABLE ISOTOPES IN ARCHAEOLOGY

The isotopic composition of naturally occurring elements appears to be independent of their geochemical source. In a few cases, however, samples from different sources show slight variations in the relative abundance of various isotopes in them.

Lead Isotopes

Lead is unique in that it occurs with a large variety of isotope mixtures, whose relative proportions vary between ores from different geographical areas. These variations reflect differences in the geological ages of the ore deposits and stem from the fact that lead-206, lead-207, and lead-208 are formed as end products of the radioactive decay series of uranium and thorium (see Fig. 5.3).

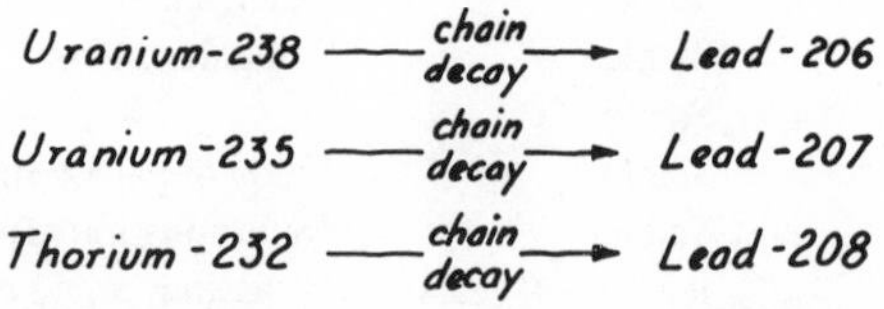

$$Uranium\text{-}238 \xrightarrow[\text{decay}]{\text{chain}} Lead\text{-}206$$

$$Uranium\text{-}235 \xrightarrow[\text{decay}]{\text{chain}} Lead\text{-}207$$

$$Thorium\text{-}232 \xrightarrow[\text{decay}]{\text{chain}} Lead\text{-}208$$

Fig. 5.3.   Lead is the end product of heavy-metal radioactive transformations. Its isotopes include the stable end products of three ''chain-decay'' series.

The measurement of the isotopic composition of lead contained in archaeological objects is thus helpful in tracing the location of the mines from which the lead could have been taken. The technique has been applied to a wide variety of archaeological objects containing lead in different forms such as metallic lead, leaded bronzes, and lead glasses (see Chapters 9 and 11).

Oxygen Isotopes

Two stable isotopes of oxygen of interest to the archaeologist are oxygen-16 and oxygen-18. Oxygen-16 is the more abundant, whereas the heavier isotope, oxygen-18, accounts for only 0.2 percent of all the oxygen in nature. Because the difference in mass between them—two atomic units—is a rather large percentage of the masses 16 and 18, and because oxygen occurs in forms that readily interchange with each other (e.g., water, ice, water vapor, and atmospheric oxygen), the oxygen-18 contents of various occurrences of oxygen are often quite different. Seawater, for example, is richer in the heavier isotopes than are fresh or atomospheric waters, a fact used in the determination of paleotemperatures (see Chapter 22).

The oxygen-18 contents of many natural minerals have been measured for archaeological purposes. Oxygen-18/Oxygen-16 ratios have permitted the identification of Greek marbles from several sources. The oxygen-18 content of glasses also reflects the contents of the ingredients from which they have been made and provides and independent means for their classification.

RADIOACTIVITY

Not all atoms are stable. The nuclei of the atoms of certain isotopes break down naturally; they are said to *decay* or to *disintegrate*, and the process of nuclear decay is called *radioactivity*. Unstable isotopes are also called *radioactive isotopes*.

Radioacitvity may be either natural or artificial. *Natural radioactivity* is a spontaneous process over which man has no control; it is exhibited by all heavy elements, that is, those with atomic numbers greater than 83 (among them uranium and thorium)—and to a lesser extent, by lighter elements such as potassium (atomic number 19) and carbon (atomic number 6). Artificial or *induced radio-activity* can be produced by nuclear reactions under controlled conditions: it involves unstable nuclei produced by the bombardment of stable nuclei with highly energetic particles (e.g., neutrons and protons) in specially designed nuclear devices.

The process of radioactivity can be neither seen, felt, or heard directly; it can, however, be recognized because radiation is emitted during nuclear disintegration. The radiation emitted can be detected and measured by special instruments such as the Geiger counter, or the ionization chamber. Because the emission of radiation can be easily detected with these instruments, very small amounts of radioactive substances can be detected and measured among large quantities of stable ones (see Fig. 5.4).

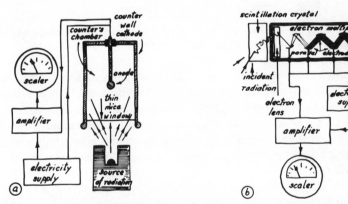

Fig. 5.4. Two types of radioactive radiation detectors: (*a*) Geiger-Muller counter (schematic): radioactive particles or radiation entering the Geiger-Muller tube produce a series of pulses of electric current; the number of such pulses is a measure of the intensity of radiation); (*b*) scintillation counter with secondary electron multiplier (schematic): radiation falling on the scintillation crystal produces flashes of light called "luminescence flashes"; these flashes are used to generate minute electric impulses that are "multiplied" and further ampliified prior to being measured).

RADIOACTIVE RADIATION

Radioactive substances possess in common the property of emitting energetic radiation. Three types of radiation are emitted: (1) alpha particles, (2) beta particles, and (3) gamma radiation (see Table 5.3 and Fig. 5.5).

TABLE 5.3   Types of Radioactive Radiation

Type and Symbol	Particle Emittted	Electric Charge	Mass (Atomic Mass Units)
Alpha α, He_2^4	Helium nucleus	+2	−4
Beta β, e^{-1}	Electron	−1	1/1840
Gamma γ	None	0	0

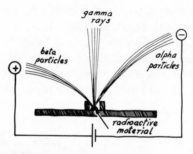

Fig. 5.5.   The three types of radiation emitted from radioactive materials. Alpha and beta particles are deflected by an electric field. Gamma rays are not.

Alpha particles consist of two neutrons and two protons packed together; each particle weighs four times the atomic mass unit. These particles are, in fact, nuclei of helium atoms that have been stripped of their electrons.

Beta particles are actually elecrons that originate in the nucleus from the destruction of a neutron: when a beta−emitting nucleus decays, a neutron in that nucleus changes into a proton and an electron; the latter is shot out of the nucleus as a beta particle.

Gamma rays have no charge; they are a form of energy, a wave motion similar to heat, light, or radiowaves. Gamma rays are similar to X-rays except that they have a shorter wavelength.

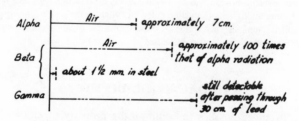

Fig. 5.6.   Penetrating power of alpha, beta, and gamma radiation.

The penetrating power of these three types of radiation is shown in Fig. 5.6. Radioactive radiation provides efficient, nondestructive means of investigation ideal for the study of antiquities.

Backscattering of Beta Particles

Backscattering of beta particles has been put to interesting uses, such as the determination of the composition of ancient glass and bronze or the thickness of layers of gilding (Asahina, Yamasaki, et al. 1958). The technique is very simple; a radioactive emitter of beta–particles is mounted in front of the surface to be analyzed and the *backscattered* (i.e., deflected) beta particles are determined with a suitable detector (see Fig. 5.7). Some of the beta particles striking the surface are

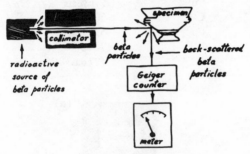

Fig. 5.7.   Backscattering of beta particles.

absorbed, whereas others are backscattered due to their interaction with the nuclei of atoms; the number of backscattered beta particles is a function of the atomic number of the elements in the surface. The most efficient backscatterers are elements of high atomic number.

The penetration of the particles into a particular surface is dependent on the energy of the beta particles of the source used: the higher the energy, the more penetrant the beta particles. Sulfur-35, thallium-204, and phosphorus-32 have been used as radiation sources, and the penetration of their particles varies in the range 0.1-1 mm (or more) in glass or enamel. Probably the most useful applications of beta backscattering so far have been in determining the lead content of leaded glass and in differentiating between soda-lime glasses and lead glasses (Emeleus 1960).

Gamma-ray Radiography

The high penetrating power of gamma-rays makes radioactive gamma-emmitters — such as those listed in Table 5.4—suitable complements of X-rays for radiographic purposes. The small size of gamma sources (only a few centimeters in diameter) makes them ideal for radiographing hollow or circular objects. The gamma source can be centrally placed inside the object, which in turn is surrounded by photographic film (see Table 5.4 and Fig. 5.8). This technique makes possible a complete radiograph on a single film, whereas for a complete X-ray examination, a large number of exposures would be required.

TABLE 5.4   Gamma-ray Emitting Isotopes Used as Radiography Sources

Isotope	Symbol	Half-life	Energy of Gamma Radiation (million eV)
Cesium-134	Co^{134}	2.1 years	0.48; 1.37
Cesium-137	Cs^{137}	30 years	0.66
Cobalt-60	Co^{060}	5.3 years	1.17; 1.33
Iridium-192	Ir^{192}	74 days	0.29; 0.61
Thulium-170	Tl^{170}	127 days	0.05; 0.08

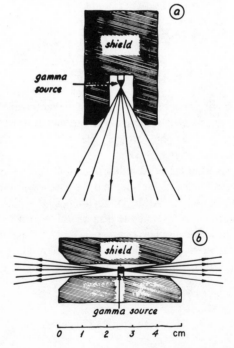

Fig. 5.8. Gamma radiation sources and shields: (*a*) for directional radiography; (*b*) for panoramic radiography.

Radiography of Large Stones

An unusual application of radiography using radioisotopes has been the examination for cracks of a fallen lintel stone forming part of a trillithon (= three stones) in the megalithic ring at Stonhenge (in Salisbury Plane, England). In 1957 the lintel fell from the top of two upright stones. Although replacement in its original position, was desired, it was feared that the stone might have been too badly cracked as a result of the fall. Some major defects were indeed visible, but it was not clear how deep they were.

Owing to the gigantic size of the stone (the shortest side measured some 1.5 metres and the longest, about 5.0 metres), breakage of the stone during reerection might have caused disastrous collapse. Ultrasonic flaw detectors failed to produce any results, and the great thickness of the stone prevented the use of portable X-ray apparatus. A strong sodium-24 radioactive source was used to provide enough high-energy gamma rays to radiograph the stone and provide a useful guide to the presence of cracks (Hinsley 1959). The gamma rays penetrated through the stone, blackened an X-ray film, and produced a sufficiently definite picture. After examination of the film it was concluded that the stone was sound enough to lift. This conclusion was vindicated by the eventual safe completion of the operation.

Radiography of the Pyramid of Chephren at Giza

The usefulness of nondestructive testing in the investigation of large solid buildings is illustrated by the search for chambers in the Egyptian pyramids. There are two chambers in the pyramid of Chephren's father (Cheops) and the same number in the pyramid of his grandfather (Sveferu). The presence of these chambers

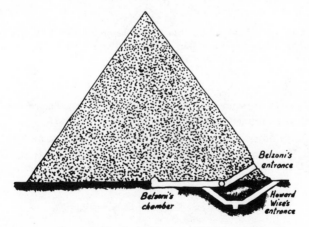

Fig. 5.9.   The Pyramid of Chephren. Section looking west. The present height of the pyramid is 137m (originally 144m). Each side meaures 213m (originally 217m).

suggested that similar structures might also be present in the second pyramid, of Chephren, at Giza (Fig. 5.9), even though none had actually been found in the apparently solid structure.

In an attempt to detect and locate known chambers, use was made of *cosmic ray* mapping (Alvarez, Anderson, et al. 1970). Cosmic rays are a highly penetrating type of radiation reaching the earth from outer space; detectors of cosmic rays were installed in the Belzoni chamber just below the base of the Chephren pyramid near its center, and cosmic-ray measurements were taken over a period of several months. In this way 19 percent of the pyramid's volume was examined and the results obtained were mapped by two different methods: (1) use of simultated X-ray photographs and (2) numerical analysis of the detected radiation.

The mapping of the pyramid structure by the simulated X-ray photograph technique is identical to what would be obtained by X-raying a small model of the pyramid with an X-ray source located in the Belzoni Chamber and with an X-ray film touching the peak of the model pyramid, as shown in Fig. 5.10.

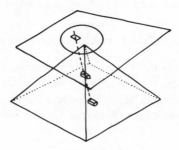

Fig. 5.10. Cosmic-ray search of the Pyramid of Chephren. Diagram showing the projection technique used to produce a simulated X-ray photograph. The plane on the top of the pyramid can be thought of as the plane of an hypothetical photographic film.

The cosmic-ray measurement clearly showed the four diagonal ridges of the pyramid and also outlined, through more than 100 m of limestone, the shape of the cap of original limestone facing blocks that gives the pyramid its destinctive appearance. However, no chamber with volumes similar to the known chambers in the pyramids of Cheops and Sveferu were detected in the mass of limestone.

Gamma-ray-Induced Polymerization

Penetrating gamma radiation can be used to initiate chemical reactions. Small *monomer* (liquid plastic materials) molecules, for example, can be made to join together into large stable ones. In this process, called *polymerization,* the molecules of the monomer, induced by irradiation with gamma rays "weld" together into large molecules called *polymers* or *plastic materials,* familiar to everyone.

Decaying stone and wood can be inpregnated with liquid monomers, and the water in waterlogged wood can be exchanged with suitable monomer solutions (Stoia, Paun et al. 1976). Irradiating the stone-monomer of wood-monomer combinations with gamma-rays causes the monomer molecules to polymerize within the ancient materials. The results are stone-polymer and wood-polymer combinations, much stronger and much less vulnerable to environmental changes than the original material (de Nadaillac 1972; Eymery and de Nadiallac 1972).

Mössbauer Spectroscopy

Mössbauer spectroscopy is a technique that has in recent years found application in archaeological research. It provides information on the effect of the immediate chemical and physical environment on the nuclear properties of certain chemical elements in solid materials and is based on a group of complex nuclear phenomena known collectively as the *Mössbauer effect.* It is not necessary to discuss the Mössbauer effect in detail here. Those aspected of it that are of interest to archaeologists are concerned with the absorption of gamma-rays by the atomic nuclei or certain elements.

Forty-five of the more than 90 naturally occurring elements have isotopes (see Chapter 2) whose atomic nuclei are able to absorb energy in the form of gamma radiation becoming "excited" in the process. Only gamma rays of exactly the right energy can be absorbed by a given isotope. This is the basis of Mössbauer spectroscopy, the instrumentation for which is shown in Fig. 5.11.

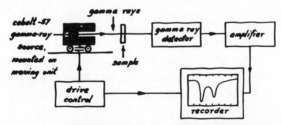

Fig. 5.11. Mössbauer effect measurements. Block diagram of the instrumental setup.

Gamma rays from a specially prepared radioactive source are allowed to pass through a thin layer of sample. The rays that are not absorbed by the sample fall upon a radiation detector. Naturally occurring isotopes in the sample absorb gamma radiation; the energy of the radiation absorbed is very precisely specific for each respective isotope. The slightest variation in the energy of the impingent radiation is enough to upset the absorption condition (called the *resonance absorption*). Any motion of the radiation source or of the absorbing sample relative to each other, produces a minute dispersion of the energy of the gamma rays, which is enough to upset the resonance conditions. Only at certain very specific relative velocities will the energy of the source exactly equal the absorption energy of the sample; at these velocities, resonance will occur.

During testing, the radiation source and the absorbing sample are moved at varying velocities relative to each other, and the degree of energy absorption is measured and recorded as a function of velocity. The positions of resonance absorption on the resulting spectrum can then be ascertained. Figure 5.12 illustrates Mössbauer spectra obtained from medieval pottery from Cheam (in Surrey, England) (Cousins and Dharmawardena 1969).

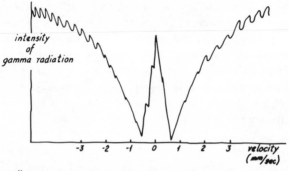

Fig. 5.12.   Mössbauer spectra obtained from pottery. The intensity of gamma-radiation transmitted varies with the relative velocities of sample and radiation source.

The information derived from Mössbauer spectroscopy measurements is based on nuclear phenomena. Hence the technique is restricted to materials in which there are elements that have naturally occurring isotopes with specific nuclear properties. Only 45 elements have such isotopes. Of these, iron is the most suited to Mössbauer studies in archaeological materials, although tin, zinc, gold and nickel may also be useful.

Two factors limit the wide application of Mössbauer spectroscopy: (1) the rather small number of elements whose isotopes can be studied without special facilities and (2) the paucity of information obtained from the spectra, which cannot be properly interpreted in the absence of other data about the compound concerned. However, studies such as those of iron in ceramics (see Chapter 8) can be used to gain information that would not readily be obtained by other methods. (Tominaka Takeda et. al 1977). It can also be of use in the study of pigments, paintings and terracotta statuary (Keisch 1976).

Radiation Damage

The interaction of radioactive radiation with matter causes changes in the properties of the matter. The energy carried by alpha, beta, and gamma rays passing through solids is expended in inducing electronic excitations, causing dislocations of electrons and leading to the increased chemical reactivity of atoms. Heavy particles moving through insulating solids, for example, leave behind narrow, continuous trails of damage that can readily be made visible by selective chemical treatment.

Radiation damage resulting from the interaction of nuclear radiation with insulating solids—minerals, pottery, enamel, and glass—is of interest to the archaeologist, for it provides method for dating events.

Fission Tracks and Alpha-Recoil Tracks Dating

α-Decay of uranium and thorium is accompanied by the emission of α-particles and the recoil of decaying nuclei. Both the alpha particles emitted and the recoiling nuclei from impurities of these elements in insulating solids produce very short *tracks* that may be useful for dating glass, ceramics and other materials of archaeological importance (see Chapter 20).

Thermoluminescence Dating

Radiation damage can also cause the freeing of electrons from ions in insulating solids. In this case the energy of the radiation is partly converted into electron displacements that are stored as potential energy in the solid's lattice. When these insulating solids are heated to a temperature below that of incandescence, they release the stored energy by emitting visible light, called *thermoluminescent glow*. The intensity of the thermoluninescent glow can be measured, and it has been shown to be a function of the age of the solid and of the rate of emission of nuclear radiation (see Fig. 5.13). It is also dependent on other factors, such as the thermal history of the material.

In natural clays the most likely sources of radiation to cause damage are radioactive elements—uranium, thorium, and potasisum-40—which occur in the clay as natural impurities; in effect, natural clay exhibits thermoluminescent properties, acquired over millions of years. When clay is heated above 400°, the thermoluminescence is removed. The firing of clay into pottery—at temperatures of 700-1000°C—suffices to remove the naturally acquired thermoluminescence. Thus firing sets back the "thermoluminesce clock" to zero time. Thereafter, the thermoluminescence of archaeological ceramics begins once more to grow with time. Measurement of its intensity makes the determination of age, that is, of the time elapsed since the firing date, possible. The absolute accuracy of thermo-

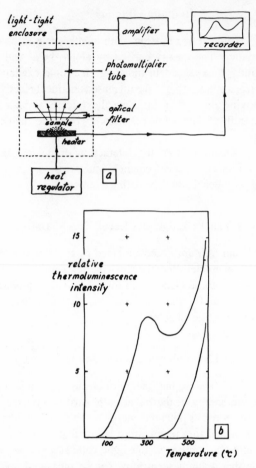

Fig. 5.13. Thermoluminescence: (a) block diagram of the instrumentation used for measuring thermoluminescence; (b) intensity of thermoluminescence emitted by a sample, shown as a function of increasing temperature.

luminescence dating has not yet been fully evaluated. It should be possible however, to use thermoluminescence for dating of archaeological ceramics, with an accuracy better than ∓5 percent (Aitken, Zimmerman, et al. 1968; Mejdahl 1972; Seeley 1975; Cairns 1976). Thermoluminescence measurements of ceramics can be used to establish whether objects are genuine or merely fakes or later copies (see Chapter 23). The thermoluminescence of marbles is also of value in restoration work: it makes it possible to ascertain whether two marble samples originated from the same block. Thermoluminescence measurements have been used to reassemble statues from fragments (Aforkados, Kessar, et al. 1974; Aforkados 1975).

DECAY OF UNSTABLE ATOMS

Unstable atoms *decay;* that is, they emit radioactive radiation and in so doing change into other kinds of atoms. If the new atoms are stable (e.g., when carbon-14 decays to nitrogen-14), the process is known as *simple decay.* The products of decay may, however, be themselves radioactive and subsequently decay in their turn. Thus radioactive atoms may decay in stages—a process called *chain decay*—until they finally end up as stable atoms (see Fig. 5.14).

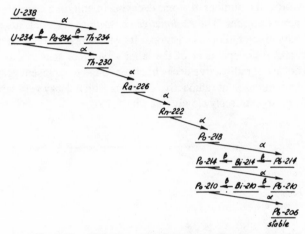

Fig. 5.14.   Chain decay. The decay series of uranium 238.

The product of the radioactive decay of a particular isotope is always the same. This means that all the disintegrating atoms undergo a similar loss of alpha or beta particles and release of gamma radiation. In some cases, however, radioactive isotopes decay through a process known as *branching decay,* to produce two different *daughter* isotopes: potassium-40, for example, decays mostly to calcium-40, but some of it decays to argon-40, a form of decay that is the basis of the potassium-argon method of dating.

LAW OF RADIOACTIVE DECAY

Using methods devised to measure the radiation emitted and the number of particles given off during radioactive decay, it has been established that all radioactive isotopes decay at a characteristic, steady rate, whereby the intensity of the radiation decreases in geometrical progression as time increases arithmetically. Since the intensity of radioactive radiation is a function of the number of atoms disintegrating, it follows that the number of radioactive atoms decreases in geometrical progression with increasing time. In mathematical terms the rate of radioactive decay is expressed by the equation:

$$N_t = N_0 e^{-\lambda t}$$

Where N is the number of atoms present at any arbitrary zero time; N_t is the number of atoms remaining after the lapse of a time interval t; e is a numerical constant $=$ 2.7183; λ is a constant called the *radioactive constant*, characteristic for each isotope; and t is a measured time interval.

In other words, the number of atoms decaying in unit time is proportional to the number of atoms present. The *radioactive constant* (λ) is a definite and specific property of any given radioactive isotope. Its value depends only on the nature of the isotope and is independent of the latter's physical state or chemical combination. The law of radioactive decay has been found to represent very accurately the rate of disintegration of radioactive elements which decay very rapidly as well as those that do so extremely slowly (see Fig. 5.15).

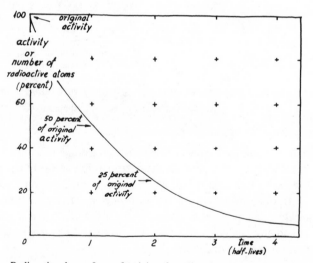

Fig. 5.15. Radioactive decay. Loss of activity of a radioactive substance with time. All radioactive substances decay in the same manner; only their half-lives differ.

HALF-LIFE

A convenient way of expressing the rate of disintegration of radioactive substances is the *half-life* ($t_{1/2}$). It is based on the time required for one-half the number of unstable atoms in a sample to disintegrate. The half-life is the same regardless of the size of the sample and is characteristic for each element. Since radioactive decay is unaffected by chemical or physical conditions, the half-life of a radioactive substance under any physical condition or in whatever chemical state remains the same. Different radioactive isotopes have different half-lifes; some are measured in millions of years, others are only a fraction of a second. Potassium-40 (half-life 1.31×10^9 years), which is present in ordinary stable potassium in the amount of 0.01 percent, has presumably been left over from the time when the primordial elements were formed. Carbon-14 (half-life 5730 years), on the other hand, of which there is only about one atom to every million-million (10^{12}) atoms of stable carbon-12, is continuously being formed at a steady rate in the upper atmosphere, where cosmic rays cause the transmutation of nitrogen into carbon-14.

Nuclear Chronometry

One of the most interesting applications of the law of radioactive decay is in the determination of the age of geological and archaeological events (Giletti and Lambert 1959). The rate of decay of radioactive isotopes is independent of environmental conditions. Therefore, if the products of radioactive decay or the damage caused by the decay are preserved in a closed system, it is possible to calculate the time elapsed since the beginning of the decay process. Methods of time measurement dependent on the decay of naturally occurring isotopes, or on the damage they cause to solids, provide absolute dates of events in the distant past. The basic techniques of nuclear dating are summarized in Table 5.5. They are discussed later in the section on dating.

TABLE 5.5   Nuclear Dating

Measurement	Dating Technique
Radio of daughter element to residual parent— radioactive isotope still present	Potassium-argon dating
Present radioactivity of an isotope	Radiocarbon dating
Radioactive damage to solids	Thermoluminescence fission tracks, alpha recoil tracks

NUCLEAR FISSION

Most radioactive elements decay, through the emission of nuclear particles, into others, the so-called *daughter* elements. The nuclei of some very heavy atoms, however, undergo the process of *nuclear fission*. This is the disintegration of a heavy nucleus into two lighter nuclei of roughly equal size, accompanied by the liberation of a considerable amount of energy. Nuclear fission is a very rare natural occurrence, but it can readily be brought about artificially by neutron induction (see Fig. 5.16).

Fig. 5.16.   Model of an atomic nucleus undergoing fission. The energy present in a fissionable atom is sufficient to cause the nucleus (*a*) to become unstable. The unstable nucleus begins to change in shape (*b*) and (*c*), until the mass splits into two fragments (*d*). Not all nuclei divide in the same way, but in all cases the total number of protons and neutrons is unchanged before and after division.

Spontaneous nuclear fission is of rare natural occurrence because there is a very small probability of its happening: the two fission fragments are strongly bound to each other, more so, for example, than in alpha-decay. The relative likelihood of spontaneous fission and of alpha-decay can be appreciated from the fact that in one gram of uranium-238 there occur about 750,000 alpha-decays every minute, but only one spontaneous fission every two minutes! Despite the rarity of its occurrence, the spontaneous fission (i.e., fission without excitation from external sources) of uranium-238 has found application as a dating technique (fission-track dating, discussed in Chapter 20).

Nuclear fission is brought about artificially by causing the nuclei of heavy atoms—such as uranium-235 or uranium-238—to be struck by neutrons traveling at speed. A nucleus struck in this way will split into two or more fragments. The splitting of the nucleus is accompanied by radioactive emission and by the release of additional neutrons, which can themselves cause further fission and produce more neutrons. The continuation of this process is known as a *chain reaction* (Fig. 5.17). *Nuclear reactors* are devices in which self-sustained nuclear-fission chain reaction take place under controlled conditions. The energy generated by nuclear reactors is used for industrial processes, and the neutrons generated in them may find application in scientific research techniques such as neutron radiography and activation analysis.

Neutron Radiography

Neutrons have special usefulness in radiograhpic techniques because they penetrate material barriers in a manner completely different from X-rays or other forms of electromagnetic radiation. Elements with high atomic numbers and high

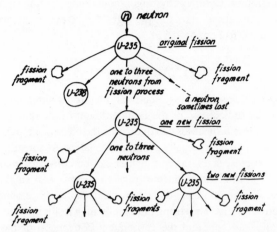

Fig. 5.17. A neutron-fission chain reaction. The collision of a neutron with a fissionable atom causes fission of the latter, producing a number of neutrons that in turn cause further fissions.

densities, such as lead or iron, arrest the penetration of X-rays much more effectively than do elements with low atomic numbers and low densities, such as hydrogen, oxygen, and carbon. Neutrons, on the other hand, are easily transmitted through materials containing heavy elements but are strongly absorbed by some light elements (e.g., boron or hydrogen). A layer of water, for instance, is a more effective barrier to a beam of neutrons than a sheet of lead of similar thickness. Boron has such a high neutron absorbing power that if even traces of this element are present in the glaze of a particular piece of pottery, its shards can be readily identified.

For radiographic work the useful properties of neutron beams are:

1. penetration through metals, particularly durable artifacts of lead or lead-bronze
2. sensitivity to organic materials of all kinds
3. high sensitivity to common materials such as water or oil, and to the presence of boron.

The utility of neutron radiography lies also in the ability of neutron beams to detect differences between materials that have practically identical X-ray-absorption characteristics.

Neutron radiography provides a useful complementary technique to X-rays for studying the internal features of archaeological specimens. The detection of hydrocarbon substances (e.g., oil) inside a metallic container could be difficult, if not impossible, by any method other than neutron radiography. Inaccessible internal passages, crevices, or dislocations in a solid mass can be easily examined by neutron radiography. Common hydrogen-containing liquids, such as water or paraffin, can be introduced into the passages and crevices to give increased radiographic contrast (Robertson 1975).

NUCLEAR REACTIONS

Energy-carrying particles—protons, neutrons, electrons, alpha-particles, and so on—can be used as projectiles for the bombardment of atomic nuclei. If given sufficient energy, these projectiles can overcome the electrical repulsion forces of the target nuclei and can then actually penetrate the latter. As a result of interaction with energetic particles, stable atomic nuclei can be converted into radioactive nuclei. Many different kinds of nuclear reactions have been studied. Neutrons carry no electric charge and can, therefore, easily approach atomic nuclei and interact with them strongly. The strong interaction may result in the absorption of a neutron by a stable nucleus and the consequent conversion of the latter to an unstable one (i.e., radioactive), which will disintegrate at a given rate. In nature, neutrons produced by the collision of cosmic particles in the upper atmosphere themselves collide with nuclei of nitrogen. The following reaction takes place:

$$\text{Nitrogen-14} + \text{neutron} \rightarrow \text{Carbon-14} + \text{proton}$$

The nitrogen-14 nucleus absorbs the neutron, and a proton is ejected as a result of this strong interaction. The new resulting nucleus is that of radioactive carbon-14, an isotope that constitutes the base of a whole branch of archaeology: *radiocarbon dating*.

Neutron Activation

The bombardment of nuclei with neutrons may be carried out artificially in the laboratory. In a nuclear reactor, for example, neutrons may be made to strike a target and penetrate it; as a result of the capture of neutrons by the target nuclei, new radioactive nuclei of higher atomic weight are obtained. Artificially produced radioactive nuclei decay and emit radiation in unique ways. Decay almost always takes the form of simultaneous emission of beta-particles and gamma-rays. The gamma-radiation emitted by neutron-activated nuclei can be measured, and the kind and the number of radioactive nuclei present can be determined from its characteristics. This method of activation and measurement is called *neutron-activation analysis*. It takes advantage of the extreme sensitivity of detection associated with radioactive measurements and, as far as archaeology is concerned, provides a method for the determination of elements at trace level in coins, obsidian, pottery, and other materials.

The sensitivity of the detection depends on the intensity of the neutron source, the ability of the elements analyzed to capture neutrons, and the half-life of the induced radioactivity. In a nuclear reactor, as little as 10^{-10} g of an element can be detected in favorable cases. Some elements, on the other hand, are not better detected by neutron activation than by other methods of chemical analysis (Gilmore 1976).

Neutron Activation Autoradiography

Information concerning the structure and composition of pigments used in paintings can be obtained through the use of *neutron-activation autoradiography* (Sayre and Lechtman 1968; Lechtman, 1966; Cotter and Sayre 1971; Sayre 1976).

Moderate radioactivation induced by neutron bombardment of paintings produces temporary radioactivity sufficiently strong to expose photographic films; the results are called *autoradiographs* and resemble X-ray radiographs. However, since the radioactivities induced are characteristic of the different elements in the pigments in the paintings, the decay occurs in different manners and at different rates, characteristic of each element. Thus a series of distinct autoradiographs can be obtained at varying times following the original activation. These autoradiographs not only permit the identification of the pigments used, but also yield information on the painting technique.

REFERENCES

Aforkados G., Kessar, A., and Demetrios, M., *Nature,* **250,** 48 (1974).

Aitken, M. J., Zimmerman, D. W., and Fleming, S. D., *Nature,* **219,** 442 (1968).

Aitken, M. J., *Phil. Transact. Roy. Soc. Lond.,* **A269,** 77 (1970).

Alvarez, L. W., Anderson, J. A., El Bedwei, F., Burkhard, J., Fakhry, A., Girgis, A., Goneid, A., Hasan, F., Iverson, D., Lyneh, G., Miligi, Z., Hilmi, A. M., Sharkawi, M., and Yazolino, L., *Science,* **167,** 832 (1970).

Asahina, T., Yamasaki, F., and Yamasaki, K., in R. C. Exterman, ed., *International Conference on Isotopes in Scientific Research,* Vol. 2, Pergamon, London, 1958, p. 528.

Cairns, T. *Anal. Chem.,* **48,** 3 (1976).

Cotter, M. J., and Sayre, E. V., *Bull. Internat. Inst. Conserv. Am. Soc.,* **11**(2), 91 (1971).

Cousins, D. R., and Dharmawardena, K.6, *Nature,* **223,** 732 (1969).

de Nadaillac, L., Commissariat a l'Energie Atomique, France, Report No. SAR-6/72-71/LN, 1972.

Detanger, B., and de Nadaillac, L., Commissariat a l'Energie Atomique, France, Report No. SAR-6/72-20/LN, 1972.

Emeleus, V. M., *Archaeometry*, **3,** 5 (1960).

Eymery, R., and de Nadaillac, L., Commissariat a l'Energie Atomique, France, Report No. SAR-6/72-22, 1972.

Fleming, S. D.,*Naturwissenschaften*, **58,**333 (1971).

Giletti, B. J., and Lambert, R. S. J., *Research*, **12,**368 (1959).

Gilmore, G. R., *Proc. Anal. Div. Chem. Soc.*, 99 (1976).

Hinsley, J. F., *Non Destructive Testing*, Vol. I, Macdonald and Evans, London, 1959.

Keisch, B., *Appl. Mossbauer Spectros.*, **1**, 263 (1976).

Lechtman, H. N., M.A. Thesis, Institute of Fine Arts, New York University, N. Y., 1966.

Mejdahl, V., *Archaeometry*, **14,** 245 (1972).

Ralph, E. K., and Han, M. C., in R. H. Brill, ed., *Science and Archaeology*, MIT Press, Cambridge, Mass., 1971, p. 244.

Robertson T. J. M., *Non Destructive Testing*, Vol. II, Macdonald and Evans, London, 1975, p. 17.

Sayre, E. V., *Technol. and Conserv. of Art, Architecture and Antiquities*, **1**(3), 26 (1976).

Sayre, E. V. and Lechtman, H. N., *Studies in Conservation*, **13,** 161, 1968.

Seeley, M. A., *J. Archeol. Sci.*, **2,**17 (1975).

Stoia, N., Paun, J., and Vultureanu, M., *Comun. Stiint. Simp. Biodeterior. Clim.*, 6th, **1,** 49 (1976).

Toishi, K., *Radioisotopes (Tokyo)*, **11,** 337 (1962).

Tominaka, T., Takeda, M., Mabuchi, H., and Emoto, Y., *Radiochem. Radional. Lett.*, **28,** 221 (1977).

Zimmerman, D. W., *Archaeometry*, **13,** 29 (1971).

Recommended Reading

Schultz, S. L., and Schultz, V., *Nuclear Technology in Archaeology*, Technical Information Centre, TID 3920, Oak Ridge, Tenn., 1975.

Various Authors, *Application of Nuclear Methods in the Field of Works of Art*; Proceedings of an Internationsl Congress, Roma-Venice, May 1973, Acc. Nazionale Dei Lincei, Italy, 1976.

PART

II

MATERIALS AND TECHNOLOGIES

Man has been, from the beginning, a maker of artifacts. This has involved him in an unending search for materials from which to make them. Stones, minerals, wood, skins, oil, and many other materials have served him from earliest times in their naturally available forms.

Techniques for modifying natural materials—by purifying, refining, and mixing them, or by causing them to interact with each other and thereby improving their properties—were evolved through the ages. These led in their turn to the discovery of synthetic materials: ceramics, glass, metal alloys, and many others offered advantages over naturally occurring ones in terms of performance, workability, durability and appearance.

STONE

Rocks are among the most plentiful of natural materials. Rocks are classed according to their mode of formation into three groups: (1) *igneous* or *primary rocks*, formed by solidification of molten masses from within the earth, (2) *sedimentary* rocks, originating in sediments of eroded materials, deposited under water, and (3) *Metamorphic* rocks, that is, igneous or sedimentary rocks that have been changed in character by the movement of the earth's crust, by heat, or by the chemical action of liquids and gasses.

All rocks are composed of one or several kinds of minerals, which determine their respective chemical and physical properties. Any book on physical geology will provide ample information on the properties of rocks.

Rock that has been cut, broken, or otherwise shaped to serve some human purpose is known as *stone*. For hundreds of thousands of years, before the discovery of metals, men worked stones into tools and artifacts. In time, men also learned to use stones for building, flooring, and roofing, construction of monuments, carving statuary, polishing into attractive gems, and many other uses. Hence it is not surprising that stone is one of the materials most frequently encountered in archaeological research.

PROVENANCE OF ARCHAEOLOGICAL STONE

Since the end of the nineteenth century, archaeologists have been interested in determining the geographical sources of stone used in ancient times in building, statuary, and the production of artifacts. (Lepsius 1890; Washington 1898). This information can be of great help in the study of ancient trade patterns.

However elaborate the processes of cutting, sizing, shaping, and polishing that may be applied to stones, the basic physical and chemical properties of the material remain unchanged and may be used to characterize the material. A large number of physical and chemical methods are available for the characterization of rocks and stones. In one well-known case, geological evidence was brought forward to prove that in Neolithic times, large stones from Pembroke in Wales were transported a distance of about 200 miles to Wiltshire in England, where they were used to build the Stonhenge Megalithic Ring (Newall 1959). The morphology and petrography of marbles have, at times, provided useful results. From the petrofabric analysis of samples from Greek quarries it was concluded that most marbles have distinctive fabric pictures (Weiss 1954).

If the distribution of trace elements throughout a rock deposit is uniform, it is reasonable to expect that chemical ''fingerprinting'' will make it possible to trace the provenance of stone-made implements. Emission spectrography, neutron activation analysis, and X-ray fluorescence have been successfully used for this purpose (see later this chapter).

OBSIDIAN

Obsidian is a natural glass of volcanic origin. Its chemical composition is similar to that of granite; the typical composition of an obsidian sample is listed in Table 6.1. Obsidian is hard and brittle and breaks with a conchoidal fracture like glass; hence

TABLE 6.1   Typical Composition
of an Obsidian Sample

Element	Percent
Silicon	43.0
Oxygen	42.0
Aluminum	7.3
Potassium	3.7
Sodium	2.9
Iron	2.3
Calcium	1.0
Titanium	0.3
Magnesium	0.3

it can be chipped like flint and fashioned into sharp-edged tools. Obsidian is usually black, due to the presence of microscopic embrionic crystallites. Colored varieties, although rare, are also known; brown is due to iron impurities and light grey may be due to the presence of tiny gas bubbles (see Fig. 6.1).

Owing to the ease with which smooth curved surfaces and sharp edges can be obtained, obsidian knives and scrapers were used by men as early as 30,000 years ago. Other tools, artifacts, and weapons were also made by primitive men from obsidian.

Fig. 6.1. A core of Obsidian.

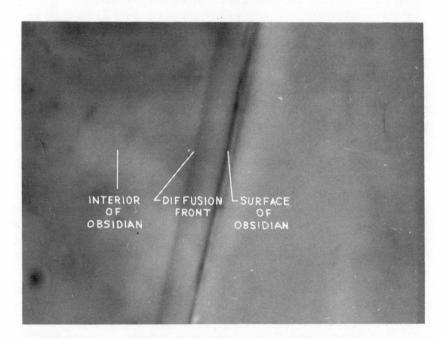

Fig. 6.2. Hydration of obsidian. Photomicrograph of hydration rim.

Hydration of Obsidian

Obsidian usually contains less than 0.3 percent water. Exposed surfaces however, undergo cortical alteration through hydration; water from the surroundings diffuses into the interior. This is a weathering process to which all glasses, whether natural or artificial, are subject. The hydrated zone on the obsidian surface contains over 10 times more water than the interior. It is referred to as the *hydration layer*.

The absorption of water causes a change in volume, accompanied by physical strains and by changes in the index of refraction of the material. Because of this, the interface between the hydrated and nonhydrated regions stands out sharply when viewed under a microscope and the depth of the hydration layer can be measured. (Friedman and Smith 1963; Freidman and Trembour 1978) (see Fig. 6.2).

Hydration commences when obsidian is chipped or flaked. It continues at a measurable rate that is dependent on a number of factors: composition of the obsidian, temperature, soil chemistry (if the obsidian is buried), and solar radiation (if exposed). Surface-collected obsidian artifacts seem to hydrate at a much faster rate than do excavated artifacts (Layton 1973). Surprisingly, relative humidity does not seem to affect the rate of hydration appreciably.

Hydration Dating of Obsidian

The thickness of the hydration layer increases at a rate given by the equation

$$T = kt^{1/2}$$

Where T is the thickness of the hydrated layer, t is time and k is a constant. Thus it is possible to calculate hydration rates: (1) if the ground temperature where the observed sample has been buried is known or can be estimated; (2) if the silica content or refractive index of the obsidian are known (Friedman and Long 1976).

Measurement of the thickness of the hydration layer of obsidian provides an estimate of the time elapsed since the surface was exposed by working. The technique may be used for relative dating of artifacts. It can also be applied to establish the reuse of artifacts, for testing stratigraphic sequences, or for the construction of artifact assemblages (Michels and Bebrich 1971) (see Table 6.2 and Fig. 6.3).

A nuclear reaction has been developed to measure the penetration of water into obsidian. This allow the measurement of much thinner hydration layers than can be measured optically (Lee, Leich et al., 1974).

Obsidian dating using hydration layers is not appicable to all chronological problems. In some regions the method is of little value. But in sites where no other

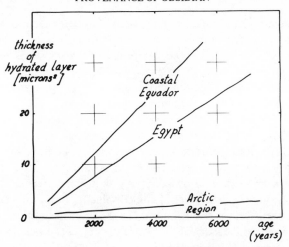

Fig. 6.3. Hydration of obsidian: the rate of hydration of obsidian of known age. Measurements were made on obsidian artifacts originating in areas of different climatic conditions.

TABLE 6.2   Correlation of Obsidian Hydration and Radio carbon Dates from West Mexico

Sample	Carbon-14 Date (Years Ago)	Hydration-layer Date (Years Ago)
1	1950	1732
2	1500	1647
3	1600	1647
4	2025	1846
5	2128	2272

evidence of age is available it may provide a significant aid in dating (Ericson 1976).

PROVENANCE OF OBSIDIAN

Obsidian—usually in the form of rather large cores—was a widely traded raw material in prehistoric times. The final purchaser worked the cores into objects. Obsidian tools and artifacts have been found at many early sites in the Middle East, Europe, and America, although only a few geological sources of obsidian are known; these are areas of volcanic activity containing obsidian flows that form the upper part of lava beds.

The investigation of obsidian trading patterns has long been a preoccupation of archaeologists and prehistorians. In obsidian from a given source, the chemical properties have been found to be unaffected by the presence of striation or banding. Geochemical studies have shown that obsidian flows are usually uniform in chemical composition. Neutron-activation analysis of a group of samples from a flow in California, USA, provided evidence of remarkable chemical homogeneity. In another group of samples there were considerable compositional variations, though the coherence in these variations was extraordinary: the assignment of provenance from these results was as definitive as it would have been if the flow were homogeneous (Bowman, Asaro, et al., 1972, 1973).

The physical properties of obsidian vary in a more irregular way. Density measurements, for example, were of limited value for identifying archaeological obsidian samples (Reeves and Armitage 1973). Attempts to identify obsidian on the basis of the refractive index of New Zealand samples showed extensive overlap and inconclusive results (Green 1962).

It follows, therefore that to identify obsidian with a particular geological source, two assumptions should first be made: (1) different flows are distinguishable from one another by their chemical composition; and (2) the chemical composition of a single obsidian flow is uniform throughout. If these assumptions hold, it should be possible to relate tools that have been distributed by trade over wide areas, to their sources.

Methods of Obsidian Analysis

Obsidian is an almost ideal material to study provenance. Except for a very thin surface layer (the hydration layer; see earlier), the chemical composition of obsidian is unaltered by weathering or burial. The most unambiguous identification is obtained by determining the trace-element composition. Emission spectroscopy (Renfrew, Cann, et. al. 1965; Green, Brook, et al. 1967), X-ray fluroescence (Parks and Tieh 1966), (Bowman, Asaro, et al. 1972, 1973) and neutron activation analysis (Stross, Weaver et al. 1968; Gordus, 1967; Gordus, Wright, et al. 1968; Gordus, Griffin et al. 1971) have all been successfully used for obsidian identifications.

Obsidian Trade in the Mediterranean Sea Area

In the Near East and the Mediterranean Sea area, obsidian flows are found only around Italy, in some islands in the Aegean Sea, and in certain localities in Turkey, Iran, and Armenia. Obsidian from these sources, and obsidian artifacts from many sites in the Near East and Mediterranean Sea region were analyzed spectrographically (Cahn and Renfrew 1964; Dixon, Cann, et al. 1968; Renfrew, Dixon, et al. 1968; Durrani, Khan, et al. 1971). A large number of trace elements was determined in each sample. Zirconium and barium showed the greatest quantita-

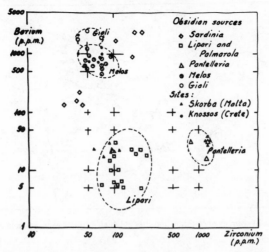

Fig. 6.4.   Relative concentrations of barium and zirconium in obsidian from different sources in the Mediterranean region. The origin of obsidian used to make artifacts found at various sites can be deduced from the graph.

tive variation, between samples from different sources. Their relative concentrations are good indicators of source origin, as can be seen in Fig. 6.4. The findings reveal quite clearly the distribution of obsidian tools in relation to the sources from which cores were obtained. Obsidian from archaeological sites in Turkey, Palestine, and Cyprus came from sources in central Turkey. Obsidian found in Mesopotamian and Persian sites originated in Armenian sources. In Syrian sites, obsidian from both sources was found.

In the Mediterranean sea area, Maltese settlers in the Neolithic period were accomplished seafarers, traveling to Sicily, Lipari, and Pantelleria to obtain raw obsidian for the manufacture of tools. No tools made from Aegean obsidian were found in Malta, indicating that the latter island's inhabitants probably had little or no contact with settlements in the Aegean area (see Fig. 6.5).

In some groups of tools the contents of barium and zirconium are similar, and the evidence for their provenance is therefore not conclusive. Samples from Palmarola, for example, are similar to those from Lipari, as can be seen on the barium-zirconium graph (Fig. 6.4). In these cases, the presence of another element, such as cesium in Palmarola obsidian, serves additionally to distinguish the particular source of the material.

Difficulties arise when specimens of obsidian from several sources show convergent compositions. This is observed in Greece, where obsidian from sources in the island of Melos resembles in composition that from the island of Giali, nearly 400 km further east. As an additional complication, both resemble

Fig. 6.5. Location of sources of obsidian and sites at which artifacts were found in the Mediterranean and Near East regions.

TABLE 6.3 Fision-track Ages of Some European Obsidians

Sample	Age (Million Years)
Obsidian tools found at Franchthi, Southern Greece	
Sample 1	8.48 ± 0.55
Sample 2	8.82 ± 0.57
Sample 3	9.33 ± 0.60
Obsidian sources	
Hungary	
Borsod	3.86 ± 0.24
Borsod	3.37 ± 0.27
Central Anatolia	
Bor	2.29 ± 0.32
Acigol	1.95 ± 0.33
Acigol	8.14 ± 0.59
Aegean	
Giali	2.01 ± 0.26
Giali	8.04 ± 0.65
Melos, Adhamas	8.95 ± 0.94
Melos, Adhamas	8.54 ± 0.73
Melos, Dhemenegaki	8.35 ± 0.72
Melos, Dhemenegaki	2.36 ± 0.53

obsidian from Hungary and from the Anatolian source at Acigol, both more than 1000 km distant. To overcome these difficulties, the date at which the obsidian was formed by volcanic eruption and its uranium content, were determined by fission-track analysis (Durrani, Kahn, et al. 1971). (The date of formation of the obsidian is not to be confused with the date of manufacture of the artifacts, which were made by chipping, not melting, the rock). Table 6.3 records the ages of obsidian samples from sources in Hungary, Central Anatolia, and the Aegean Islands, together with those of obsidian artifacts found in a grave at Franchthi in the Argive Peninsula in southern Greece.

The results shown in Table 6.3 are consistent with the tools having originated in the source at Melos. The possibility of unrecorded obsidian sources in mainland Greece cannot be excluded, although it is most unlikely that the latter should match the Melian source in appearance, trace-element composition, age, and uranium content.

The Melian origin of the obsidian found in Franchthi is of great interest. The oldest strata in which obsidian was found at Franchthi were dated by radiocarbon to the period 7530-6790 B.C. They belong, therefore, to the Mesolithic, pre-agricultural period and constitute the earliest positive evidence available anywhere of the transport of goods by sea.

Obsidian Trade in Central America

Among the ancient inhabitants of Central America the Olmec civilization is recognized as the earliest advanced culture. It flourished in San Lorenzo Tenochtitlan, on the Mexican Gulf Coast, between 1150 and 400 B.C. The Olmecs manufactured most of their tools and weapons either from obsidian or with its help. The importance of obsidian to their economy was probably similar in magnitude to that of steel to the economic of modern industrial nations. Obsidian does not, however, occur naturally in Olmec territory. Even so, the large number of obsidian artifacts recovered at San Lorenzo indicates its importance to the ancient inhabitants, who must have imported it either through trade or by direct procurement.

X-Ray-fluorescence analysis of obsidian artifacts found at San Lorenzo has thrown light on obsidian-procurement systems and ancient trade in that part of the world. In conjunction with the analyis of the artifacts, sources of obsidian in Central America were also sampled and analyzed (Cobean, Perry, et al. 1971) (see Fig. 6.6).

The concentrations of iron, manganese, rubidium, strontium, and zirconium were determined. Of these, rubidium and zirconium are the significant elements in the determination of compositional groups. Obsidians from some sources, such as from the Guadalupe, Victoria area, contain only 50 ppm of zirconium, whereas

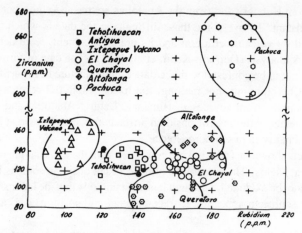

Fig. 6.6. Concentrations of zirconium and rubidium in obsidian from different sources in Central America. Raw material sources for artifacts found at San Lorenzo, Tenochtitlan can be deduced from the graph.

others at Pachuca, Hidalgo, for example, contain more than 600 ppm (see Fig. 6.7).

Out of 25 sources of obsidian sampled in the course of the investigation, only eight are represented in the artifacts at San Lorenzo; approximately 83 percent of the artifacts can be identified with one or other of these eight sources. The remaining 17 percent are from as yet unidentified sources.

Difficulties arose when it was found that obsidians from different sources had similar zirconium content. Obsidian from Teotihuacan in Mexico and from San

Fig. 6.7. Sources of obsidian and sites where artifacts were found in Mexico and Central America.

TABLE 6.4   Concentration of Trace Elements in Obsidian from San Lorenzo
Artifacts and Idenified Sources

	Range of concentration (ppm)	
Source—Artifact Group	Zirconium	Rubidium
Guadalupe Victoria, Puebla	30-50	95-130
Pico de Orizaba, Veracruz	30-50	125-135
Pachuca, Hidalgo	600-870	175-210
Ixtepeque Volcano, Guatemala	130-185	95-120
Teotihuacan, Mexico	115-160	120-150
El Chayal, Guatemala	100-135	145-190
El Paraiso, Queretaro	85-130	140-195
Altotonga, Veracruz	120-175	155-185

Bartolome in Guatemala both contain roughly 120-150 ppm of zirconium. The high ratios of mangangese and sodium in the Teotihuacan samples, as determined by neutron activation, made it possible to distinguish between the sources.

These studies on the provenance of obsidian have revealed that trade played a major part in the expansion of Olmec influence. The Olmecs were constantly establishing new commerical relationships with peoples in other areas of Central America. The importation of obsidian from El Chayal, Guatemala, at an early period in the Olmecs civilization development seems to have taken place several centuries earlier than any known occupation of the latter site. Also, the great distance from San Lorenzo to Pachuca, Hidalgo—more than 800 km by air and much farther by land or river—makes it highly unlikely that the Olmecs acquired their obsidian through mining expeditions. They must have participated in a trade network with intermediaries who supplied them with obsidian and other materials.

Prehistoric Obsidian Tools from Alaska

Artifacts and chippings of obsidian have been found at several archaeological sites in Alaska, notably in Onion Portage and at Cape Denbigh. The findings at Onion Portage are of particular interest because of the discovery at the site of other stone tools, similar to those made in distant Japan and Siberia. For a long time no natural sources of obsidian were reported throughout the vast northwestern region of Alaska. Some archaeologists suggested that obsidian found in archaeological sites in Alaska had been carried from natural sources as far away as the Aleutian Islands. Clearly, the source of obsidian was an important key to the elucidation of cultural ties and trading routes in ancient Alaska, especially since the variation in

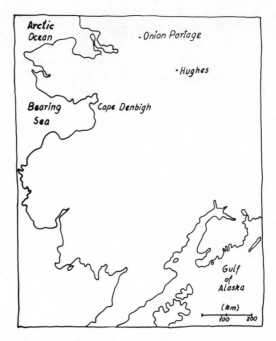

Fig. 6.8. Sources of obsidian and sites where artifacts were found in Alaska.

abundance of obsidian finds at Onion Portage was said to be related to the ebb and flow of trade relation between coastal cultures and inland cultures (see Fig. 6.8).

An end to speculation came with the discovery of obsidian deposits near Hughes, only 180 km from the Onion Portage site. The mere proximity of obsidian sources to ancient settlements does not, however, necessarily prove a physical connection. But most of the excavated and indigenous obsidians from Alaska compare well in color, opacity, and the absence of devitrification or of hydration layers. More important in this connection is the sodium:manganese ratio, which has been found to vary widely between obsidian sources. The sodium:manganese ratios in the obsidian from the Onion Portage artifacts and at the Hugh sources are similar (Patton and Miller 1970).

The circumstancial and scientific evidence for the connection between the two sites is thus very strong. Furthermore, artificially chipped obsidian fragments were found in the countryside surrounding the Hughes sources. There is still an open question as to whether this source provided all or only part of the obsidian required by the ancient communities in northwestern Alaska. But the findings undoubtedly help in establishing the prehistory of one of the more significant sites in a very important migration route.

In conclusion, it can be said that chemical analysis of the trace elements present in obsidian premits definite identification of artifacts with known sources of the material. By tracing the varieties of obsidian from their sources to the sites where they occur in the form of manufactured objects, the trade routes of ancient man can be reconstructed.

QUARTZITE

Quartzite is a very dense and exceptionally hard metamorphic rock. It is derived from sandstone and consists largely of quartz (silicon dioxide, SiO_2). Most quartzites contain 90 percent quartz or more, and in some cases the content exceeds 95 percent. The color of quartzite is usually white or light yellow, but it may also be red if iron oxides are present. The rock is so hard that quarrying is costly; nevertheless, due to its hardness, strength, and durability, quartzite has occasionally been used for ornamental construction and for statuary.

PROVENANCE OF QUARTZITE

Colossi from Memmnon

Ferruginous quartzite was used by the Egyptians in the fourteenth century B.C. in making a famous group of monumental statuary: the Colossi of Memmnon near Thebes (see Fig. 6.9). The statues were each carved from single blocks of quartzite and are of impressive dimensions; the southern statue measures over 14 meter high, 10 meter from front to back, and over 5 meters wide and has a calculated volume of 275 cubic meters and a weight of 720 tons. The northern one is of similar size.

The large dimensions of the Colossi, the technology involved in moving them to their present location, and the desire to know the exact source of the stone prompted an investigation in which neutron activation analysis was used to determine the provenance of the stone (Heizer, Stross, et al. 1973). The chemical composition of rock from the Colossi and from eight ancient quartzite quarries was determined. Ideally, several elements in the stone would have different relative concentrations for each quarry of possible interest. On the other hand, the amounts of these elements in the status would match the samples from one quarry and none of the others. The situation was not ideal in the present instance, but there was coherence between the abundance of some elements (europium, iron, etc) in the stone of the Colossi and in that from a quarry at Gebel el Ahmar near Cairo, 700 kilometers down the Nile from Thebes. Petrographic analyses also provided evidence of similarity between the samples from the Colossi and from the quarry at Gebel el Ahmar, establishing the source of the stone at the latter site beyond any reasonable doubt (see Fig. 6.10).

Fig. 6.9. The Colossi of Memmnon.

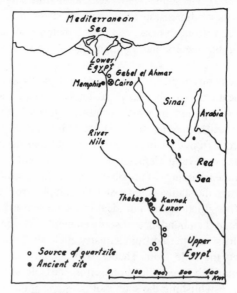

Fig. 6.10. Sources of quartzite in Egypt.

MARBLE

Marble is the term usually applied of those forms of limestone (carbonate rocks whose chief component is calcium carbonate, $CaCO_3$) or dolomite [the carbonate of calcium and magnesium, $CaMg(CO_3)_2$], which are sufficiently compact to take a high polish. The color of marble is usually white, but markings of many hues and patterns are produced by the presence of trace amounts of compounds of metals that are either constituents of the original limestone or are later intrusions.

Marble has been used for building and statuary. Many types were known to the ancients, of which Pentelic marble (from Mount Pentelikus in Attica, Greece) and Parian marble (from the Aegean Island of Paros) are the most famous.

PROVENANCE OF MARBLE

Neutron activation analysis of marble from the Aegean Islands, Attica (Greece) and Anatolia (Turkey) made it possible to characterize certain marbles as being low in manganese (at Marwara and Paros),and others as high in the same element (at Pentelikus and Vresthena) (Rybach and Nissen 1965). The conclusions do not seem to be broadly applicable since the calcium carbonate lattice apparently offers a hostile environment for the incorporation of trace elements, which may result in considerable compositional differences between specimens from a single quarry.

The determination of *isotope ratios* of carbon and oxygen has shown that marble from different sources in the Aegean Island and in Greece can be identified (Craig and Craig 1972). The ratios of carbon-13 to carbon-12 and oxygen-18 to oxygen-16, respectively, were found to be different and specific for several marble sources. On the carbon-13-oxygen-18 diagram shown in Fig. 6-11, the marbles fall into well defined groups, with the exception of the Naxian and Pentelic varieties. The analysis of ten archaeological marbles (Table 6.5) revealed that five of these could be related to the identified sources, whereas the other five had no counterparts among the sources analyzed. It appears that the isotope-ratio method, especially when used in conjunction with other techniques, provides the most useful test for determining the origin of Greek marble.

The natural thermoluminescence of marble may be of assistance in the of restoration of archaeological objects when the fragments to be joined have to be selected from large number of pieces (Afokardos, Alexopoulos, et al. 1974). Samples of marble from different quarries have different points of disturbance (called "centers") that are the source of natural thermoluminescence. But the distribution of the centers is not uniform over the whole of a single quarry, and the attribution of provenance by this method is not always reliable. On the other hand, if samples are taken from adjacent points on the same block of marble, there are no differences in the thermoluminescence centers. This provides the possibility of

TABLE 6.5   Isotopic Analyses of Archaeological Marble Samples

No.	Sample Description	$\delta^{13}O$ (per Mil)	$\delta^{18}O$ (per Mil)	Probable Provenance
1A	Athens: "Theseion"	2.63	−7.90	Pentelic
1B	Athens: basal slab	2.63	−8.03	Pentelic
2A	Delphi: carved block	3.89	−3.10	Parian
2B	Delphi: drum of column	1.96	−4.15	?
3A	Epidarus: carved block		−1.12	?
3B	Epidaurus: Corinthian capital	2.16	−0.78	?
3C	Epidaurus: cut slab, white marble	2.58	−3.96	?
4A	Naxos: Apollon Gate block	1.67	−5.51	Naxian
4B	Naxos: Apollon Gate, large block	1.64	−5.53	Naxian
5	Caesarea, Israel: Corinthian capital, grey marble	3.51	−2.64	?

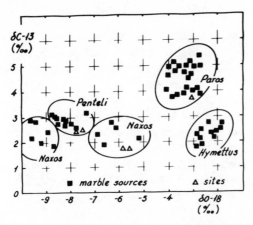

Fig. 6.11. Carbon-13 and oxygen-18 in marble samples from ancient Greek quarries and archaeological sites.

determining whether two fragments of marble belong to different objects or are part of the same piece.

GEMSTONES

Gems are natural stones and minerals (and a few materials of organic origin) that have sufficient beauty and durability to make them sought after for use as personal adornment. More than 100 kinds of natural stone have been used at one time or another for ornamental purposes. Only a very few materials of organic origin are suitable for gem making. These include pearl, coral, jet, and amber; of these, only amber has thus far attracted the attention of archaeological chemists.

AMBER

Amber has been used for ornamental purposes since prehistoric times, when man believed that amber was sunlight solidified by the sea waves. Amber is, in fact, a fossil resin derived from coniferous trees. It is soft and has a refractive index of 1.55 and a hardness from 2-3 on the Mohs scale. Amber occurs as irregular rods or in droplike shapes in all shades of yellow, with nuances of orange, brown, and, rarely, red.

The largest and most significant European deposits of amber, known since prehistoric time, occur along the shores of the Baltic Sea. Southern European deposits were also known in antiquity, especially in what are today France, Rumania, and Italy.

PROVENANCE OF AMBER

Amber has been found at many archaeological sites: in neolithic dwellings in southern France, at Stonehenge in England, and in Mycenaean tombs, to mention only a few. Until the late nineteenth century amber found anywhere in Europe or in the Mediterranean region was assumed to be of Baltic origin. This assumption was fortified by the knowledge that a regular amber trade existed across Europe during the Bronze Age and in Greek and Roman times. However, it may be important to know whether specimens of archaeological interest were made of, say, Baltic or Mediterranean amber.

The problem of the geographic origin of archaeological amber was first brought up in the late nineteenth century (Capellini 1872); since then, systematic investigations of amber have been made to determine its provenance.

Succinic-acid Content

It has been known since the sixteenth century that Baltic amber contains small quantities of succinic acid, an organic compound of formula $(CH_3)_2NNHCOCH_2CH_2COOH$. Sublimes when the fossil resin is destructively distilled. The concentration of succinic acid in amber was used by early workers to identify some important archaeological finds (Helm 1877, 1891; Olhausen and Rathgen, 1904). The succinic-acid content of some European ambers is given in Table 6.6. For identification purposes, the only conclusion that can be drawn from these results in a negative one—samples containing less than 3 percent succinic acid cannot be of Baltic origin. On the other hand, samples with appreciable amounts of the acid could be of either Italian, French, Portugese, or Baltic origin.

Amber samples of archaeological interest are usually small and mostly unique objects. The succinic-acid method for the determination of amber provenance requires the destruction of most of the specimen and is hence, unacceptable by archaeologists. Methods to be used for analysis and identification of amber should call for very small samples or, preferably, be nondestructive. X-Ray diffraction, mass spectrometry (Mischev, Eichhoff, et al. 1972), fission-track dating (Uzgiris and Fleischer 1971), infrared-absorption spectroscopy, and neutron activation analysis may help in the determination of the provenance of amber. Infrared-absorption spectroscopy and activation analysis have already provided some positive results.

Trace Elements

Instrumental neutron-activation analyses were used to determine the presence of trace elements in amber samples from Sicily and from the Baltic Sea (Dass 1969). Data obtained from both groups are given in Table 6.7.

TABLE 6.6 Succinic-acid Contents of European Ambers

Origin of Sample	Succinic-acid Content (Percent)
Italy	—
Sicily	—
Sicily	Traces
Sicily	0.6
Baltic Sea	5.6
Baltic Sea	7.2
Baltic Sea	5.8
Occidental France	—
Portugal	9.1
Southern Russia	5.7

TABLE 6.7 Analysis of Amber from the Mediterranean and Baltic Regions

Sample		Composition					
		Sodium	Manganese	Scandium	Zinc	Cobalt	Gold
Origin and Description	Number of Samples	ppm			ppb		
Baltic amber							
Light bright	12	13	0.9				23
Dull	13	10	1.0		1-3	0.01-0.1	50
Scaly	12	11	1.1	0.001-0.01			30
Flamed	12	18	2.9				26
Dark bright	10	18	2.0		1-10	0.01-0.1	46
Flamed	5	22	2.0				17
Silcilian amber							
Light bright	9	30-60	3-80	0.001-0.01	0.01-0.5	0.01-0.5	30-600
Dark dull	27	100-1600	0.05-150		0.1-2	0.01-0.2	50-200
Scaly	5	500-1500	1-5				

The elements showing greatest variation, dependent on place of origin, are sodium and gold. The sodium and the gold content of the Baltic Sea samples are low and almost constant. The Sicilian samples, on the other hand, usually have higher concentrations of both elements, which vary from sample to sample of the same origin. From these results it can be concluded that the determination of sodium and gold contents can give a good indication as to the origin of amber.

Infrared-absorption Spectra

The infrared spectra of geographically widely separated types of amber show greater similarities than differences. Infrared spectra over the spectral range of 2.5-16 micron, show the manner in which atoms are linked to each other. In pure compounds, identity of infrared spectra establishes identity of chemical composition. For amber, which is a complex mixture of organic compounds, a similarity of spectra can be expected (see Fig. 6.12).

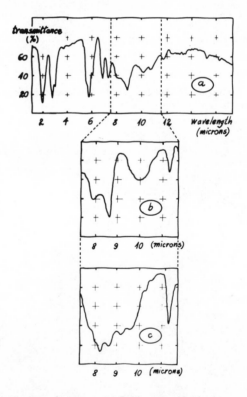

Fig. 6.12. Infra-red spectrum of amber: (a) Baltic amber, 2-16 microns.   (b) Baltic amber, 7-11 microns.   (c) Sicilian amber, 7-11 microns.

The provenance of Japanese amber beads was determined by comparing and contrasting their full infrared spectra (Fujinaga, Takenaka, et al. 1974). Baltic amber has an easily recognizable band at 8.6-8.7 microns which is never found in any other amber. A very simple mathematical analysis of the slope of the spectrum of amber, (8-9 microns), serves to characterize Baltic amber and to distinguish it from others (Beck 1970, 1972; Beck, Adams, et al. 1971; Beck, Wilbut et al. 1964). This permits the positive identification of specimens originating in the Baltic region and allows the direct recognition of imports from the north among the large number of amber fossils in European archaeology. On the other hand, infrared spectrum analysis does not distinguish unequivocally between the many non-Baltic European amber resins. The absorption bands of these are very variable and cannot be related to a geographic origin. This is probably because many of them are from the same botanical sources.

TURQUOISE

Turquoise is a secondary mineral found in arid regions. It is formed during the alteration of rocks containing a high proportion of aluminum. The mineral consists of hydrated phosphate of aluminum and copper. It has the formula $CuAl_6(PO_4)_4(OH)_8 \cdot 4H_2O$. The color ranges from blue (due to the presence of copper) through various shades of green (which results from the presence of iron) to greenish and yellowish grey. Turquoise has been extensively used and prized as a semiprecious gem for over 8000 years.

Turquoise is found at many archaeological sites. The distribution of turquoise sources, however, is limited to relatively few regions. The Sinai Peninsula sources were mined by the Egyptians who extracted the stone from seams in the sandstone rock. Much fine material of gemstone quality has come from Iran. Pre-Columbian turquoise sources in the southern United States were apparently confined to the margins of the Colorado Plateau.

PROVENANCE OF TURQUOISE

With the aid of instrumental neutron-activation analysis to detect trace elements, it has been possible to tell which turquoise specimen comes from which mine (Sigleo 1975). Turquoise beads, dating from 800-2000 years ago, were found at a pre-Columbian site at Snaketown, Arizona (USA). Beads found at the site and samples from each of 24 turquoise mines in the southwestern United States were analyzed. The trace-element pattern from the beads was found to correlate with that of samples from a group of mines near Hallorn Springs, California, but not with the turquoise samples from the other 23 mines; thus the turquoise of the beads must have been mined near Hallorn Springs.

The results were unexpected since there are several mines with prehistoric workings closer to the Snaketown site. But it is possible that other turquoise sources were being used concurrently, because the beads examined represented only a portion of the turquoise from the archaeological site.

REFERENCES

Afokardos, G., Alexopoulus, K., and Milotis, D., *Nature*, **250**, 48 (1974).

Beck, C. W., *Archaeology*, **23**,7 (1970).

Beck, C. W., *Naturwissenschaften*, **59**, 294 (1972).

Beck, C. W., Adams, A. B., Gretchen, C., and Fellows, C., in R. H. Brill, ed., *Science and Archaeology*, MIT Press, Cambridge, Mass., 1971, p. 235.

Beck, C. W., Wilbut, E., and Meret, S., *Nature*, **201**, 256 (1964).

Bowman, H. R., Asaro, F., and Perlman, I., Lawrence Berkeley Lab. Rep., LBL 661, 1972.

Bowman, H. R., Asaro, F., and Perlman, I., *Archaeometry*, **15**, 123 (1973).

Cahn, J. R., Dixon, J. E., and Renfrew, C., in D. Brothwell, and E. S. Higgs, eds., *Science and Archaeology*, Thames and Hudson, London, 1969, p. 578.

Cahn, J. R., and Renfrew, C., *Proc. Prehist. Soc.*, **30**, 111 (1964).

Capellini, G., *Zeitsch. fur Ethnologie*, Verhandlungen, 198, 1872.

Cobean, R. H., Perry, E. A., Turekian, K. K., and Kharkar, D. P., *Science*, **174**, 666 (1971).

Craig, H., and Craig, V., *Science*, **176**, 401 (1972).

Dass, H. A., *Radiochem. Radioanal. Lett.*, **1**, 289 (1969).

Dixon, J. E., Cann, J. R., and Renfrew, C., *Sci. Am.*, **218**, (3), 38 (1968).

Durrani, D. A., Kahn, H. A., Taj, M., and Renfrew, C., *Nature*, **223**, 242 (1971).

Ericson, G. E., *World Archaeol.*, **7** (2), 151 (1976).

Friedman, I., and Smith, R. L., in D. Brothwell and E. Higgs, eds., *Science and Archaeology*, Thames and Hudson, London, 1963, p. 47.

Friedman, I., and Long, W., *Science*, **191**, 348 (1976).

Friedman, I., Smith, R. L., and Long, W. D., *Geol. Soc. Am. Bull.*, 323 (1966).

Friedman, I., and Trembour, F. W., *Am. Sci.*, **66**, 44 (1978).

Fujinaga, T., Takenaka, T., and Muroga, T., *Nippon Kugaku Kaishi*, **9**, 1653 (1974).

Gordus, A. A., *Archaeometry*, **10**, 87 (1967).

Gordus, A. A., Griffin, J. B., and Wright, G. A., in R. H. Brill, ed., *Science in Archaeology*, MIT Press, Cambridge, Mass., 1971, p. 222.

Gordus, A. A., Wright, G. A., and Griffin, J. B., *Science*, **161**, 382 (1968).

Green, R. C., *Archaeol. Assoc. Newslett.*, **5**, 8 (1962).

Green, R. C., Brook, R. R., and Reeves, R. D., *N. Z. J. Sci.*, **10**, 675 (1967).

Heizer, R. F., Stross, F., Hester, T. R. Albee, A., Perlman, I., Asaro, F., and Bowman, H., *Science*, **182**, 1219 (1973).

Helm, O., *Arch. Pharm.*, **211**, 246 (1877).

Helm, O., *Schriften der Naturforschenden Gesellschaft, Danzig, N. F.*, **7**, 189 (1891).

Layton, T. N., *Archaeometry*, **15**, 129 (1973).

Lee, R. R., Leich, D. A., Tombrello, T. A., Ericson, J. E., and Friedman, I., *Nature*, **250**, 44 (1974).

Lepsius, R., *Abh. Akad. Wiss. Berlin* (1890).

Meighan, C. W., Foote, L. J., and Aiello, P. V., *Science*, **160**, 1069 (1968).

Michels, J. W., and Bebrich, C. A., in H. N. Michael and E. K. Ralph, eds., *Dating Techniques for the Archaeologist*, MIT Press, Cambridge, Mass., 1971.

Mischev G., Eichhoff, H. J., and Haevernick, T. E., *Jahrb. Rom. Germ. Zentral., Mainz,* **17**, 111 (1972).

Newall, R. S., *Stonehenge*, HM SO, London, 1959.

Olshausen, O., and Rathgen, F., *Zeitsch. fur Ethnologie, Verhendlungen*, 153 (1904).

Parks, G. A., and Tieh, T. T., *Nature*, **211**, 289 (1966).

Patton, W. W., and Miller, T. P., *Science*, **169**, 760 (1970).

Reeves, R. D., and Armitage, G. C., *N. Z. J. Sci.*, **16**, 521 (1973).

Renfrew, C., Cann, J. R. and Dixon, J. E., *Ann. Br. School Archaeol. Athens*, **60**, 225 (1965).

Renfrew, C., Dixon, J. E., and Cann, J. R., *Proc. Prehist. Soc.*, **34**, 325 (1968).

Rybach, L., and Nissen, H. U., in *Radiochemical Methods of Analysis, Proceedings of Salzburg Symposium*, 1965, p. 105.

Shakleton, N., and Renfrew, C., *Nature*, **228**, 1062 (1970).

Sigleo, M., *Science*, **189**, 459 (1975).

Stross, F. H., Weaver, J. R., Wyld, J. E. A., Heizer, R. F., and Graham, J. A., *Contributions University of California Archaeological Research Facility*, No. 5, 1968.

Uzgiris, E. E., and Fleischer, R. L., *Nature*, **234**, 28 (1971).

Washington, H. S., *Am. J. Archaeol.*, **11**, 71 (1898).

Weaver, J. R., and Stross, F. H., *Contributions University of California Archaeological Research Facility*, No. 1, 1965.

Weisss, L. E., *Am. J. Sci.*, **252**, 641 (1954).

BUILDING MATERIALS

There is no reliable evidence about the earliest prehistoric building techniques. Probably the first man-made dwellings were tents made of animal skins. Later rushes, straw, and reeds would be used, as well as timber. Changes in the form of human settlement would have brought about efforts to make buildings stronger and more lasting, eventually leading to the use of masonry. The convenient general heading of *building materials* embraces a number of natural and artificial materials that find their principal application in building. These include natural stone, discussed in the preceeding chapter, and brick and mortar, to be discussed here.

BRICK

Man has been making bricks for at least 6000 years. In fact, one of the oldest technologies is that of brickmaking, for which almsot every kind of clay has been used at one time or other. In Mesopotamia, solid earth was compressed in molds, whereas in Egypt a mixture of Nile mud, fine gravel, and small amounts of impurities were shaped into regular blocks.

Brickmaking

Digging the raw materials, mixing, molding, and finally drying are the steps involved in brickmaking. After the clay was dug out, it was exposed to the atomsphere to weather it and make it fit for shaping. If the raw clay was too rich, the finished bricks would dry too slowly, cracking and losing their shape in the process; therefore, sand or marl were added to prevent this, whereas the addition of animal dung gave strength and plasticity.

Primitive bricks were always sun dried and could only be used in dry climates; humidity could bring about their softening, and rain might dissolve them completely. The discovery of firing (baking bricks in an oven at high temperatures) was a major technical advance. Firing gives bricks greater strength and makes them weatherproof. The chemical changes that it causes in clay are discussed in Chapter 8.

CEMENT, MORTAR, AND PLASTER

Several terms are used by archaeologists to refer to the various materials used in building as bonding agents: the names *cement, concrete, mortar, plaster,* and *stucco* have often been applied indiscriminately to all sorts of materials used for joining or coating masonry. It is probably best, therefore, to begin by defining broadly each of these terms. *Cement* is a general term used to refer to substances that act as bonding or coating agents in masonry. *Hydraulic cements,* sometimes also called *concretes,* are cements capable of setting and hardening under water. *Mortars* are cements used to bond masonry units (bricks or stones) together. When cements are used to cover and conceal masonry, they are known as *plaster.* *Stuccos* are fine plasters of high quality, used to give a smooth coating to interior walls. It can be concluded that mortars, plasters, and stuccos are forms of cement; the latter word is used in the discussion that follows as a generic term.

Cements of Antiquity

Clays, asphalt, gypsum, and lime have all been used at different times as cementing materials. Clay mortar was used by the Egyptains for cementing sun-dried bricks. It consisted of ordinary clay, to which sand was sometimes added. For use, the mixture was brought to the required consistency by the addition of water.

Brick walls of ancient Egyptian buildings were made of bricks dried in the sun without baking, and each course was covered with a moist layer of the Nile mud used for making the bricks with or without the addition of chopped straw. The drying of this layer made the wall a solid mass of dry clay as it is found, for example, at Giza and in the step pyramid at Saqqara. In samples from the latter, the mortar contained a relatively high proportion of powdered limestone varying in the range from 3-55 percent (Lucas 1962). Such a mode of construction was only possible in a rainless climate, as the unburned material possesses little power of resistance to water. (Clifton 1977)

Burned bricks and alabaster slabs were used by the Babylonians and the Assyrians and were cemented together with bitumen. Stones joined by asphalt were found in the ruins of Nineveh. The asphalt, which was applied hot, penetrated into the cracks and holes in the stones and bound them strongly, in addition to keeping out the weather very effectively.

For the great majority of the buildings of antiquity, two principal cements were used: one based on gypsum and the other on lime; however, mixtures of both were sometimes used. (Furlan and Bissegger 1975).

The basic ingredients of building cements are: (1) a *binder*—gypsum or lime— and (2) a *filler,* usually sand or crushed limestone, but also occasionally consisting of crushed ceramic materials, crushed shells, or similar. *Hydraulic components,* if present, include volcanic earths, clay, and similar materials.

GYPSUM CEMENT (PLASTER OF PARIS)

Gypsum plasters and mortars employed in building are prepared by calcination of gypsum (hydrated calcium sulfate, $CaSO_4 \cdot 2H_2O$), the commonest of all sulfate minerals, found in abundance throughout the world.

When gypsum is calcined at a temperature of 150-200°C it loses part of its water of cyrystallization, and calcium hemihydrate ($CaSO_4 \cdot \frac{1}{2}H_2O$)—*Plaster of Paris*) is obtained:

$$CaSO_4 \cdot 2H_2O \xrightarrow{150°} CaSO_4 \cdot \frac{1}{2}H_2O + 3/2H_2O$$

If the temperature is raised above 200°C all the water of crystallization is lost. The product is then anhydrite (anhydrous calcium sulfate, $CaSO_4$), which has poor setting qualities:

$$CaSO_4 \cdot \frac{1}{2}H_2O \xrightarrow{>200°} CaSO_4 + \frac{1}{2}H_2O$$

After calcination Plaster of Paris is ground and finally slaked before use.

Setting

Plaster of Paris sets in 6-8 minutes after addition of water. The process of setting is one of hydration and crystallization: calcium sulfate recombines with water to form the hydrate, $CaSO_4 \cdot 2H_2O$. The crystals formed during setting grow together, become entangled and interlocked, and embed any gains of filler present. The result is a rigid structure of crystalline plaster and filler, with occasional voids. During setting the cement undergoes a linear expansion of about 0.3 percent, which enables sharp casts to be made.

Gypsum cement was used in ancient Egypt as a plaster and as a mortar for stone (Lucas 1962). The reason for using gypsum instead of lime, although limestone was more abundant and more accessible than gypsum, was the scarcity of fuel; lime requires a much higher temperature, and consequently more fuel, for calcination.

LIME CEMENT

Lime cement were used at a very early date by the Greeks and earlier still in Crete. Later the Romans must have borrowed their use since they, like the ancient Greeks, used them. Lime cements were also used by the Japanese (Yasuda, Iuchi, et al. 1976).

Lime cements used in masonry and plastering are made by mixing lime with sand and water. *Lime* is a general term for the various products of calcined limestone; quicklime and hydrated lime are examples.

Manufacture

Lime consists essentially of calcium oxide, CaO, either alone or together with a small proportion of magnesium oxide, MgO. It is made by heating a natural rock, usually limestone, to a high temperature; the product disintegrates on contact with water. The manufacturing process consists of burning the limestone in a kiln to a temperature above 900°C so that it will decompose into *quicklime*, CaO, with a corresponding release of carbon dioxide:

$$CaCO_3 \xrightarrow{\ >900°\ } CaO \ + \ CO_2 \uparrow$$

Before use, quicklime must undergo *slaking* (or hydration); the quicklime is mixed with water to obtain *slaked lime*, $Ca(OH)_2$:

$$CaO \ + \ H_2O \rightarrow Ca(OH)_2$$

Setting

Slaked lime sets into a hard material by the evaporation of excess water. Setting is completed by a process of fixation of carbon dioxide from the air, with conversion of slaked lime back into calcium carbonate:

$$Ca(OH)_2 \ + \ CO_2 \ = \ CaCO_3 \ + \ H_2O$$

The fixation of carbon dioxide is extremely slow and often incomplete. Minute traces of quicklime were found in the pure lime plaster in the walls of the palace at Knossos (in Crete), which has been exposed freely to the atmosphere for about 3500 years. Free lime has also been found in plaster used in the Egyptian pyramids. Still, most of the lime is converted into calcium carbonate in a few weeks following the use of the cement. It has, therefore, been suggested that lime mortars could be dated by the radiocarbon method, a procedure discussed in Chapter 18. The ultimate result of the setting process is the formation of a coherent mass of calcium carbonate crystals, practically limestone. To this is generally attributed the strength of old plaster.

Lime Mortars and Plasters

Lime is a very good binder, but by itself it has little strength. The addition of fillers increases its strength. In *lime cements*, lime is used as a binder and sand as a *filler*. Occasionally crushed ceramic material is added instead of sand. *Carbonate cements* are lime mortars and plasters in which crushed limestone is used as a filler. They can be easily detected by their low content of sand, which may even be absent entirely.

Ancient Lime Cements

Lime is one of the oldest known chemicals manufactured by men: the process of lime burning seems to have been known during the Stone Age (Gourdin and Kingery (1975). Lime plaster in good condition is found in Egyptian pyramids built 1500 years B.C.. Lime is mentioned in the Bible several times. Lime cements were used by the Greeks, Etruscans, Romans, Arabs, and Moors. The composition of Roman mortar is described by Vitruvius, an architect and engineer of the first century B.C., as 3 parts of pit sand to 1 part of slaked lime. When river sand or sea sand were employed, 2 parts of sand and 1 part of crushed and sifted shells were mixed with 1 part of lime. These porportions are in accordance with accepted modern practice.

It is interesting to examine the results of the analysis of Greek and Roman mortars listed in Table 7.1. The average quantity of sand in all the samples is about 25 percent, whereas that of lime is about 36 percent. This corresponds to about 48 percent slaked lime, giving a ratio of 1 part of sand to 2 parts of slaked lime, which is not in agreement with Vitruvius's prescription. The similarity in composition of the stucco from the Greek fountain house and that from the Roman tomb, separated from each other in time by about eight centuries suggests that the Romans may have learned from the Greeks the technique of preparing lime cements (Foster 1934).

TABLE 7.1   Composition (in Percent) of Greek and Roman Mortars

Sample		Composition (percent)	
Source	Date (Century)	Sand (SiO_2)	Lime (CaO)
Lining of reservoir	5-6th B.C.	14.87	36.80
Floor of draw-basin	4-5th B.C.	36.22	22.62
Small aqueduct	3rd B.C.	26.34	32.84
Greek fountain house	4th B.C.	25.33	33.22
Roman stucco from chamber	4th C.E.	25.58	35.14
Roman repair of Peirene	1st C.E.	30.41	27.77
Wall of Chamber II, Peirene	2nd C.E.	31.70	24.27
Painted wall chamber of tomb	2nd C.E.	12.50	43.98

Plasters from houses in Rome and Pompeii were composed of three layers. The deepest layer consisted of plaster with a lime to filler ratio of 1:3. This ratio fell to 1:2.5 in the middle layer and surface plaster; the grain size of the middle layer was, however, much smaller (Muller-Skjold 1940).

HYDRAULIC CEMENTS

To build in water, one must have waterproof cement that will harden under water. When limestone is heated at high temperature, decarbonation occurs and lime is formed. If the limestone contains a considerable amount of clay; or if clay or other siliceous materials are added to the limestone prior to heating, the high temperature will cause chemical reaction between the lime formed and active silica and aluminum oxide in the additives. The products are silicates and aluminosilicates of calcium, compounds that are water proof and harden under water. Such products are called *hydraulic cements*.

In Mesopotamia, ashes were added to lime to produce hydraulic cements for the building of wells. A water-tight cement was produced by the Romans from pure lime and oil. But the most important Roman hydraulic cement was made by adding a volcanic ash—*pozzolana*—to lime. For building in water, one part of ordinary lime was mixed wih two parts of pozzolana. A study of ancient Roman hydraulic cements in Israel confirmed their high strength and durability under severe conditions in harbor quays at Caesarea and lakeshore and hot mineral structures in Lake Tiberias (Malinovsky, Slatkine, et al. 1961).

REFERENCES

Clifton, J. R., *Technology and Conservation of Art, Architecture and Antiquities*, **77**, 30 (1977).

Foster, W., *J. Chem. Ed.*, 223 (1934).

Furlan, V., and Bissegger, P., *Zeitsch. fur Schweitzerische Archaeologie und Kunstgeschichte*, **32**, (2), 166 (1975).

Gourdin, W. H. and Kingery, W. D. *J. Reld Archaeol.*, **2**, 133 (1975).

Lucas, A., *Ancient Egyptian Materials and Industries*, E. Arnold, London, 1962.

Malinovsky, R., Slatkine, A., and Ben-Yair, M., in *International Symposium on the Reliability of Concrete*, RILEM, 1961, p. 531.

Muller-Skjold, F., *Angew. Chem.*, **53**, 139 (1940).

Yasuda, H., Iuchi, I., and Mukai, K., *J. Archaeol. Soc. Japan*, **61**, 277 (1976).

POTTERY

"Pottery" is a general name for artifacts made entirely or partly of clay and hardened by heat. Sometimes the term pottery is applied to the lower grades of ceramic wares as distinct from porcelain. In this book, however, the term is used in the accepted technical sense to include all products made of fired clay. The craft aspects of pottery, fascinating in themselves, are outside the scope of this book. Of greater interest to archaeological chemistry are the raw materials from which pottery is made, their properties, and the reactions and processes that they undergo during manufacture. These are discussed here, together with the chemical methods available today to obtain information about the artifacts and working methods of ancient potters and the provenance of their wares.

PRECERAMICS

Puddles formed in loess soils are known to fissure on drying; the natural sedimentation of the soil produces irregular, dish-like concave crusts. The upper, fine layer, contracts more quickly than the rest and the pieces break away from the soil beneath. The dish-like soil crusts may attain an area of 100 cm^2 and a thickness ranging from 1 mm to 1 cm. (Fig. 8.1).

Firing these relatively soft crusts in small bonfires produces hard, dish-like pieces, which can be broken by hand only with difficulty. Their physical properties (hardness, porosity, and strength) are similar to those of primitive pottery (de Korosy 1975). Primitive man may have accidentally obtained this type of dish when making fire for warming himself or for cooking a meal. Could such accidentally fired dishes have given to prehistoric man the idea of molding clay by hand and then firing it?

Any such hypothesis is impossible to prove, but there is a way of distinguishing between accidental "preceramic" dishes and intentionally made ones. The outer surface of man-made pottery is always smooth, as its inner face. The outer face of accidentally formed soils crusts, not having been smoothed, is normally coarse. If ceramic dishes with coarse outer surfaces were to be found in ancient habitation sites, they would provide evidence of "preceramic" pottery having been used by man.

Fig. 8.1. Preceramics: cracked loess crusts found in a gully: *(a)* Inside unfired; *(b)* outside unfired: *(c)* inside fired; *(d)* outside fired.

CERAMIC MATERIALS

The main constituent of pottery is *clay*. Pottery makers have, from earliest times, made use of two main groups of properties of clay: plastic properties and firing properties.

A mixture of clay and water has *plastic properties;* that is, it can be pressed into a shape that it will retain when the pressure is released. Plasticity is used by the potter for shaping different wares. Once the shaped mixture is dry it loses its plasticity and the clay becomes hard and brittle.

On firing, chemical changes take place in the clay over a wide range of temperature. Some types of clay will dry and be fired satisfactorily. Most will dry unevenly and crack while doing so. For this reason and others explained later, pottery is seldom made from clay alone. Other ingredients are added to the clay mixture: *nonplastic materials* or *fillers*. Their function is to facilitate uniform drying and to counteract excessive shrinkage of clay mixtures on drying and firing. The combination of clay and nonplastic filler is usually called the *ceramic body*.

Clay

The term *"clay"* has been applied to a variety of earthy materials of different origin and composition. It is difficult to define clays precisely, the definition varying between different fields of interest. For the archaeologist interested in pottery, clay can be defined as any earthy aggregate that when pulverized and mixed with water exhibits plasticity, becomes rigid on subsequent drying, and develops hardness and strength when heated to a sufficiently high temperature.

Geologically speaking, clays are fine-grained weathering products of silicate rocks. The weathering of rocks may be caused by a variety of factors. Atmospheric oxygen, carbon dioxide, and surface water are the chief agents causing rocks to break down. Volcanic gases and organic acids dissolved in water may also contribute to the conversion of rocks into clay. The final result of weathering is a mixture of clay, quartz, ferric oxide, and micaceous material, characterized by the extreme fineness of particles (usually below 2 microns in diameter) and varying widely in chemical composition and physical properties.

Primary clays are those formed by natural chemical action on felspathic rocks. They are quite pure and are to be found in their place of origin. *Kaolin* is the best known; others are listed in Table 8.1.

TABLE 8.1   Chemical Composition (in Precent) of Clays

Clay	Silica (SiO_2)	Alumina (Al_2O_3)	Iron Oxides (Fe_2O_3)	Lime (CaO)	Others	Water H_2O
Primary Minerals						
Kaolinite	45-49	38-35	0.2-2	0.1-1	5-1	12-15
Montmorillonite	50-55	20-16	0.06-6	0.5-3	4-6	8-23
"Pure" clay	46.5	39.5	—	—	—	14.0
Secondary:						
"Red" clay	57.5	25.6	7.8	1.7	5.4	20.0
"Ball" clay	50.5	32.1	2.3	0.4	3.5	11.7

Secondary clays are "transported" or "sedimentary" clays, which have been carried by water to a site more or less distant from their place of origin. In their travels they have picked up various impurities, each of which affects the way in which the clay has to be tempered, fashioned, dried, and fired. Secondary clays are generally more plastic than primary clays due to greater fineness of particles, resulting from mechanical friction during transportation and from the addition of decayed vegetable matter.

Red clay is the commonest of all types of clay. It is a secondary clay that at proper temperatures fires to a rich red color. When extracted from the earth its color may be red, yellow, green, gray, black, and so on. The rich red color of the fired clay will be modified if other components of the clay are varied.

Chemical Composition and Structure of Clay

The chief constituents of clay are the oxides of silicon and aluminum. The "assumed" composition of pure clay is:

$$Al_2O_3 \cdot 2SiO_2 \cdot 2H_2O$$

This formula expresses the composition of a theoretical, ideal clay. In practice, clay minerals are composite materials that may also contain quartz, organic compounds, talc, mica, iron compounds, and others. Most raw clays contain over 50 percent impurities and are only slightly plastic. To serve for pottery making, they must be modified before use.

Chemical composition alone is not much use in characterizing a clay, since clays of similar chemical composition may have very different physical and technological properties. Clays are better characterized by their structure than by their composition. The structure of clays has been studied in considerable detail, and modern crystallographic techniques have made it possible to describe the arrangement of the atoms in clay particles. In most clays, the crystals are built up from two basic, sheet-like arrangements of atoms: one is a two-dimensional network of tetrahedrons, where silicon atoms are surrounded by oxygen atoms or hydroxyl ions (Fig. 8.2). The other consists of aluminum, iron, or magnesium atoms embedded between two sheets of closely packed oxygen atoms or hydroxyl ions.

The most important ceramic clays, kaolinite, montmorillonite, illite, and halloysite, are built up of combinations of these basic structural units (see Fig.

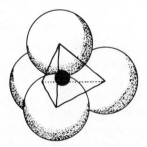

Fig. 8.2.   Silicate tetrahedron. Each silicon atom is surrounded by four oxygen atoms and each oxygen atom, by four silicon atoms (only one of the latter is shown in the figure). Silicate tetrahedra are the basic structural units of all clays.

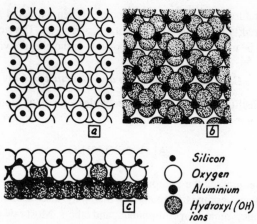

Fig. 8.3.   The structure of clays: (*a*) silicate sheet formed by silicate tetrahedra (each tetrahedron shares three of its oxygen atoms with three other tetrahedrons);   (*b*) aluminate sheet, built up from aluminum ions and hydroxyl (OH) ions

When the two sheets are superposed they mesh, forming kaolinite; in kaolinite each of the aluminum ions is surrounded by six close neighbors: oxygen and hydroxyl ions); (c) kaolinite sheet (cross-section) is a laminate of silicate and aluminate sheets.

8.3). Because of the sheet-like nature of the crystalline arrangement, the particles of most clays have the shape of thin, flat platelets. Some clays, however, are composed of structural units different from those noted in the preceding paragraphs, and their particles are fibrous or tubular, whereas still others are amorphous, due to the random arrangement of the atoms, without any symmetry.

Properties of Clay

During each stage in the manufacture of pottery a different group of properties of clay become particularly important. These groups are as follows: (1) plastic properties, (2) drying properties, (3) firing properties, and (4) others (color, etc.).

Plastic Properties.   The addition of water to dry clay produces a plastic mixture, readily shaped without rupturing, that will keep the form given to it. The plasticity of clay is essentially due to three factors: (a) the plate-like structure of clay particles, (b) the small size of these particles, and (c) the presence of water—called *water of formation*—between the clay particles. The water of formation acts as a lubricant between the small, plate-like particles that slide readily across each other when a shearing force is applied (see Fig. 8.4).

Plasticity develops only within a narrow range of water content: dry clay is nonplastic, and too much water makes the mass too fluid to retain a shape. Table 8.2 gives some idea of the amount of water that will produce a workable, plastic clay.

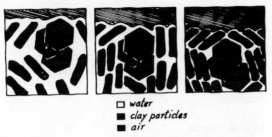

☐ *water*
■ *clay particles*
■ *air*

Fig. 8.4. Enlarged cross section of a clay mass at three different levels of water content. Loss of water results in reduced plasticity.

TABLE 8.2   Consistency of Clay-water Mixtures

Clay	Water (Percent)
Dry	5-18
Semidry or stiff	10-20
Stiff plastic	12-25
Plastic	15-30
Oversoft	15-35
Liquid	20-50

Drying properties.   When wet clay is dried, the elimination of water takes place in two phases. During the first phase *water of formation* evaporates freely and uniformly. This water is present in the clay as an envelope surrounding each individual clay particle. Layers of "water of formation" molecules adhere to the surfaces, edges, and pores of the clay particles. As evaporation proceeds and the water molecules around the clay particles are removed, the latter draw closer to

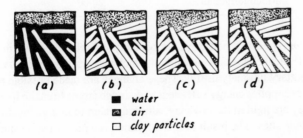

(a)   *(b)*   *(c)*   *(d)*

■ *water*
▨ *air*
☐ *clay particles*

Fig. 8.5. Stages in the drying of moist clay. Water is lost from between the clay particles until the latter come so close that they touch (b) and can consolidate no further. From this point onward the bulk volume does not change; water is lost from the pores (c) until the clay is quite dry (d).

POTTERY

TABLE 8.3 Linear Drying Shrinkage of Clays

Clay Mineral	Shrinkage[a] (Percent)
Kaolinite	3-10
Illite	4-11
Montmorillonite	12-23

[a]Calculated as percent of the dry length after drying at 105°C.

each other until they eventually touch and adhere. Figure 8.5 illustrates the stages of drying of a moist clay.

If water evaporates at too fast a rate during drying, internal stresses will be set up that may crack, break, or even burst the shaped clay. The drying shrinkage and dry strength of the clay are therefore very important properties. *Drying shrinkage* is the reduction in size that takes place when a mass of clay is dried. Table 8.3 gives values of drying shrinkage from some clays. *Dry strength* is the strength needed to break clay after water of formation has been removed; it is important when handling the clay before firing. Different clays vary greatly in dry strength, as illustrated in Table 8.4.

TABLE 8.4 Dry Strength of Clays

Clay Mineral	Strength (kg/cm^2)
Kaolinite	0.5-50
Illite	15-75
Montmorillonite	20-60

Firing properties. When clays are heated to a high temperture they undergo profound chemical changes, as a result of which they acquire many of the qualities that make pottery uniquely useful. These changes are made possible because the firing process supplies energy sufficient to dislodge atoms from the fixed positions in which they are held in the raw clay, allowing them to migrate by diffusion into more favorable sites. The result of this migration is a product more stable than the original material.

When clay is heated to 450-500°C (the exact temperature varies for different clays), chemically combined water is almost completely lost. On further heating the internal structure of the clay mineral is distorted and begins to break down above 800°C. Breakdown of the clay's structure continues as the temperature increases up to 1800°, a temperature far beyond anything within the reach of ancient technology; at this point the clay fuses. Table 8.5 summarizes the chemical changes taking place in clay undergoing heat treatment.

TABLE 8.5   Thermochemical Changes Induced by Heating Clay

Temperature (°C)	Reaction
100-150	Loss of water of plasticity
450-600	Modification of clay's structure brought about by loss of chemically combined water (water of constitution)
600-950	Brakedown of clay's structure and incipient vitrification
900-1750	Development of new crystalline phases
>1750	Fusion

Color of clay

The color of clays is due to the presence of impurities. Clays relatively free from impurities are white. The most frequently found impurities in clay and the colors that they impart to it are listed in Table 8.6.

TABLE 8.6   Impurities Affecting the Color of Clay

Impurity	Color
Ferric oxides (hematite, limonite, goethite, etc.)	Red, brown, buff, or yellow
Ferrous oxides (magnetite, siderite, etc.)	Gray
Ferrous silicates (glauconite)	Green
Organic materials	Gray, brown, or black

Nonplastic Materials

Materials that do not become plastic when mixed with water and withstand high temperatures are added to clay by the potter; these are called *nonplastics, fillers,* or *temper materials.*

The addition of nonplastic materials weakens the body of pottery wares. However, these materials improve other physical properties: they reduce the stickiness of clay, prevent shrinkage and undue cracking, and assist in the escape of water during drying. During drying, water evaporates from the suface of clay objects and is replaced by water diffusing outward from the interior. The interstitial spaces between the minute clay particles are very narrow and do not allow rapid diffusion. As a result, there is a tendency for the clay near the surface to become progressively dryer than the interior as drying proceeds, and consequently to shrink more than the latter. Stresses are set up, and cracking may occur. The filler particles, coarser than those of clay, leave larger interparticle spaces through which water can migrate more rapidly, decreasing strains and preventing cracking.

Nonplastic Fillers

Potters have used as fillers a wide range of materials differing in their composition, reactions on heating, an effect on strength. Even a short list of fillers would include such varied materials as volcanic ash, sand, crushed limestone and dolomite, pulverized igneous rock, twigs, chopped feathers, straw, dung, and even crushed sherds.

Some fillers undergo thermal changes on firing, such as decomposition or oxidation. Carbonates, for example (limestone, shells, etc), begin to decompose above 750°C, giving off carbon dioxide and leaving behind calcium dioxide. After a period of time the calcium dioxide recombines with atmospheric carbon dioxide and returns to its original composition. The calcium carbonate so formed, however, differs from the original in crystal structure. The change in structure of carbonate fillers can be used as a rough indicator of the temperature attained by ancient potters during firing. Observed under a microscope, limestone filler particles will appear powdery and soft if firing took place above 750° but consolidated and crystalline if firing was performed at lower temperatures.

When organic materials—dung, twigs, feathers, and so on—are used as fillers, they may be either partially oxidized (charred) or completely burned away during firing, depending on the conditions prevailing in the kiln. In the former case a carbonaceous residue is left within the body of the pottery, which thereby acquires a gray or black color. The complete oxidation of organic materials yields gaseous products that escape from the body of pottery, leaving it porous.

POTTERY MAKING

There are four main stages in the making of pottery: (1) preparation of the clay paste, (2) forming and shaping, (3) drying, and (4) firing. Of these, all but the second—which involves mainly stylistic considerations—are of importance to the archaeological chemist and are considered in some detail in the paragraphs that follow.

Preparation of the Clay Paste

Clay pastes—usually called the *ceramic body*—vary widely in composition, depending on the nature of the local clay, the intended use of the wares, and the ceramic traditon of the potter.

Raw clay usually requires preliminary separation of the coarser fractions, followed by grinding, to obtain a very fine-particled material. Foreign matter is at the same time removed and the clay is then brought to a uniform consistency by prolonged kneading with water. Tempering (the addition of nonplastics, mentioned earlier) is done at this stage, and the material is then left to dry to the proper consistency required for fashioning into pottery.

Smoothing and Slipping

After an object has been modeled and while it is still wet, the surface is smoothed, usually with a wet hand. This operation improved appearance and also makes the material less permeable to liquids by filling up pores with fine particles of clay.

"*Slipping*" is usually done after the object has dried. A "*slip*" is a thin paste of clay and water with or without the addition of a pigment. It is applied to molded wares for smoothing the surfaces and rendering them impervious to liquids. A slip makes an excellent ground for painting and decorating and, if a pigment is added to it, changes the color of the surface to which it is applied.

Drying

During drying, water held mechanically in the body—water of plasticity—is given off. The water is lost at low temperatures (below 150°C) and is not to be confused with chemically held water—water of constitution—which is driven off during firing.

The drying of clay draws the clay particles together and closes the pores. Most ceramic bodies shrink during this stage. Differential shrinkage (when the outer layers shrink faster than the rest) may lead to distortion, cracking or bursting of wares. One of the objects of adding nonplastic fillers to the clay is to obviate this. The coarse particles of the filler not only facilitate the escape of water, but also reduce the amount of water used, by displacing some of the clay particles and their accompaning water film.

Firing

Firing is in some ways the most critical stage in the process of making pottery. It is during this stage that the ceramic body acquires the hardness, toughness, and durability essential to the finished product. Three major changes take place during firing: oxidation, dehydration, and vitrification.

As the temperature rises to 200-500°C, organic materials oxidize. Organic materials in the form of straw, twigs, dung, and so on constitute an integral part of primitive ceramic bodies. When heated in an oxidizing atmosphere (i.e., one in which oxygen is present in abundance), the carbon in organic materials combines with oxygen to form carbon dioxide, which is then given off:

$$C + O_2 = CO_2 \uparrow$$

As a result, the space occupied in the wet body by organic compounds is left vacant, and porous ware is obtained. If heated under reducing conditions (when oxygen is not available), organic compounds are not oxydized but are charred (i.e., partly burned), leaving a residue of elemental carbon. Gray-black ware is obtained. The temperature and time required for oxidation may vary greatly for different bodies. As a rule, however, firing is timed so as to allow oxidation to proceed to completion before dehydration commences. In this way organic matter is not trapped in the fired clay.

Dehydration, the loss of water of constitution of clays, occurs and 450-600°C. During dehydration, structural water, chemically held in the clay molecules in the form of hydroxyl ions, is given off. The atomic structure of the clay is thus irreversibly changed.

TABLE 8.7   **Thermochemical Changes Induced by Firing
Pottery**

Temperature (°C)	Chemical Change
100-150	Evaporation of water of constitution
200-500	Decomposition of organic matter
450-600	Decomposition of clayey substance (loss of water of constitution)
600-800	Incipient vitrification
800-900	Decomposition of calcium carbonates (limestone fillers)
>850	Recombination of oxides and vitrification
>900	Development of new crystalline phases
>1500	Fusion

Above 600°C the *baking* stage begins, when vitrification (formation of glass) is started by the more fusible (fluxing) impurities present in clay. Above 600°C enough vitrification takes place to cement the clay particles together and give the ware a moderate degree of hardness while remaining highly porous. At temperatures above 700° the clay can be considered baked.

The full firing or *burning* stage begins at about 850°C: vitrification proceeds further and the pottery becomes progressively more dense and less porous and acquires a high strength. If two baked vessels are struck together, they emit a dull sound; if two "burned" objects are struck together, they emit a good ring. Above 1000°C—a temperature usually unobtainable by the primitive potter—the component particles of pottery melt and coalesce into a glass.

After firing, pottery is normally held for some time at its finishing temperature and then cooled slowly. If allowed to cool too quickly, it may crack or spall. A rapid change in temperature is too great a shock for fired clay to stand. (Table 8.7).

KILNS FOR POTTERY

Primitive pottery, from earliest times until the present, was (and still is) fired in *periodic kilns* (some very primitive pottery was fired in domestic "open" fires). In the periodic kiln the firing process consists of setting up the kiln, firing, cooling, and finally withdrawing the pottery.

The design of periodic kilns varies widely. In most, called "open-flame" kilns, the heat is used directly: the fire box is usually below the ware, the flames surround it, and the gases of combustion move upward by natural convection (Fig. 8.6).

Temperature uniformity in open flame kilns is poor; the bottom of the kiln is hotter than the top, and because of the gas flow, there are also horizontal temperatures gradients. This results in irregular firing, some articles reaching higher temperatures than others.

FIRING ATMOSPHERE AND COLOR

The color of fired pottery, as that of clay, is due to impurities; white clay will produce white fired ware. But the kiln's atomsphere will affect the color of fired pottery.

When secondary clays are used, clear colors—reds, buffs, or yellows—are the result of an oxidizing atmosphere. The best reds are produced by slow firing in a strongly oxidizing atmosphere, at temperatures in the range 700-900°C. A temperature of 800°C is enough for most secondary clays to "mature," but some require 950°C and others, 1000°C. Overfiring will turn a rich red to a muddy brown. Underfiring will yield buff, grey, or mottled colors.

High temperatures and a good draft—to ensure abundance of oxygen—will facilitate the oxidation of organic matter. If the draft is shut off and the kiln is filled

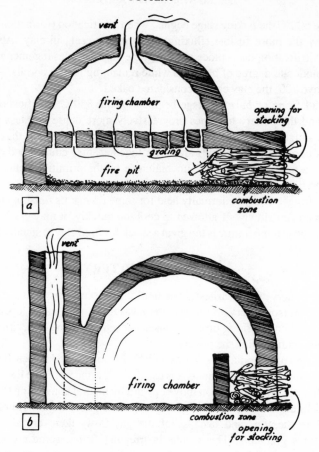

Fig. 8.6. Wood-fired ceramic kilns: (*a*) updraught kiln; (*b*) downdraught kiln.

with reducing gases (e.g., carbon monoxide or hydrocarbons released in the early stages of combustion), oxidation is prevented and the pottery is partially or totally blackened by particles of unburned carbon, from wood smoke, for example, depositing within the clay's fabric. Red ferric oxide (Fe_2O_3) is also reduced by carbon monoxide to black ferrous oxide (FeO) and magnetite (Fe_3O_4), which contribute to the darkening of ware.

A strong reducing atmosphere is often used deliberately to produce black pottery. Some ancient objects owe their black color to oil or bitumen, which was allowed to permeate through the walls of vessels that were afterwards heated to a low temperature in a reducing atmosphere. The oily substance would be carbonized, yielding a black that could readily be burnished to give a sheen to the surface and that also made the pottery impervious to thick liquids. Other gray or black pottery owes its color to reduced iron obtained by firing the kiln under smoky conditions; that is, with excess combustion gases and lack of oxygen.

It has been suggested that the atmosphere in which a given type of ancient pottery was fired can be determined by refiring shards in air: a shard is heated in a temperature-controlled muffle furnace to a temperature sufficiently high to burn out carbon and to bring iron oxide to the ferric state (700-750°C is usually satisfactory for low-fired ware). If the shard originally contained carbon or partially burned organic material—indicative of a reducing atmosphere in the original kiln—it will usually become lighter within a few minutes. If there is no change of color on refiring, the article was originally fully oxidized. It must be borne in mind, however, that a light-colored, fully oxidized shard may differ from a dark, unoxidized one only in the length of time it was fired. The dark one may have been fired in an oxidizing atmosphere for a short time only: under such conditions, not all organic material would have had time to be fully oxidized. Therefore, it is more accurate to describe dark or gray pottery as unoxidized, rather than as having been fired in a reducing atmosphere.

GLAZING

Fired pottery is porous, absorbs moisture readily and has a rough surface. For protection against moisture absorption, for concealment of rough surfaces and for decoration, ceramic ware is often glazed. Glazing is the application of a vitreous coating to fired pottery, followed by further firing to a temperature at which the glaze melts and adheres to th underlying surface. The subject of pottery glazes is dealt with in Chapter 9.

GRADES OF POTTERY

A number of physical properties are used to qualify different grades of pottery. These are strength, color, translucency, and, most important, porosity. *Translucency*—the converse of opacity—is the capacity to transmit light. In fired pottery it depends on the degree of vitrification and on the nature of the impurities present in the body. *Porosity* is measured by the amount of water absorbed by a ceramic body and is expressed as a percentage of the original weight. The porosity of pottery depends on the raw materials used and on the degree of vitrification. The quality of pottery is usually related directly to the degree of translucence and inversely to the porosity.

Terracotta

Terracotta is a general term comprising all brown, red, or yellow porous pottery not covered by a glaze. Fired at temperatures below 900°C, it undergoes very little vitrification. Hence it is soft—it can be scratched with a hard steel—and very porous (5-15 percent). Terracotta is not translucent. Most primitive pottery falls into this category. When the same type of ceramic body as is used for making terracotta is fired at temperatures of 1000-1200°C, it is referred to as *red earthenware*.

Earthenware

Earthenware is stronger than terracotta. It is not translucent and has a porosity of 3-10 percent. Earthenware is usually glazed, with a red, buff, or cream ceramic body. Red earthenware is fired at 1000-1100°C, and buff or cream, at 1100-1200°C.

Majolica and Faience

Two particular types of earthenware are *majolica* and *faience*. These terms were originally used during the middle ages to refer to glazed earthenware made in Majorca (Spain) and Faenza (Italy). Today these two terms are used in many countries to designate different types of locally made earthenware. The body is relatively weak, usually red or gray, not translucent, and having a porosity of about 15 percent.

In archaeology the name ''faience'' has become associated, quite incorrectly, with ancient Egyptian glazed ware made of quartz frit.

Egyptian Faience

Egyptian faience, which is not pottery, should not be confused with neither *true faience* (see preceding paragraph), or with other ceramics such as porcelain, nor with the pigment ''Egyptian blue'' discussed elsewhere in this book. Egyptian faience is a pseudo-ceramic material containing practically no clay at all. It consists of a siliceous body made up of powdered quartz (ca. 90 percent), lime (1-3 percent) and some felspar, held together by a bonding substance composed of soda, lime, alumina, and copper and coated with an alkaline glaze. Powdered quartz has little rigidity. The strength of Egyptian faience depends chiefly on the glaze. As soon as the very hard glaze (usually blue) surface coating is removed, the material powders easily.

Chemical analysis of Egyptian faience has yielded the results shown in Table 8.8. These indicate quite clearly that a very limited range of raw materials was used in making up the body. In a white body, for example, powdered white quartz rock or powdered white quartz pebbles would be used; yellowish brown or gray bodies would be made up with powdered stone, sandstone, or flint. These assumptions are confirmed by microscopic examination; both fine- and coarse-

TABLE 8.8   Composition of Ancient Faience

Faience	Silica	Alumina	Iron Oxide	Copper Oxide	Lime	Magnesia	Alkalies
Egyptian body material	90-99	0.5-15	1	—	0.5-3	0.5-2	0.4-3
Non-Egyptian body material	95	0.5	0.5	0.3	0.2	0.2	0.2
Egyptian Glaze	80	0.5	0.5	2	4	0.6	15

body faience consist of sharp, angular grains of quartz without any visible admixture of other materials (Lucas 1962). X-Ray-diffraction analysis offers further confirmation that only varieties of powdered quarz were used to make Egyptian faience (Kiefer and Allibert 1968).

The glazes are essentially soda-lime or potassium-lime glasses, usually termed *alkaline glazes*. They are usually blue, green, or blue-green, although sometimes violet, white, and even occasionally two or more colors are found. It has been suggested that the body glazed itself during firing. If so, is it possible that a flux was applied over the entire surface to facilitate the process of self-glazing. By refiring ancient Egyptian faience it has been possible to show that the original firing was carried out at temperatures ranging from 890° up to (but never over) 920°C.

Faience was fashioned into small objects such as beads, scarabs, amulets, bowls, and tiles. Its use can be shown to date from as early as 3200 B.C. in Mesopotamia (Mallowan 1947). This is more than 1600 years before its use became widespread in ancient Egypt, during the eighteenth dynasty (1550 B.C.). From that time onward it was used for about 1500 years in an unchanged form. Variations of Egyptian faience continued to be made until the fourteenth century C.E.

Stoneware

Stoneware is the name commonly applied to highly vitrified ceramic bodies—red, cream, brown, or gray in color. Stoneware is strong, nontranslucent, and has a porosity of 0-5 percent. It may be either glazed or unglazed, the body being fired at 1200-1300°C.

Porcelain

Thin porcelain ware is translucent. The translucency is due to a high degree of vitrification. Porcelain is very strong, with less than 2 percent porosity. It is always white and always glazed. Firing temperatures of porcelain are in the range 1300-1450°C.

CHEMICAL STUDIES ON ANCIENT POTTERY

Pottery making is one of the most ancient technologies. Because of the durability of pottery the universality of its use and its total lack of value after breakage, pottery fragments are by far the most abundant of all the material vestiges of former cultures.

Archaeologists have evolved systems of classification of pottery based on style. Chemists have also been interested for a long time in the composition and properties of ancient ceramics and in the techniques of pottery production. During the nineteenth-century speciemens of ancient pottery were already being chemically analyzed. However, interest in the subject was sporadic, and no systematic investigations were attempted. Early chemical work was concerned with the description of the matrix elements of pottery clay and tempering materials, which are very similar throughout the world. It was only with the advent of modern sensitive techniques of examination that the study of ancient pottery came into its own. The availability of rapid and sensitive analytical methods has promoted investigation into the raw materials used by ancient potters, the composition of pottery, the geographical provenance of ceramic artifacts, and the techniques of manufacturing and decorating pottery. (Peacock 1969).

Attic Vase Painting: Black-and-Red Glaze

The technique used by the Greeks to produce their characteristic black-and-red glazed "Attic" ware has long intrigued scholars. Only in the past three decades, however, has important progress been made in analyzing and reproducing this technique.

In some ways the term "glaze," used in reference to the surface of Attic ware, is a misnomer. A glaze is actually a glass, with coloring matter dissolved in it, which fuses and produces a glassy surface. This is not true of Greek glaze, which does not melt to a glass but only *sinters,* that is, partly melts, becoming a coherent solid. Perhaps it would be more accurate to refer to it as a *sintered engobe.* The terms "Greek glaze" and "black glaze" have been so often used, however, that it seems easiest to retain them here.

"Greek glaze" is an actual part of the clay from which the ware itself was made. It is not an applied material, nor is it a true glaze. Spectrographic analysis has revealed that the composition of the black glaze and of the clay body beneath it are basically the same; there is no element found in substantial quantities in the black glaze that is not also found in the clay body (Noble 1960).

To produce the black-and-red ware, advantage was taken of the peculiar properties of the compounds of a single element, iron, which are red when oxidized and black when reduced. Iron compounds, suspended in the glaze in the form of finely divided particles, were either oxidized or reduced during firing, to produce red and black colors, respectively. The red pigment is always ferric oxide

Fig. 8.7. Black-and-red glazed Attic ceramics.

(Fe_2O_3) and the black, ferrous oxide (FeO) and magnetite (Fe_3O_4) (Oberlies and Koppen 1953, 1954, 1962; Oberlies 1968). Both came from the iron naturally present in the carefully chosen clay that was used in making the pottery (Schumann 1942, 1943; Rijken and Farejee 1941). If the entire firing were done under oxidizing conditions, both the ware and the glaze would turn ,and remain, red.

The production of Attic Greek ware involved, as a first step, the *elutriation* of the clay, that is, the separation of smaller particles from larger ones by suspending the powdered clay in water and allowing it to settle (see Fig. 8.8). The rapid-settling, coarse particles were discarded; the finer, slow-settling fraction of the clay was used for making pottery, whereas the finest part of the suspension, which did not settle out, yielded material for the paint slip (Winter 1959).

The paint slip was prepared by modifying the sintering properties of the ellutriated clay. To do this the clay was first peptized, that is, colloidally dispersed

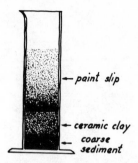

Fig. 8.8.   The elutriation process (schematic).

in water by the addition of dispersing agents such as tannins (q.v.) (Hofmann 1962). Further modification of the properties of the slip was achieved through the addition of alkali (probably extracted from wood ash). Alkali acts as a flux, lowering the temperature at which sintering takes place. Its addition in the proper amount to the slip, followed by a careful regulation of the firing conditions, yielded the dense, glaze-like oxidation-resistant black layer over a porous red body, which is characteristic of this type of Greek ware (Farnsworth 1959).

After the vessels had been shaped and air-dried, their surfaces were polished and a very thin slip layer, probably not more than 50 microns thick, was then applied over the desired black areas. Firing was carried out in three stages. The first stage was under oxidizing conditions, that is, in an excess of air, at about 850°C. After this stage the entire vessel—body and slip-painted areas—would be colored red. To develop the black color, the firing was continued under reducing conditions: the air vents of the kiln were closed and the furnace fueled with wood rich in resin or wet wood, or water was poured over hot glowing charcoal. Under these conditions, hydrogen and carbon monoxide—both good reducing agents—would be produced. Carbon monoxide reduced the red ferric oxide to black ferrous oxide:

$$Fe_2O_3 + CO = 2FeO + CO_2$$

In the presence of water vapor, the reduction would also yield magnetite (Fe_3O_4), "blacker" than ferrous oxide:

$$3Fe_2O_3 + CO = 2Fe_3O_4 + CO_2$$

Heating under reducing conditions was continued for a short while (probably 5-10 minutes) until the slip sintered and became resistant to oxidation. After this second stage the entire surface of the pottery was black. In the third, and last, firing stage the air supply was restored and oxidizing conditions reestablished. At the same time the temperature was probably raised to nearly 900°C. During this reoxidation stage the sintered slip remained black and the underlying body once again became red. The kiln was finally allowed to cool in an oxidizing atmosphere.

Electron microscopy has shown that the red surfaces are rough and porous; oxygen was thus able to penetrate the pores during the last stage and oxidize the black ferrous oxide and magnetite back to red ferric oxide. The black surface, on the other hand, is smooth and very finely grained.

Intentional Red Pottery

The red color in Attic pottery is that of the body and is termed *occasional* or *accessory red*. Some Greek pottery has bright coral-red surfaces that appear to be painted-on, like the black. This red is usually called *intentional red* glaze. The obvious points of difference between "intentional red" and "accessory red" are the gloss and the color; that of the former being often spoken of as "sealing-wax" or "coral" red.

Intentional red was made by adding ocher (i.e., ferric oxide) to a good "black" slip. The amount of ocher added was not always the same. It would depend on the quality of the clay used. Analysis indicates that enough ocher was added to bring the iron content of the paint slip to about 11 percent. In the three-stage firing-cycle already described (oxidation-reduction-oxidation) the black glaze slip would remain black, whereas that with added ocher turned an intense red (Farnsworth and Wisely 1958).

Manganese Black Decoration

A manganese-black enriched clay used as a coloring slip, made possible the simultaneous production of black-and-red colored pottery fired under oxidizing conditions only. The technique of manganese-black decoration was reconstructed from analytical data obtained by X-ray diffraction, X-ray microanalysis, and scanning electron microscopy (Noll, Holm, et al. 1973). Maganese ore, containing manganese dioxide (MnO_2) and iron oxide (Fe_2O_3), was added to the clay slip used for decorating vessels. Firing at 900-1000°C resulted in very black manganese compounds. Because of its simplicity, this technique of making black decorations became very popular and once discovered, was used whenever the raw materials were available.

Chinese *"Sang-de-boeuf"* Porcelain

Reduction firing was successfully employed by the Chinese to make their *celadon* and *sang-de-boeuf* ware. The reduction of finely divided copper pigments to copper metal produced a very beautiful red color. Sometimes the copper pigment dissolved in the glaze, whereupon the color would become light green (Farnsworth 1959).

Mössbauer Spectroscopy of Pottery

Iron is often present in pottery materials. Relatively small variations in its chemical state seem to bring about a rich variety of colors in the latter. Iron is one of the elements with isotopes suitable for Mossbauer spectroscopic measurements. The results of these measurements yield information about the chemical state and the physical environment of iron atoms in ceramic materials. The information may be useful in providing a basic understanding of the chemistry of iron in clays and of the chemical changes occurring when the clays are fired under controlled conditions. Such understanding may help the archaeologist in a number of ways:

1. Clays from different areas contain different minerals in different proportions. The treatment each clay receives during firing differs from kiln to kiln. Hence it is reasonable to assume that pottery from one particular kiln (or at least area) should have Mossbauer spectra characteristic of the mineral composition of the fired clay (Eissa, Sallam, et al., 1976). An exploratory study of ancient pottery from northwestern Greece revealed that differences in Mössbauer spectra correlated well with a classification of the samples based on archaeological criteria. The spectra are sufficiently different to be used as supporting evidence for the classification of pottery (Gangas, Kostikas, et al. 1971).

2. Useful deductions can be made as to why particular clays were selected for making pottery.

3. By comparing Mössbauer spectra of pottery with those of clays and baked clays from known sources, it may be possible to obtain an explanation for their chromophoric effect. One interesting experiment was concerned with the study of red clay from Tel Ashdod (Israel). Mössbauer spectroscopy revealed the chemical form of the iron in the clay and the changes that resulted from firing under different conditions. A very delicate control of firing was required to produce ware of specific color from local clay (Hess and Perlman 1974).

PROVENANCE OF POTTERY

When investigating the provenance of pottery, special attention is given to the detection and measurement of minor and especially of trace constituents. Since potsherds consist of fired clay, their classification and identification are very similar to those of mineral samples.

Classification and Identification of Pottery

Given a series of pottery samples having a number of elements in common, each present in a different concentration, a system of groups must first be defined, into which the samples—usually potsherds—can be divided and that can serve as a basis for later classification. If full prior information is available regarding the origin of the samples, only identification will be required—that is, the assignment of each individual potsherd to one of a set of possible groups. In many instances, however, there is only limited knowledge available concerning the possible origin of a set of samples. In this case classification is called for, that is, the establishment and definition of groups among the available samples (Op de Beck and Hoste 1974). The nature of the groups that can be expected is illustrated in Fig. 8.10, a model

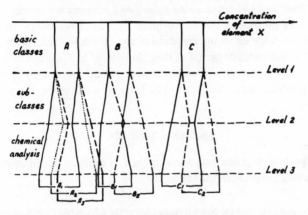

Fig. 8.9. Provenance of pottery. The model illustrates the predictable differences of pottery classes that can be expected from the analysis of a single element and also the lack of precision inherent in the classification of pottery on the basis of analysis of a single chemical element.

that summarizes the predictable differences for a single element. The model is ordered in three levels. In the first level are depicted the ranges of concentration of the element in different clay beds. The existence of considerable differences in trace-element concentration between different clay beds is well established, although at present little is known about the variation of trace-element concentration within a single clay bed.

The composition of a clay bed may differ considerably from that of the pottery made from it. This can be due to a number of factors: refining of the clay (removal of coarse material), addition of tempers and salts during the preparation of the ceramic body, and even addition of clay from other clay beds, may all enhance or decrease the concentration of any trace constituent. As a consequence of these operations, the original ranges of concentration tend to separate into a number of subgroups, as illustrated in the second level in Fig. 8.10. It follows that there is little point in trying to establish a relationship between a given sherd and any particular clay bed; instead, each sherd should be assigned to an identified group, with the objective of establishing compositional uniformity within groups.

The imprecision of chemical analysis at trace level may be considerable; errors may be as high as 30-35 percent. This leads to a spreading of the ranges of concentration within groups, as indicated in the third level in the figure, where classes are observed after chemical analyis.

Several methods have been suggested for classifying a group of sherds. For example, a group of well-known sherds, sharing a common stylistic origin, may be selected and analyzed and correlations established between the concentrations of various elements present in them. (Perlman, Asaro, et al. 1972). In this way different types of trace element composition can be established, each one characteristic of a distinct area or period of time. Individual samples can then be fitted into one or another group with greater or lesser certainty, depending on the extent to which the results of their analysis deviate from the mean of the group.

Analytical Methods

The most important requirements for a method of analysis suited to the identification of pottery are perhaps the following:

1. The method should be sensitive to elements which may be present in very small (trace) amounts.
2. The number of elements to be determined should be as large as possible. The more elements are measured, the greater will be the degree of certainty in identifying the source of the pottery.
3. A large number of samples may have to be analyzed.

Given these requirements, it emerges that the ideal analytical method for use in the identification of pottery is a rapid—preferably nondestructive—instrumental method capable of determining as many as 10-20 trace elements simultaneously. *Optical emision spectroscopy* and *neutron-activation analysis* are techniques that fulfill these conditions. (Kostikas and Simopoulos 1976, Widemann 1976). Advances in the development of X-ray fluorescence have recently added the later techniques to the list of methods suitable for provenance studies (Giauque and Jaklevic 1972) (Gautier 1976).

Minoan and Mycenaean Pottery

Articles of pottery from Mycenaean Greece and Minoan Crete were widely used not only in the lands of their origin, but also in Asia minor, Syria, Palestine, Egypt, Cyprus, Sicily, and Italy. Stylistically, the ceramic ware produced in the Aegean region is remarkably uniform. Consequently, there is aways some uncertainty about the source of Aegean pottery found in distant regions. In addition, it is difficult to distinguish Mycenaean and Minoan pottery from the imitations made by local potters. Chemical analysis of trace and minor elements present in sherds was used to obtain a more detailed view of Greek ceramic provenance (Cattling 1963; Cattling, Blin Stoyle, et al. 1961; Millett and Cattling 1967; Richards, Blin-Stoyle, et al. 1963).

In one research project, more than 1000 Minoan and Mycenaean shards, found at 50 sites in the Greek mainland, the Peloponnesian Islands and Cyprus were analyzed by emission spectroscopy. The most obvious gain arising from this study was the possibility of distinguishing between: (1) Mycenaean and Peloponnesian pottery and (2) Minoan and Cretian pottery. The results indicate that Mycenaean pottery was produced only in the Peloponnese and was exported to the Greek Islands and to Cyprus. It was not produced by Mycenaean migrant potters. Local workshops, using local clay unique to those sites, imitated the original imported pottery styles.

An idea of the magnitude of the research project can be gained by examining just one of the 17 distinct compositional types of pottery identified (arbitrarily named A to Q): out of 60 Cypriot shards of Mycenaean type analyzed, 40 specimens, stylistically Mycenaean, shared the compositional characteristics of group A, irrespective of the sites at which they were collected. The 20 others, which appear to be imitations of Mycenaean, were divided into two groups that were nowhere else recorded.

The analytical similarities within groups, as well as the differences from one group to another, were confirmed by analysis of trace elements using neutron activation (Harbottle, Hall, et al. 1969; Bieber, Brooks, et al. 1976). The use of this technique, which allows the determination of concentrations of elements of less than 1 ppm confirmed the homogeneity of the groups that had been established by emission spectroscopy. Neutron activation provided a useful complement to emission spectroscopy, both as a check on the latter and as a means of extending the range of elements determinable at low concentrations.

Pottery Analysis by Neutron Activation

A detailed evaluation of the accuracy obtainable when using neutron-activation analysis was conducted. Pottery samples were irradiated in a nuclear reactor for a definite period of time and the gamma radiation that they emitted after irradiation was measured using silicon-and-germanium γ-ray spectrometers. In this way more than 40 elements could be determined in each sample (Perlman and Asaro 1971).

From the data obtained, ''trial pottery groups'' were established, disregarding any archaeological information available about the groups. To do this, a computer was employed to make computations for about 20 chemical elements in each sample and to calculate the probability of any particular sample belonging to one or another of the "trial pottery groups." Using a computer program especially developed for this study, groups were later analyzed and the results statistically tested to establish where single samples fitted into any of the groups. (The probability with which a sample fits into any group depends on the dispersion of the analytical results in the group.) From several thousand analyses made, it was learned that there is no difficulty in distinguishing pottery groups. In many instances, samples that were statistically different proved to have a common origin of source materials. Others were labeled as imports on both stylistic and chemical grounds.

Pottery Groups from Upper Egypt.   The results of analyses of material from Upper Egypt revealed a complex pattern of pottery composition. Categories of shards from Nag-e-Deir in Upper Egypt, established by compositional analysis, were also found to be easily distinguishable by visual examination. Such a clear correlation is, however, not universal, and typologically similar specimens some-time differ in composition.

An interesting demonstration of the accuracy of the results obtainable from neutron activation of pottery is provided by the analysis of a clay plug used for sealing a predynastic Egyptian jar. The composition of the clay plug was found to agree closely with that of pottery found in the vicinity (Table 8.9).

TABLE 8.9   Composition of "Nile Mud" Pottery and a Clay Plug, Probably Inserted at a Later Date

Composition	"Nile Mud" Ware (Mean Value, 32 Sherds) (ppm)	Fired-clay Plug
Antimony	0.29 ± 0.07	0.29
Arsenic	0.88 ± 1.14	1.42
Cobalt	34.96 ± 1.60	39.64
Cesium	1.39 ± 0.21	1.66
Chronium	180.8 ± 15.6	194.5
Hafnium	8.67 ± 0.75	8.66
Lanthanum	32.77 ± 1.20	33.92
Lutetium	0.512 ± 0.027	0.520
Manganese	1204 ± 68	1277
Scandium	23.11 ± 0.96	25.46
Tantalium	1.445 ± 0.106	1.372
Uranium	2.26 ± 0.41	2.05

For progress to be made in neutron activation of pottery, a combination of factors is required: a powerful nuclear reactor, analytical instruments of high sensitivity, good computation facilities, and the accumulation of background data. For instance, to provide a reliable analysis of an unknown Cypriot sample, it was estimated that about 10,000 pieces of pottery would have to be analyzed. An adequate study of Mediterranean pottery would require the analysis of about 100,000 pieces.

GLOSSARY—CERAMICS

Ball clay. A secondary clay with highly plastic characterisics, usually high in organic matter.

Bisque. Unglazed, once-fired pottery, also called *bisquit*.

Body. A mixture of ceramic raw materials (clay and nonplastics) having properties suitable for forming, drying, and firing. Also called *clay paste*.

Ceramic materials. Inorganic materials that contain metallic and nonmetallic oxides (e.g., clays, glasses) and are subjected to high temperatures during manufacture.

Clay minerals. Fine-grained hydrated aluminosilicate minerals formed by natural chemical action of felspathic rocks.

Common Clays

Name	General Formula
Halloysite	$[Al_2Si_2O_5(OH)_4 \cdot 2H_2O]_2$
Kaolinite	$[Al_2Si_2O_5(OH)_4]_2$
Montmorillonite	$[(Al(MgR^+))_2Si_4O_{10}(H_2O)]_2$
Serpentine	$[Mg_3Si_2O_5(OH)_4]_2$

Clay, primary. Clay formed *in situ* by natural chemical reactions of igneous rocks. Also called *residual clay*.

Clay, secondary. Clay moved from its point of origin by natural agencies, before deposition at a site. Also called *sedimentary clay*.

Colloid. A mixture consisting of very finely divided solid particles dispersed in a liquid, usually water, and exhibiting some of the gross properties of solutions.

Crazing. The cracking of glazes as a result of internal tensile stresses.

Earthenware. Low-fired, colored-body ceramic ware covered with a glaze.

Efflorescence.   The appearance of a white powdered deposit on the surface of pottery, resulting from leaching of one or more of the components from the interior to the surface.

Elutriation.   The separation of small particles from larger ones by suspending a powder material in water and allowing it to settle.

Engobe.   A slip used as a pottery coating.

Faience.   A name loosely given to quartz-cored objects of Egyptian origin, as well as to Moorish pottery ware covered with a glaze.

Fire clay.   Clay having approximately the composition of kaolite.

Firing.   The high-temperature baking of clay materials that sets the particles into a coherent product with desired properties.

Flocculation.   The aggregation of colloidal particles within a suspension.

Green ware.   Formed and hardened, but as yet unfired, clay products.

Kiln.   A furnace for the production of ceramic ware.

Mullite.   Crystalline phase formed when clay is fired over 1000°C.

Peptization.   The liquefaction or transformation of thin particle clay into a clay colloid.

Porcelain.   A fired and highly vitrified clay product that has almost no apparent porosity and is translucent.

Refractory.   A material capable of withstanding high temperatures, (e.g., ceramics).

Sintering.   The agglomeration of material particles caused by high-temperature treatment.

Slip.   A fluid suspension of colloidal ceramic materials.

Stoneware.   A highly vitrified clay product, coarser than porcelain.

Tannins.   Substances found in many plants and chemically related to phenols.

Temper.   Inert materials added to ceramic bodies to alter the plasticity of clay. Also called *filler* and *nonplastic* material.

Terracotta.   Pottery made from clay and coarse tempers and fired at temperatures well below 1000°C.

Vitrification.   The reduction of porosity through the formation of glass in a ceramic product. The initial stage in the formation of glass within a ceramic body is termed *incipient vitrification*.

REFERENCES

Bieber, A. M., Brooks, D. W., Harbottle, G., and Sayre, E. V., in *Application of Nuclear Methods in the Field of Works of Art,* 111, 1976.

Cattling, H. W., *Archaeometry,* **6,** 1, (1963).

Cattling, H. W., Blin Stoyle A. E., and Richards, E. E., *Archaeometry,* **4,** 33 (1961).

de Korosy, F., *Naturwissenshaften,* **62,** 484 (1975).

Eissa, N. A., Sallam, H. A., and Negm, S. M., *J. Phys. (Paris), Colloq. No. 6,* **873,** (1976).

Farnsworth, M., *Archaeology,* 242 (1959).

Farnsworth, M., and Wisely, H., *Am. J. Archaeol.,* **62,** 165 (1958).

Gangas, N. H. J., Kostikas, A., Simopoulos, A., and Vocotopoulon, *Nature,* **229,** 1971 (1971).

Gautier, J., *Ann. du Lab. de Res. des Musees de France,* 58, 1976.

Giauque, R. D., and Jaklevic, J. M., *Adv. X-ray Anal.,* **15,** 164 (1972).

Harbottle, G., Hall, E. T., and Cattling, H. W., *BNL* 13740 (1969).

Hess, J., and Perlman, I., *Archaeometry,* **16,** 137 (1974).

Hofmann, U., *Angewandte Chemie, Internat. Ed.,* **1,** 341 (1962).

Kiefer, Ch., and Allibert, A., *Ind. Ceramique,* **607,** 395 (1968).

Kostikas, A., and Simopoulos, A., *European Spectrosc, News,* **1**(6), 10 (1976).

Lucas, A., *Ancient Egyptian Materials and Industries,* Arnold, London, 1962, p. 158.

Mallowan, M. E. L., *Bazar Iraq,* 9 (1947).

Millett A., and Cattling, H. W., *Archaeometry,* **10,** 70 (1967).

Noble, J. V., *Am. J. Archaeol.,* **64,** 307 (1960).

Noll, W., Holm, R., and Born, H., *Jahrbuch der Staatlichen Kunstsammlungen in Baden-Wurttenberg,* **10,** 103 (1973).

Noll, W., Holm, R., and Born, H., *Ber. Deutsch. Keram. Geselt.,* **50,** 328 (1973).

Oberlies, F., *Naturwissenschaften,* **55,** 277 (1968).

Oberlies, F., and Koppen, N., *Ber. Deutsch. Keram. Geselt.,* **30,** 102 (1953).

Oberlies, F., and Koppen, N., *Ber. Deutsch. Keram. Geselt.,* **31,** 287 (1954).

Oberlies, F., and Koppen, N., *Ber. Deutsch. Keram. Geselt.,* **39,** 19 (1962).

Op de Beek M. and Hoste, L., *Analyst,* **99,** 978 (1974).

Peacock, D. S. P., *World Archaeol.,* **1,** 295 (1969).

Perlman, I., and Asaro F., *Archaeometry,* **11,** 21 (1969).

Perlman, I., and Asaro F., *The Application of the Physical Sciences to Archaeology,* Vol. 12, California U. P., Berkeley, 1971, p. 21.

Perlman, I., and Asaro F., in R. H. Brill, ed., *Science in Archaeology,* MIT Press, Cambridge, Mass., 1971 p. 183.

Perlman, I., Asaro, F., and Michel, H. V., *Annu. Rev. Nucl. Sci.,* **22,**383 (1972).

Richards, E. E., Blin-Stoyle, A. E., and Millett, A., *Annu. Brit. School at Athens,* **58,** 94 (1963).

Rijken, A. J., and Farejee, J. Ch. L., *Chem. Weekbl.,* **38,** 262 (1941).

Schumann Th., *Ber. Deutsch. Keram. Ges.,* **23,** 408 (1942).

Schumann Th., *Forsch. und Fortschr.,* **19,** 356 (1943).

Widermann F., Report CSNSM-R-76-1, France, 1976.

Winter, A., *Technische Bertrage sur Archaeologie,* Romisch-Germanischer Zentral-museum, Mainz, 1959.

Glass, Glaze, and Enamel

Glass, glaze, and enamel have essentially the same chemical composition; the difference between them lies in the way in which they are used. *Glazes* and *enamels* are coatings applied to the surfaces of artifacts for decorative effects or to make them impervious to chemical attack or impermeable to liquids. Glazes are used to coat pottery, whereas enamels are fused to metallic surfaces, *Glasses,* on the other hand, are fashioned into objects unsupported by a backing or a body of other material. In the making of some very ancient glass a temporary supporting core was used, but this was removed when the object was finished; the glass was not intended to adhere to it.

The differences in use determine a number of technical differences between glass, glaze, and enamel. For example, a glass is usually required to have a low melting point and to exhibit plasticity over a wide temperature range. For glaze and enamel these requirements are irrelevant. On the other hand, the *coefficient* of *expansion* (the change in size brought about by a change in temperature) of a glass is relatively unimportant, whereas in a glaze or enamel, differences between the

TABLE 9.1   Qualities Required from Glass, Glaze, and Enamel

Property	Material	
	Glass	Glaze or Enamel
Ease of work	Low melting point; plasticity over a wide range of temperature	Properties determined by the characteristics of the underlying body
Transparency	Desirable	Not desired; opaque material will cover defects of the body
Coefficient of expansion	Relatively unimportant	Differences between body and coating lead to peel and "craze"

coefficients of expansion of coating and body, respectively, may lead to cracking or even to detachment of the coating. (Shafer 1976). Table 9.1 summarizes some of the qualities called for in glass, glaze and enamel.

Glass, glaze, and enamel all have in common the fact that they are in a particular physical condition called the "glassy state." The terms *glass* or *glasses* are often used as generic names to describe all these materials. It is in this wider sense that these terms are used in the following discussion.

GLASS

Glass is one of the oldest of man-made materials and one of the most valuable and versatile. Yet glassmaking remains, even today, one of the most empirical and least understood of technologies.

The oldest known glassmade articles have been estimated to be 9000 years old. But it is accepted that glass manufacture as an industry began about 3000 B.C. By 1500 B.C. a fairly high standard of glassmaking had been reached. Early work was crude. Glass vessels were formed around a core of sand or clay which was scraped out after the object was finished (Schuler 1959). The discovery of glassblowing, about the year 30 B.C., revolutionized the industry. In Roman times glassmaking developed into a large-scale industry, akin in many ways to the one we know today.

THE NATURE OF GLASS: THE GLASSY STATE

Glasses are materials obtained by cooling molten solids in such a way that crystallization does not occur and the product remains amorphous. Though glass

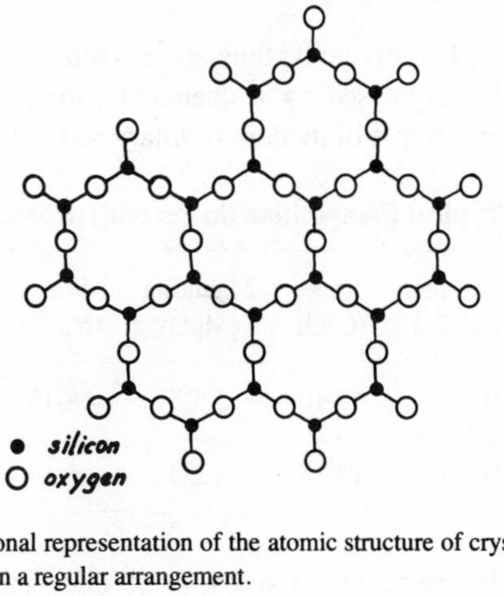

Fig. 9.1. Two-dimensional representation of the atomic structure of crystalline quartz. Silicon and oxygen atoms are linked in a regular arrangement.

has several of the properties of crystalline materials—for example, hardness and brittleness—it belongs to a class of materials known as *supercooled liquids,* which differ from crystalline solids in a fundamental way. In *crystalline solids* the constituent atoms occupy definite compositions in a perfectly ordered three-dimensional lattice. In glasses the atoms are also arranged in extended three-dimensional networks; but X-ray diffraction and other analyses have shown that these networks lack the precise periodicity and symmetry found in the structure of crystals (Zachariansen 1932) (see Figs. 9-1 and 9-2).

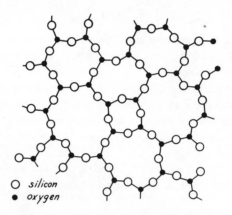

Fig. 9.2.   Two-dimensional representation of the atomic structure of glassy (amorphous) quartz. Note the disordered array of atoms characteristic of a supercooled liquid.

THE COMPOSITION OF GLASS

Glasses are not chemical compounds; they are mixtures of oxides. Usually their composition cannot be expressed by a chemical formula. Table 9.2 lists the components of a typical sample of modern ordinary soda-lime glass.

TABLE 9.2   Typical Composition (in Percent) of Modern Glasses

Glass	Silica (SiO_2)	Alumina (Al_2O_3)	Lime (CaO)	Magnesia (MgO)	Soda (Na_2O)	Potash (K_2O)	Lead Oxide (PbO)
Soda-lime	72-73	0.5-2	9-10	2.5-3	13-15	0.05	—
Lead	52-70	0.1-0.5	0-4	—	0-2	5-13	10-35
Potash	70-80	0.5-1	12-18	2-3	5-10	6-10	—

Glasses are obtained by the fusion of silica, or of silica and some metal oxides, followed by suffcently rapid cooling to prevent crystallization. With these components an infinite number of formulations is possible: one present-day glass manufacturer has more than 65.000 formulations on file!

The main component of ordinary glass is silica, (silicon dioxide, SiO_2), in the form of sand. Most glasses are based on this material, although certain kinds can be made without it. Silica sand fused and cooled on its own, under the proper conditions will result in a glass of pure silica. This is very difficult to work with; the temperature needed to fuse it is very high (over 1700°C), far beyond temperatures attainable by ancient craftmen, and the resulting melt is so viscous that entrapped air bubbles can not escape. The addition of certain metal oxides to silica sand lowers the melting temperature of the mixture to a workable level (700-900°C); it also makes the melt less viscous and more easy flowing. The most commonly added oxides are those of sodium and potassium, usually in the form of salts (carbonate, nitrate, etc.) that yield a glass of adequate quality for many purposes.

Formers and Modifiers

Silica, the main component of ordinary glasses, is known technically as the *glass former*. The flux of metallic compounds (soda or potash), which lowers the melting point of the silica by changing its structure, is called the *modifier*. The terms *glass former* and *modifier* define the respective roles of the main components of glass. The glass former has a three-dimensional crystal lattice. When heated, the crystal lattice breaks down into a melt and takes on the random

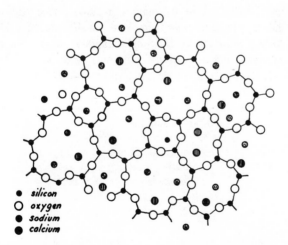

Fig. 9.3. Two-dimensional representation of the atomic structure of soda-lime glass. Sodium and calcium (modifier) ions are entrapped within the disordered array of glassy quartz.

configuration of a liquid. If the melt is cooled slowly, the molecules crystallize back into the same three-dimensional network in which they were arranged before melting. However, if a *modifier* is added and the melt is cooled rapidly without allowing crystallization to occur, the molecules of the melt set into a glass. The product then has an irregular, three-dimensional network of silicon and oxygen atoms, within which the modifier atoms are entrapped (see Fig. 9.3). The modifier atoms, usually sodium and potassium, lower the melting temperature of the glass: a high content of soda or potash reduces the melting point of glass to about 700°C. The modifier also reduces the viscosity of the glass when liquid, and alters the coefficient of expansion of the solid. It also diminishes the chemical durability of the glass by increasing its solubility in water. *Stabilizing compounds* are thus added to the melt in the form of alkaline-earth oxides: lime or magnesia. These decrease the solubility of soda and potash glasses and improve their chemical durablity. The alkaline-earth oxide content also determines the hardening range, and hence the period during which glass can be shaped.

Soda-Lime Glass

Lime (calcium oxide, CaO) is an effective glass stabilizer. It is added to glass in the form of limestone. The resulting product is a basic *soda-lime-silica glass*. Most common glasses are of this type. Soda-lime glasses have been used with only slight variations throughout history all over the world. They are easily molded, can be blown thin, and can be drawn into fine threads. The usual composition of soda-lime glasses is approximately 75 percent sand, 15 percent soda, and 10 percent lime.

Although soda, lime, and silica are the main ingredients, other oxides (e.g., aluminum oxide and magnesium oxide) are also usually present. Alumina (aluminum oxide, Al_2O_3) improves the chemical durability and reduces the tendency of glass to crystallize during forming operations. Magnesia (magnesium oxide, MgO) decreases the solubility of glass in a similar way to lime. Table 9.3

TABLE 9.3 Soda-Lime Glasses: Main Constituents (in Percent) of Ancient and Modern Samples

Glass	Silica (SiO_2)	Alumina (Al_2O_3)	Lime (CaO)	Magnesia (MgO)	Soda (Na_2O)
Egyptian	60.7	2.6	3.6	3.0	29.8
Mesopotamian	65.9	3.6	7.9	5.5	15.7
Roman	66.5	3.5	7.1	0.3	21.1
Japanese	69.5	4.0	6.7	0.5	17.9
Modern	~70	~1	~10	~3	~15

THE COMPOSITION OF GLASS

gives typical analysis of modern and ancient soda-lime glasses. A balance of silica and modifiers is required. Too much soda makes the glass easily corroded, even by water. Too much lime gives it a tendency to devitrify (crystallize) and makes it susceptible to weathering. Excess alumina raises the softening temperature of the glass.

Potash Glass

The use of potash instead of soda in the making of glass results in a product that hardens more rapidly and at a higher temperature than does soda glass. For this reason, potash glass was less suitable than soda glass for the molding techniques of the ancient glassmakers (Table 9.4).

TABLE 9.4   Potash glass: Main Constituents (in Percent) of Ancient Samples

Glass	Silica (SiO_2)	Alumina (Al_2O_3)	Lime (CaO)	Magnesia (MgO)	Soda (Na_2O)	Potash (K_2O)
Egyptian	69.7	1.7	9.6	1.9	2.6	12.6
Roman	62.1	1.8	8.6	2.9	4.3	20.4
Indian	73.6	2.9	3.9	2.4	3.1	13.4

Lead Glass

Lead oxide is a good modifier that yields glasses of low softening temperature and high density. Lead gives glass a peculiar brillance that was very much appreciated in Roman times (Table 9.5).

TABLE 9.5   Lead Glass: Main Constituents of Ancient Samples

Glass	Silica (SiO_2)	Alumina (Al_2O_3)	Lime (CaO)	Soda (Na_2O)	Lead Oxide (PbO)
Egyptian	71.0	2.5	1.5	4.5	13.0
Mesopotamian	39.5	4.3	4.4	9.7	22.8
Roman-Egyptian	32.2	2.1	8.1	10.3	31.1

Minor Constituents of Glass: Coloring Materials

The main properties of glass are governed by the major constituents already described. Certain properties, the most important of which is color, are determined by minor constituents; these rarely exceed 10 percent of the total mass and are often present in concentrations of less than 5 percent.

Mixtures of silica, soda, potash, and lime yield glasses that are essentially colorless. The color of glass is determined by the presence—whether intentional or otherwise—of various metallic oxides, usually in small amounts. Iron is the almost universal coloring agent of glasses, ancient and modern, even though the color it produces is often unintentional. When yellow sand is used for glass-making, the iron compounds always present in sand give a green tint to glass. This can easily be seen when a sheet of ordinary glass is viewed edgewise.

Iron is usually present in glass as a mixture of ferrous (Fe^{2+}) ions, which color the glass blue, and ferric (Fe^{3+}) ions, which color it yellow; the combined effect of the two ions is the familiar "bottle green." Less than 1 percent of iron is sufficient to give glass a distinct green hue. In Roman times glass makers already knew how to counteract this green color by physically masking it with manganese oxide; the color of the manganese dioxide adjusts the green hue to a less noticeable gray.

The presence of iron was sometimes turned to good use; melting glass in the mildly reducing atmosphere produced by a smoky fire, increased the proportion of ferrous ions, and blud glass was obtained. Manganese was also added as a coloring agent, yielding an amethyst color under certain conditions. Many ancient glasses were colored by the presence of oxidized copper: cupric (Cu^{2+}) ions give glass a bright blue color, whereas cuprous (Cu^+) ions color the glass red. Table 9.6 lists some of the most common metal ions used by ancient glassmakers to color glass.

TABLE 9.6   Glass-coloring Ions

Color of Glass	Metal Ions
Black	Manganous (Mn^{3+})
	Cuprous (Cu^+)
Red	Cuprous (Cu^+)
	Colloidal gold
Pink	Manganic (Mn^{4+})
Orange	Cadmium (Cd^{2+})
Yellow	Uranium (U^{4+})
	Cadmium (Cd^{2+})
Amber	Ferric (Fe^{3+})
Green	Cupric (Cu^{2+})
	Ferrous (Fe^{2+})
Blue	Cupric (Cu^{2+})
	Cobalt (Co^{2+})
Violet	Manganous (Mn^{3+})

GLASSMAKING

Glassmaking techonology has changed little over the centuries; although methods and techniques have been refined, the basic process has remained almost unaltered since its discovery. Five steps are involved in the glassmaking process: (1) mixing, (2) melting, (3) fabrication, (4) annealing, and (5) finishing.

Mixing

The mineral raw materials used in glassmaking are frequently, though not invariably, in the form of oxides. They may also be carbonates, nitrates, or other compounds that, during the glass melting process, are converted into oxides.

The mixture, or "batch," from with which the glassmaking process begins contains almost invariably the same basic ingredients: a glass former (usually sand), a modifier (soda or potash), and a stabilizer (lime). The batch may also contain "cullet," excess glass from a previous melt. If necessary, one or more of the following may also be added: (1) decolorizing agents, (2) coloring agents, and (3) opacifiers.

Melting

The mixed ingredients are heated in a furnace. The cullet will soften first, at a temperature lower than the melting point of the other ingredients. The fluid cullet helps the still solid batch granules to move around the mix with each other. At about 1400°C all the entire mixture fuses together into a viscous liquid. The temperature is then kept steady for several hours or is raised to improve mixing and remove streaks and bubbles. During this period a number of chemical reactions take place, so that the homogeneous mixture takes on the properties of glass. The temperature is then lowered (to 1100-1300°C); the object of this is to raise the viscosity of the batch to make it suitable for fabrication.

Fabrication

The fabrication of glass differs from that of other materials in that the operation must be completed in the short cooling period—ranging from a few seconds to several minutes—during which the glass changes from a highly viscous liquid to a solid. If more work is still required, the glass must be reheated.

Glass can be shaped into useful articles by a variety of techniques: casting, drawing, blowing, and pressing. Their description is outside the scope of this book, and the reader is referred to one of the many books on the subject (Kirk-Othmer 1966; Thorpe and Whiteley 1962).

Annealing

All finished glass objects need to be annealed by a heat treatment cycle that prevents harmful stresses from being fixed in the glass on solidifying. During annealing the glass is heated to a temperature near its softening point. It is held a few minutes at this temperature until internal stresses are relieved. The glass is then cooled at a controlled rate.

Finishing

Finishing operations can be divided into "cold" and "hot" operations. Cold finishing operations involve the removal of excess glass. This can be done by cutting, drilling, engraving, grinding, or polishing. Hot finishing operations include reshaping and fusion sealing.

ANALYSIS OF ANCIENT GLASS

Chemical analysis affords the most direct way of studying ancient glass. A knowledge of the chemical composition is not only of interest in itself, but may often help in the clearing up of archaeological problems, not to mention its fundamental value for the history of glass technology. However, one of the difficulties encountered when studying the results of chemical analyses of ancient glass is the similarity in composition of glass from different sources: nearly all ancient glass is—as already remarked—of the soda-lime variety. Analyses that do no more than restate this fact do not contribute new knowledge. To be of significance, analytical results should reveal differentiating characteristics or conformity to known patterns recognizable in well-defined groups of ancient glass.

From the results of chemical analyses, the ancient glassmakers' selection of raw materials can be evaluated. To appraise their technical skill, however, it is also necessary to inquire into the workability of the materials used. It is important, for example, to know the softening temperature as well as the hardening range of temperatures, both governed by chemical composition (Table 9.7).

It is only when, having discovered the composition of ancient glass, we attempt to duplicate the working methods of the ancient glassmakers using the primitive equipment available to them that we can fully appreciate their technical skill (Schuler 1959).

METHODS OF GLASS ANALYSIS

Glass usually has a large number of chemical components: more than 10 constituents can be identified, even in very early specimens. Analysis by classical

TABLE 9.7   Softening Temperature of Ancient Glass

Provenance of Sample	Period (Century)	Softening Temperature (°C)
Thebes (Egypt)	15thB.C.	880
El Amarna (Egypt)	14thB.C.	865
El Amarna (Egypt)	,,	760
Elephantine (Egypt)	2nd B.C.	760
Elephantine (Egypt)	,,	720
Alexandria (Egypt)	1st B.C.	770
Mainz (Germany) (Roman glass)	3rd C.E.	805
Mainz (Germany) (Roman glass)	,,	670
Salona (Roman tesserae)	2nd C.E.	775
Köln (Germany)	4th C.E.	770
Köln (Germany)	,,	720
Samarra (Islamic glass)	9th C.E.	850
,,	,,	790

chemical methods is likely to be long and tedious since it involves the fusion of the glass—to bring the elements into solution—and the subsequent determination of each element successively. The detailed analysis of a single sample in this way may entail several days' laboratory work. New instrumental methods have helped greatly to reduce the labor involved. Modern X-ray diffraction methods, fluorescence analysis, and microprobe techniques not only yield rapid results, but also provide information that could not have been obtained by any means available a few years ago (Brill 1968).

Analytical results of the major and minor components are usually presented as percentages of the oxides present in the sample. It is accepted practice to specify all or most of the following components:

Silica	SiO_2
Alumina	Al_2O_3
Lime	Cao
Magnesia	MgO
Soda	Na_2O
Potash	K_2O
Manganese oxide	MnO
Ferric oxide	Fe_2O_3

Lead oxide   PbO
Antimony oxide Sb_2O_5
Tin oxide SnO_2

Trace element components, on the other hand, are usually expressed in parts per million of the total mass present.

THE COMPOSITION OF ANCIENT GLASS

Silica was universally used as the glass former. The source of silica was sand or occasionally crushed pebbles. The most common modifier was soda, which could be obtained from natron lakes (as in Egypt) or, together with potash, from wood ash that also contains lime and magnesia. Lime was usually added intentionally as limestone—calcium carbonate—but could also enter glass together with soda from plant ashes, as mentioned previously, or with sand, where it is often present as shell detritus. The most common type of ancient glass is the soda-lime type. It usually contains magnesia and alumina in addition to silica, soda, and lime (Turner 1956). It has been suggested that the chemical composition of ancient glass should be recognizably related to the chemical composition of the soil of the place where it was made, since the composition of the plant ashes used as modifiers depends upon the elmentary composition of the soil where the plants were grown (Besborodov and Zandnesprovskii 1967).

Analytical Results

Some generalizations can be made about the components of ancient glass. Those related to the major elements are summarized as follows (Caley 1962):

Silica, the universal constituent of glass, is present in the 60-68-percent range. In only a few exceptional cases does it exceed 70 percent.

Alkaline oxides (soda and potash), are usually present in a higher proportion than in modern glasses. Their proportions vary widely, with an upper limit of 30%. Soda almost always predominates over potash.

Alkaline earth oxides (lime and magnesia) are almost universal components of ancient glass (i.e., up to about 10 percent). The proportion of lime is similar to or higher than that in modern glass. Magnesia is usually present in the range 2-5 percent.

Alumina was probably not always added deliberately, but rather introduced accidentally with other raw materials. It is usually present in the 1-5-percent range.

Iron was not intentionally added. It was probably introduced through being present in the sand (silica). It occurs in glass in a wide range of concentrations, from 10 percent down to trace amounts.

Egyptian Glass

Ancient Egyptian glasses are mainly of the soda-lime variety. The constituents are similar to those of modern ordinary glass, although variations in the respective proportions of the major components are much greater. Egyptian glasses generally have a high magnesia content.

A specimen from the twelfth dynasty (2000-1800 B.C) was certainly of the soda-lime type as shown by the analysis. Egyptian glasses of the Ptolemaic and Roman periods are also—with few exceptions—basically of the soda-lime type (Table 9.8).

Ancient Egyptian glass may be of the following colors: red, yellow, green, blue, violet, black, and white. The nature of the various coloring matters is listed in Table 9.9.

TABLE 9.8   Composition (in Percent) of Ancient Egyptian Glasses

Sample	Silica	Alumina	Lime	Magnesia	Soda	Potash	References
Twelfth dynasty	68.3	3.2	4.9	1.0	20.2	2.0	Parodi 1908
Eighteenth dynasty	67.8	4.4	10.6	5.5	22.7	7.4	Caley 1962
Ptolemaic period	64.7	2.8	7.1	1.5	18.8	0.5	Caley 1962

TABLE 9.9   Coloring Matter of Egyptian Glass

Color	Coloring Matter
White and opaque	Tin oxide
Black	Copper, manganese, and ion oxides
Red	Red oxide of copper
Yellow	Lead antimoniate, ion oxide
Green	Copper carbonate
Blue	Copper carbonate, cobalt oxide

Mesopotamian Glass

Glass of excellent quality was being made in Mesopotamia by the middle of the third millennium B.C. The major components of Mesopotamian samples show that these glasses were also of the soda-lime type. The magnesia content is so high, however, that perhaps they should be properly called soda-lime-magnesia glasses (Table 9.10). Although Egyptian glasses also often have a high magnesia content and thus a high magnesia:lime ratio (MgO:CaO), this ratio is much higher in Mesopotamian than in Egyptian glasses. This difference serves sometimes to distinguish between glasses from the two regions.

TABLE 9.10   Lime and Magnesia in Mesopotamian and Egyptian Glass[a]

| | Composition (Percent) | | |
Glass	Lime	Magnesia	Magnesia:Lime Ratio
Egyptian			
Thebes, 15th cent. B.C.	7.3	4.1	0.6
El Amarna, 5th cent. B.C.	9.2	4.1	0.4
Elephantine, 10th cent. B.C.	7.4	2.1	0.3
Mesopotamian			
Fifteenth cent. B.C.	5.9	4.1	0.7
Seventh cent. B.C.	6.9	3.8	0.6
Third cent. B.C.	6.6	4.7	0.7

[a]Data from Caley (1962).

TABLE 9.11   Coloring Agents (Composition in Percent) in Ancient Mesopotamian Glasses[a]

	Iron Oxide (Fe_2O_3)	Copper Oxide (CuO)	Manganese Oxide (MnO)	Cobalt Oxide (CoO)
Pale rose	1.8	0.4	0.4	—
Deep rose	0.7	0.6	0.9	—
Pale green	2.1	0.5	3.9	—
Dark green	1.8	1.0	5.3	—
Light blue	1.5	1.5	0.8	—
Light blue	1.3	2.6	—	—
Dark blue	1.4	1.2	2.5	—
Dark Blue	1.0	1.9	0.5	0.9

[a]Data from Caley (1962).

Lead glass seems to have been first manufactured in Mesopotamia, probably between the eighth and sixth centuries B.C, and possibly as early as the fifteenth century B.C. The coloring agents used in Mesopotamia are the same as those found in Egyptian glass, as can be seen from Table 9.11.

Roman Glass

Surprisingly, there is little information on the composition of Roman glass manufactured in Italy because of the few specimens that have been analyzed (Table 9.12). From the examples available, it seems that several kinds of glass were manufactured there in ancient times. Ordinary soda-lime Roman glass from Italy does not seem to bear any resemblance to Asian or Egyptian glass of the same type (Caley 1962). An opaque soda-lime-lead glass was of similar composition to lead glass from Mesopotamia. Potash glass found at Cape Polisipo (near Naples) contains uranium and differs in composition from any other ancient glass (Gunther 1912) (Table 9.13).

TABLE 9.12   Ancient Roman Glass: Main Constituents (Composition in Percent)

Analyst	Silica	Alumina	Lime	Magnesia	Soda	Ferric Oxide	Lead Oxide
Klaproth (1801)	71.0	2.5	1.5	—	4.5	1.0	13.0
Klaproth (1801)	81.5	1.5	0.3	—	6.0	10.2	—
Pettenkofer (1857)	49.9	1.2	7.2	0.9	11.5	2.1	15.5
Neumann and Kotyga (1925)	67.7	3.8	6.1	0.5	21.7	0.6	—

TABLE 9.13   Roman Glass Containing Uranium

Component	Percent
Silica	62.1
Alumina	1.8
Lime	8.9
Magnesia	2.9
Soda	—
Potash	20.4
Uranium (U_3O_8)	1.2

Pieces of colored glass (tesserae) were much used by the Romans for making mosaics. Data on the coloring components of such tesserae are given in Table

9.14. It is remarkable that in contrast to glasses produced elsewhere in the ancient world, Roman glass produced in Germany had little variety of color. The explanation advanced for this difference rests on the different end uses of the glass: whereas in Mesopotamia or Egypt glass was made into ornamental objects, in Germany it was usually made into utensils for daily use (Caley 1962).

TABLE 9.14   Coloring Matter in Roman Tesserae

Color of Tesserae	Coloring Matter
Turquoise blue	Copper oxide
Dark green	Ferric oxide + manganese oxide
Red	Feric oxide + manganese oxide + cuprous oxide

COMPARATIVE GLASS STUDIES

An analytical survey of several hundred samples of ancient glass produced in the Middle East, Europe, and Africa was conducted using spectroscopic methods of analysis (Sayre and Smith 1962). The results revealed the existence of five compositional groups (Table 9.15) as follows: (1) second millennium B.C. (2) Roman, (3) early Islamic, (4) Islamic lead glass, and (5) antimony-rich glass.

TABLE 9.15   Oxides Used to Characterize Ancient Glass (Composition in Percent)[a]

Glass Group	Magnesium oxide (MgO)	Potassium oxide (K_2O)	Manganese oxide (MnO)	Antimony oxide (Sb_2O_5)	Lead oxide (PbO)
Second millennium B.C.	3.6	1.1	0.03	0.06	0.007
Roman	1.0	0.4	0.4	0.04	0.001
Early Islamic	4.9	1.5	0.5	0.02	0.008
Islamic lead	0.3	0.03	0.02	0.08	36
Antimony rich	0.9	0.3	0.02	1.0	0.02

[a]Data from Smith (1963).

All the groups are of the soda-lime type, except the Islamic, which is of the lead type. All contain similar concentrations of the major components: silica, sodium oxide, and calcium oxide. The groups can be distinguished according to the concentration of five other oxides: magnesium (MgO), potassium (K_2O), manganese (MnO), lead (PbO), and antimony (Sb_2O_5). Variations in the concentrations of potash (K_2O) and magnesia (MgO) are probably due to the compositional variation of the raw materials used in manufacture. This is not so in the case of antimony and managanese oxides. Antimony-containing minerals were used, through only sporadically, from the time of the earliest production of hollow colored glassware. By about sixth century B.C. antimony minerals had begun to be intentionally used as decolorants and apparently became part of the standard components of glass from Persia, Asia Minor, and Greece during the sixth and fifth centuries B.C. Later, during the ascendancy of Rome, antimony-rich glass continued to be made east of the Euphrates, whereas in Syria, Asia Minor, Egypt, and Italy, what is known today as Roman glass begun to be made, using manganese instead of antimony minerals as decolorants. The distinction between antimony-rich and manganese-rich glasses suggests, therefore, the deliberate use of one or the other elements to eliminate unwanted tints and to produce colorless glass.

Early Islamic glass is similar in composition to ancient glass of the second millennium B.C. It has been postulated that this is due to the uninterrupted continuation of a glassmaking tradition. Islamic lead glasses, it is interesting to

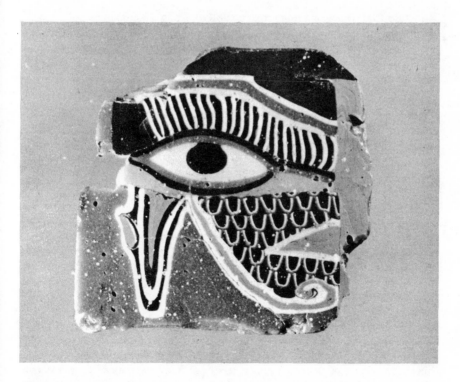

Fig. 9.4. Miniature mosaic plaque possibly made in Alexandria; first century B.C. to first century C.E. (2 cm × 1.8 cm). Myriads of crystallites suspended within the matrix cause opacity of the glass.

notice, although originating from a variety of sources, all possess a remarkably similar composition.

CRYSTALLINE GLASSES

In glassmaking operations crystallization sometimes occurs and spoils the melt. The glassmaker generally tries to avoid crystals in his products. However, glasses containing small proportions of crystals have been produced from earliest times. When one of the components of glass separates out on cooling in the form of very small crystals within the matrix, the effect on the glass may vary from a mere cloudiness or smokiness to opalescence. This effect was used by ancient glassmakers to produce *opal* and *opaque* glass. A miniature fused mosaic (magnified in Fig. 9.4) made in Alexandria between 100 B.C. and 100 C.E. represents an ancient example of a glass that contains crystals (Douglass 1966).

Glass—Opacifying Compounds

Tin oxide (SnO_2) and antimony oxide (Sb_2O_5) are two compounds traditionally used to produce opal and opaque glass. Antimony was probably added to the glass melt as the mineral stibnite (antimony sulfide, Sb_2S_3). This material was used in the East as a cosmetic for darkening the eyes. In Syria and Mesopotamia, where there were deposits of stibnite, the mineral was an important standard component of glass; it was intentionally added as a reducing agent to eliminate undesired colors. The data presented in Table 9.16 afford evidence of the widespread and deliberate use of antimony minerals, many centuries before the Christian era, for making opal glass. Stibnite reacts with metal oxides in the glass melt to form metal antimoniates; calcium antimoniate ($Ca_2Sb_2O_7$) produces white opal, and lead antimoniate ($Pb_2Sb_2O_7$) produces yellow opal glass.

TABLE 9.16   Antimony in Ancient Glass

	Opaque Light Blue Glass Knossos, Crete (ca. 1400 B.C.)	Sealing Wax Red-lead Glass (8th-6th Cents. B.C.)	Cameo Opal Glass (1st Cent. B.C.)
Antimony Oxide (percent)	0.5	4.1	8.8

Archaeologists and glass historians have often expressed the view that tin oxide (SnO_2) was also used in ancient times, as a opacifying compound. To investigate this assertion, a study was undertaken of the materials used to produce white opal or opaque colored glasses (Turner and Rooksby 1959, 1961). The problem was investigated using x-ray-diffraction methods. The results show that until the first or second century C.E. calcium antimoniate was the sole agent used for the production of white opal and opaque glasses and lead antimoniate for yellow opal and opaque glasses. Between the second and fourth centuries C.E., tin oxide began to be used as an opacifier, in addition to antimony oxide. During the fourth century C.E. a tradition of some 1800 years' standing was for some reason interrupted: antimony oxide ceased to be employed, and tin oxide became the only glass opacifier used (Table 9.17).

TABLE 9.17   Opacifiers in Ancient Glass

Sample	Color	Opacifier
Fused mosaic glass vessel from Mailik-Tepe (11-10 cent. B.C.)	White	Calcium antimoniate ($CaSb_2O_7$
Cullet from Nimrud (7th cent. B.C.)	Blue	Calcium antimoniate ($Ca_2Sb_2O_7$)
Mosaic Tessera from Shavei-Zion, (Israel) (5th cent. B.C.)	Yellow	Lead-tin oxide ($Pb_2Sn_2O_6$)
Glass from Dendereh, (Upper Egypt) (2nd cent. B.C.)	White	Tin oxide (SnO_2)
Cullet from Rhodes (2nd cent. B.C.)	Yellow	Lead antimoniate ($Pb_2Sn_2O_7$)
Glass from Koban (Russia) (9th cent. C.E.)	Yellow	Lead-tin oxide ($Pb_2Sn_2O_6$)

Microscopic examination and electron-beam-microprobe analysis of Roman *millefiore* have provided additional quantitative confirmation of these results (Brill and Moll 1963). The opacity is due to myriads of small crystallites suspended in the glassy matrix; white particles were shown to be calcium antimoniate, whereas yellow particles consist of lead antimoniate, both containing additional coprecipitated metal oxides. Opal glasses from the fourth to twelfth centuries C.E. contain tin, probably in the form of tin oxide.

Cameo Glass

Cameo glass is a glass in which a design in relief is produced by molding or by cutting away a layer of opal or opaque glass that is backed, completely or partially,

Fig. 9.5. The Portland vase.

by a more or less transparent body glass. The cameo technique was in use in the years just preceding, and the century following, the birth of Christ. The Portland vase (see Figure 9.5), one of the best known of all ancient glass objects, is made of cameo glass. Chemical and physical analysis of the glass from which the Portland vase is made yielded much valuable information, including the answers to such questions as how the color of the body glass and the opacity of the overlayer were produced and how the body glass and the overlayer were bonded, firmly enough to withstand the rough mechanical treatment unavoidable during cutting, engraving, and polishing (Turner 1959).

Chemical analysis of a splinter of the dark blue body glass yielded the results shown in the first column of Table 9.18. The composition of the blue glass conforms to that of the more stable types of ancient glasses. Soda is the dominant alkaline oxide, most probably derived from plant ash. This is indicated by the presence of a small amount of phosphorus pentoxide as well as by the contents of lime, magnesia, soda, and potash. Four constituents of the glass produce a marked color effect: cobalt, copper, iron, and manganese. All four of these were used from

TABLE 9.18 Composition (in Percent) of Cameo Glass

Component	Blue Body Glass of Portland Vase	Opal Glass of a Medallion (1st Cent. B.C.)	Relative Proportions of Basic Constituents	
			Blue Glass	Opal Glass
Silica	65.7	60.0	69.2	67.6
Alumina	2.4	2.8	2.5	3.1
Lime	9.0	8.3	9.5	9.4
Magnesia	0.6	1.1	0.6	1.2
Soda	16.2	15.8	17.1	17.8
Potash	1.0	0.7	1.0	0.8
Ferric oxide	2.3	0.5		
Manganese oxide	1.6			
Phosphoric oxide	0.2	0.1		
Copper oxide	0.5			
Cobalt oxide	0.1			
Antimony oxide		8.8		
Sulfur trioxide		1.2		
Loss in ignition at 600 °C		0.8		
Total	99.6	100.1	99.9	99.9

earliest times to produce various shades of blue. Table 9.19 lists examples where these elements were used as colorants in ancient glasses.

Melting of experimental glasses of a composition similar to that of the Portland vase led to the conclusion that the blue color is due primarily to copper and cobalt

TABLE 9.19 Cobalt, Copper, Iron, and Manganese as Colorants of Ancient Glass (Composition in Percent)

Component	Beads from Azarbaijan Persia (2400-2000 B.C.)	Glass from Enidu (200 B.C.)	Glass from Nippur (1400 B.C.)	Portland Vase (1st Cent. B.C.)
Cobalt oxide	Traces	0.2	0.9	0.1
Copper oxide	Traces	0.5	1.9	0.5
Iron oxide	Traces	2.4	1.0	2.3
Manganese oxide	Traces	0.04	0.6	1.6

oxides and is modified by iron oxide. The blue color due to cobalt is further reinforced by the presence of manganese oxide.

The composition of the opal glass of the Portland vase was determined in an indirect way. Only a minute splinter of opal glass from the vase itself was available, which was not sufficient for accurate analysis. Instead, an analysis was made of a fragment of opal glass from a medallion, which corresponded as closely as possible in date and in physical characteristics to the glass of the vase. The composition is shown in the second column of Table 9.18. The most salient feature is the presence of antimony oxide in a concentration higher, perhaps, than in any other ancient glass thus far analyzed. This high concentration is due to the fact that the opacifying agent in the glass is calcium antimoniate in the form of very small crystallites. X-Ray-diffraction analysis provided confirmation of this. Another salient feature is the presence of sulfur trioxide (SO_3), an indication that the antimony was added to the glass melt as the mineral stibnite (antimony sulfide Sb_2S_3).

It must be remembered that the data discussed in the preceding paragraphs were obtained, not from the Portland vase, but from a medallion of similar qualities. Nonetheless, the basic constituents of the two glasses are comparatively close to one another. The figures in the third column of Table 9.18 were calculated by taking the analytical results from the first and second columns of the table, excluding from that of the blue body glass the coloring oxides and from that of the opal glass the oxides of iron, antimony, and sulfur, and then recalculating the basic constituents to a 100-percent total in each case. Bearing in mind the variability in the composition of ancient glasses of the same period and provenance, the similarity in composition of the blue body glass and of the opal glass of the medallion is remarkable. Likewise, when the Portland vase was made, a common parent glass was probably used. From it the body glass was prepared by the addition of coloring agents, whereas the opal glass was derived by adding stibnite. The almost identical composition of the two glasses would explain the firm bond between the body glass and the overlayer—a bond firm enough to withstand without fracture the jar and shock caused by the penetrating action of cutting and engraving tools and the subsequent polishing treatment.

DICHROIC GLASS

Dichroism is the property some materials have of exhibiting two colors: one when viewed in reflected light and the other when viewed in transmitted light. A few pieces of ancient glass, mostly Roman, but occasionally some of Islamic origin, exhibit dichroism. The analysis of some of these glasses and the production of experimental glass with composition resembling that of ancient dichroic glass helped to clarify the source of dichroism. (Davison, Giauque et al. 1971).

Fig. 9.6.   The Lycurgus cup.

The "Lycurgus cup" (see Fig. 9.6) in the collection of the British Museum, is well known for its elaborate cutting and especially for having been made from glass with dichroic properties. The cup has a "pea-soup" green or jade-like appearance in reflected light but shows a deep magenta color by transmitted light. Analysis of a dislodged chip from the cup, only 9 mg in weight, yielded the results shown in Table 9.20 (Brill 1965; Chirnside and Proffit 1963).

The Lycurgus cup was made of a soda-lime-silica glass similar to other Roman glasses of the same period. Manganese was very probably added as a decolorizer rather than as a coloring agent. Iron, in unusually high concentration, and antimony were probably added intentionally: both metals may have served as reducing agents.

The most unusual characterisitic of the composition of the Lycurgus cup (also found in other ancient dichroic glasses) is the presence of minute quantities of silver and gold. These offer crucial evidence for explaining the dichoric effect, although the presence of traces of silver and gold in an ordinary glass matrix has no noticeable effect on color. Heat treatment of glass and the use of the reducing agents, together with a carefully controlled furnace atmosphere, cause the silver and gold to be chemically reduced and to form colloidal metal particles within the glass matrix. If the heating is continued, the colloidal particles grow to a range of

**TABLE 9.20   Chemical Analysis of
The Lycurgus Cup**

Glass Component	Percent
Silica	73.5
Alumina	2.5
Lime	6.5
Magnesia	0.6
Soda	14.0
Potash	0.9
Ferric oxide	1.5
Manganese oxide	0.5
Antimony oxide	0.3
Lead oxide	0.2
Boron oxide	0.1
Titanium oxide	0.07

sizes that causes both absorption and scattering of light and results in the dichroic effect.

The color of the glass viewed by transmitted light is a combined effect of the absorption and scattering of light by the colloidal metal particles. The color obtained by reflection is the result of intense scattering by the same particles. The color combinations of dichroic glass cannot be achieved with either gold or silver alone. The color obtained by reflection seems to be associated more with the light-scattering properties of silver and the transmitted color, with the absorption of light by gold.

PROVENANCE OF GLASS AND ITS RAW MATERIALS

Use of Natron in Ancient Soda-Lime Glassmaking

Natron, a naturally deposited mixture of sodium carbonate and bicarbonate, was used for glassmaking since early dynastic times. Deposits of this salt in Wadi Natroun in lower Egypt (between Alexandria and Cairo) were commercially exploited for a long time. Pliny mentions the use of "nitrum" in Roman glass-making. It has long been assumed that natron from Wadi Natroun was used in the manufacture of ancient glasses characterized by low concentrations of potassium and magnesium. Hence if the concentration of some element present only in trace concentrations were shown to be correlated with the concentrations of potassium and magnesium in both natron and glass, the use of natron in these glasses would be proved.

Using nondestructive neutron-activation analysis, the concentrations of europium oxide in natron from Wadi Natroun and in ancient glass were determined (Sayre 1965). The results indicate that europium oxide concentrations have a significant negative correlation with the concentrations of potassium and magnesium in glass; that is, the europium oxide concentration in low potassium and magnesium glasses is consistently higher than in high potassium and magnesium glasses. Natron samples—both ancient and modern—do not, however, contain the relatively high europium concentration characteristic of glasses having low potassium and magnesium concentrations. It is hence obvious that whatever may have been the source of a high concentration of europium in the low potassium and magnesium glasses, it was not natron.

Lead Isotopes in Glass

Lead minerals were used by ancient glassmakers to color opaque glasses. Red opaque glass owes it color to the presence of crystallites of cuprite (Cu_2O), which are reaidly formed if lead oxide is present in the glass. Many early red opaque glases contain 5-30 percent lead oxide (PbO). Opaque yellow is caused in glass by the presence of lead-antimony or lead-tin colorant-opacifier ($Pb_2Sb_2O_7$ or $PbSnO_3$). These glasses usually contain 5-10 percent lead oxide (Rooksby 1962).

In nature lead existed as a mixture of four stable isotopes: lead-204, lead-206, lead-207, and lead-208. These occur in a very large variety of isotope mixtures. The isotope ratios of lead ores vary from one mining region to another, as they are characteristic for different sources. This makes it possible to relate lead-containing glass artifacts to the place of origin of their raw materials.

According to the lead isotope ratios found in glass samples from a number of sites, artifacts have been classified into groups (Brill and Wampler 1967; Brill 1968, 1970). Isotope determinations on lead extracted from the decoration of Egyptian glass vessels prove that the pigment used was made from locally mined lead. Similar objects from Mesopotamia have a different lead isotope ratio.

Certain glass panels found in Greece are very unlikely to have been made there, since there was not much glassmaking activity in Greece at the supposed time of their manufacture. Isotope analysis of the colored inlay of the panels provided quite unexpected results: the lead from yellow and red glasses came from different respective sources. In at least one case, that of the red glass, and probably in both, the lead came from sources outside Greece.

There are certain limitations to this method of localizing the source of glass. The most important is that although the isotope ratios of lead in a mineral may be characterisitic of a given mining region, they are rarely uniquely so. Mines located in geologically similar environments occasionally have identical isotope ratios, even though the regions themselves may be widely separated.

Oxygen Isotopes in Ancient Glass

Oxygen is an elementary constituent of all the materials used in making glass. In accounts for about 45 percent of weight of most glass. Natural oxygen consists of three stable isotopes: oxygen-16 (99.76 percent), oxygen-17 (0.04 percent), and oxygen-18 (0.2 percent). Variations in the relative abundance of oxygen-18 in minerals from different sources have been used in the determination of palaeo-temperatures (see Chapter 22). The measurement and interpretation of these variations may also be used to distinguish between glasses of different origins. Analysis reveals considerable variation in the isotope composition of oxygen extracted from ancient glasses (see Fig. 9.7), even though combinations of entirely different raw materials can also produce glass with identical oxygen-18 contents. A classification system based on this may be of value and can provide answers to occasional specific problems (Brill 1968, 1970).

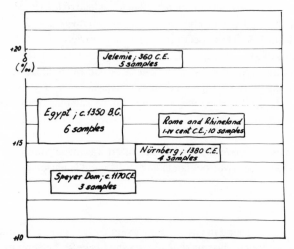

Fig. 9.7. Oxygen-18 contents of some early glasses: δ is the difference (expressed in per mil) in the oxygen-18:oxygen-16 ratio between the sample and an accepted reference value.

DECAY OF GLASS

Soda-lime glasses are usually considered to have considerable chemical resistance and durability. Exposure to the elements or burial in the soil for long periods of time does, however, cause weathering of the glass, and crusts of decomposition products are formed: the colorful iridescence usually seen on ancient glass objects is the cumulative result of slow and presistent deterioration (see Fig. 9.8).

Fig. 9.8. Weathering crusts in ancient glass.

Glass found in dry locations shows little evidence of decomposition. In general, the more moist the burial site of ancient glass, the more extensive the weathering. This is because deterioration is caused by water containing dissolved salts, which leaches away soluble substances. Some glass components are dissolved relatively rapidly, whereas others persist for longer periods. There are at least four factors that can greatly affect the rate of attack: (1) composition of the glass, (2) nature of the salts contained in the water, (3) temperature, and (4) time. Time seems to be the controlling factor that might result in a periodic effect and the consequent formation of layers. Weathering causes more or less complete removal of the alkali metals (sodium and potassium) and partial removal of the alkaline earths (calcium and magnesium). A few other components (iron, aluminum, and titanium) remain as insoluble decomposition products within the silica matrix (Geilman 1956). The final product of weathering is a residue of more less separate flaky layers of hydrated silica (Shaw 1965).

The chemical mechanism of the formation of the layers poses a most intriguing problem. Apparently, seasonal variations of temperature, moisture, or both account for their formation, although this suggestion has not been experimentally verified. The weathered crust is highly porous and very finely laminated (Raw 1955). The thickness of individual layers has been measured and found to vary in the range 0.3-15 μ. Iridescence on ancient glass is due to interference effects, arising from the reflection of light from the top and bottom surfaces of the thin decomposition layers (Browster 1863; Laubengayer 1931; Raman and Rajagopalan 1940).

DATING OF GLASS

It has been suggested that the appearance of separated layers of weathering products in ancient glass might be the consequence of cyclic changes imposed on the glass during long periods of exposure (Brill and Hood 1961). Since seasonal variations in temperature or alternating dry and rainy seasons occur in yearly cycles, the possibility arises of dating ancient glass by counting the number of layers comprising weathering crusts (Table 9.21).

TABLE 9.21   Weathering Layers and the Age of Glass

Date of Sample Based on Archaeological Evidence	Number of Weathering Layers	Date Calculated from Counting layers
Late 3rd-early 4th cent. C.E.	1582 ± 10	378 ± 10 C.E.
Ninth-thirteenth cent. C.E.	775 ± 20	1175 ± 20 C.E.
Eighth-tenth cent. C.E.	930 ± 30	1020 ± 30 C.E.
Submerged 1621 C.E.	334 ± 15	1625 ± 15 C.E.
Buried 1676 C.E.	269 ± 10	1691 ± 10 C.E.

Table 9.21 shows the dates obtained by counting the layers in ancient glass observed in cross section under a microscope. The technique does not yield a date of manufacture or use, but rather the date at which the weathering process started, presumably at the time of burial or submersion. Obviously not every sample of

ancient glass can be dated by this method. Datable samples must have the weathering crust preserved intact, a requirement seldom met. And, of course, the conditions at the site of burial must have been subject to a yearly cycle of changes.

Roman and Byzantine glasses are generally too resistant to corrosion to become heavily weathered. Egyptian glasses are usually found in arid environments and thus have not suffered much from corrosion. Mesopotamian glasses are ofen so heavily weathered that little or no glass remains, and what does remain is too fragile to be handled. The results obtained so far add up to quite a strong case for the applicability of the method. However, accelerated layering of glass has been induced artifically and it has been suggested that layering of glass could occur even where no cyclic variation of conditions is perceptible (Newton 1971).

The development of a glass dating method, based on the depth profile of the layer of hydrated glass, has been reported (Lanford 1976). The measurement of the thickness of the hyration layer—which increases with time—requires the bombardment of the sample wih a beam of Nitrogen-15 ions in an accelerator. There is, therefore, a need for further study of the complicated phenomena taking place during the weathering of glass, so as to provide a firm scientific ground for the method of dating by counting layers or measuring the depth profile of the hydration layer.

GLOSSARY—GLASS, GLAZE AND ENAMEL

Annealing (glass).   Removal of stresses within the glass by controlled cooling.

Crazing (in glazes and enamels).   Cracking that develops as a result of tensile stresses.

Crissling.   Diminished transparency of certain glasses due to very fine surface crazing (Brill 1975).

Cullet.   Broken glass from production waste, recycled in the glass-melting furnace.

Devitrification.   Crystallization of a glass.

Enamel.   A glassy coating for metals.

Flux.   A substance that facilitates the fusing of minerals.

Former (glass).   Oxides whose atoms constitute the basic structural network of glass. The former of ancient glass was silica (SiO_2).

Frit.   Finely powdered glass, used in making glazes and enamels.

Glass.   Product of fusion of materials, cooled to a rigid condition without crystallizing.

Glaze.   Glassy coating for ceramics.

Luster (glass). Glass decorated with a thin metallic film.

Milliefiore. A Roman type of glass mosaic inlay. The mosaic pieces were produced by casting either a tube or rod of glass with several layers of a differently colored glass and then reheating, rolling on a corrugated surface and cutting into short lengths. The cross section of the rod received, during the rolling, the shape of flowers. For very small flowers the rod was drawn after rolling.

Modifiers (glass). Metal oxides that, when added to silica, result in a marked change of its properties without destroying its structural identity.

Opacifier. A material that decreases the transparency of glass.

Refraction. Change in direction that light rays undergo as they pass from one medium (e.g., air) to another (e.g., glass).

Softening point. Temperature at which the viscosity of glass is reduced sufficiently for the glass to be workable (see Fig. 9.9).

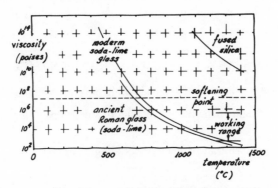

Fig. 9.9. Dependence of that viscosity of glass on temperature. Viscosity is the resistance that a liquid offeres to flow. Glasses soften gradually as their temperature is raised. At high temperatures they become quite fluid. At a viscosity of 10^7 poises glass may become deformed under its own weight. When the viscosity drops to 10^3-10^5 poises, glass will flow quite readily into a mold, or it can be blown with ease.

REFERENCES

Bezborodov, M. A., and Zadnesprovskii, J. A., in M. Levey, ed., *Archaeological Chemistry*, Philadelphia, 1967.

Brewster, D., *Transact. Roy. Soc. Edinburgh*, **23**, 193 (1863).

Brill, R. H., in *Proceedings, 7th International Congress on Glass*, paper No. 223, Brussels, 1965.

Brill, R. H., in *Proceedings, 8th International Congress on Glass,* 1968, p. 47.

Brill, R. H., *Lead Isotopes in Ancient Glass,* International Association for the History of Glass, 1969.

Brill, R. H., *Transact. Roy Soc. London,* **A269,** 143 (1970).

Brill, R. H., in *Conservation in Archaeology,* Stockholm Congress, IIC, 1975.

Brill, R. H., and Hood, H. P., *Nature,* **189,** 12 (1961).

Brill, R. H., and Moll, S., in F. R. Matson and G. E. Rindone, Eds., *Advances in Glass Technology,* Part 2, Plenum Press, New York, 1963, p. 293.

Brill, R. H. and Wampler, J. M., *Am. J. Archaeol.,* **71,** 63 (1967).

Caley, E. R., *Analysis of Ancient Glasses, 1790-1957,* Corning Museum of Glass, 1962.

Chirnside, R. C., and Proffit, P. M. C., *J. Glass Stud.,* **5,** 18 (1963).

Davison, C. C., Giauque R. D., and Clarck, D., *Man (new series),* **6**(4), 645 (1971).

Douglass, R. W., *Discovery,* **27,** (1) 40 (1966).

Farnsworth, M., and Ritchie, P. D., *Technical Studies in the Field of the Fine Arts,* **6,** 155 (1938).

Geilman, W., *Glastech. Ber.,* **29,** 145 (1956).

Gunther, R. T., *Archaeologia,* **63,** 99 (1912).

Kirk-Othmer, *Encyc. Chem. Technol.,* Vol 10, 533, 1966.

Klaproth, M. H., *Memoires de l'Academie Royale des Sciences et Belles Lettres,* Classe de Philosophie exper., Berlin, 3, 1801.

Landorf, W. A., *Science,* **196,** 975 (1977).

Laubengayer, A. W., *J. Am. Ceram. Soc.,* **14,** 833 (1931).

Lucas, A., *Ancient Egyptian Materials and Industries,* Arnold, London, 1962, p. 191.

Neumann, B., *Chem. Zeit.,* **51,** 1013 (1927).

Neumann, B., and Kotyga, G., *Angew. Chem.,* **38,** 864 (1925).

Newton, R. G., *Archaeometry,* **13,** 1 (1971).

Pettenkofer, M., *Dingler's Polytechnisches Journal,* **145,** 122 (1857).

Parodi, D. H., Thesis, Cairo University, Egypt, 1908.

Raman, C. V., and Rajagopalan, *Proc. Ind. Acad. Sci.,* **11a,** 469 (1940).

Raw, F., *J. Soc. Glass Technol.,* **39,** 128T (1955).

Rooksby, H. P., *J. Soc. Glass. Technol.,* **43,** 285T (1959).

Rooksby, H. P., *GEC Journal,* **29,** 20 (1962).

Sayre, E. V., *Proceedings, 7th International Congress on Glass,* paper No. 220 Brussels, 1965.

Sayre, E. V., and Smith, R. W., *Science,* **133,** 1824 (1962).

Schuler, F., *Archaeology,* **12,** 47 (1959).

Shafer, T., *Ceramics Monthly,* **24**(7), 44 (1976).

Shaw, G., *New Scientist,* **27,** 290 (1965).

Smith, R. W., in D. Brothwell and E. Higgs, eds., *Science in Archaeology,* London, Thames and Hudson, 1963, p. 520.

Thorpe, J. F., and Whiteley, M. A., *Dict. Appl. Chem.*, Vol. 5, 536, 1962.

Turner, W. E., *J. Soc. Glass Technol.*, **40,** 162T (1956).

Turner, W. E., *J. Soc. Glass Technol.*, **43,** 22T (1959).

Turner, W. E., and Rooksby, *Glastech. Ber.*, **32K,** 17 (1959).

Turner, W. E., and Rooksby, *Jahrbuch des Romisch-Germanischen Zentralmuseum,* Mainz, 1961.

Zachariansen, W. H., *J. Am. Chem. Soc.*, **54,** 3841 (1932).

Recommended Reading

Bezborodov, M. A., *Chemistry and Technology of Antiques and Middle Age Glasses,* Von Zaber, Mainz, 1975.

COLOR: PIGMENTS AND DYES

Colors have been used for decorative purposes since prehistoric times. Color is a psychophysical effect: it is the mental response to a stimulus consisting of light varied in its spectral characteristics by chemical or physical means. To produce decorative color effects, pigments and dyes are used.

PIGMENTS

Pigments are finely divided, insoluble colored materials used to impart color to other materials. Pigments do not usually combine, chemically or otherwise, with the material to which they impart color; they must be deliberately attached to the surface to which they are applied, by means of a *binding medium.*

The binding (or fixing) of pigments may be performed in either of three ways:

1. The binding medium is applied first, followed by the pigment, as in fresco painting.
2. The binding medium and the pigment are mixed and applied together. This is the normal method used in most kinds of painting. For this purpose the pigment is mechanically dispersed in the binding vehicle or medium to give a more or less fluid paint or coating.
3. The third possible method, is to apply the pigment and then the binding medium. An example of this method is the covering of colored pottery decorations with a transparent glaze.

The category of pigments comprises of a wide variety of chemical compounds. There are pigments of mineral, animal, and vegetable origin. Those most readily available to ancient man were pigments of vegetable origin (powdered berries, barks, flowers); these are usually *fugitive*—that is, their color soon fades when exposed to light. The burning of almost any sort of bone would provide a white pigment. Colored earths, abundant on the surface of sedimentary rock deposits, provide yellow, red, and brown ochers. The caves of Lascaux in Southern France are decorated with paintings made with natural earth pigments dating as far back as the early Stone Age. Less readily available to ancient men were the brightly colored compound of the heavy metals (oxides, sulfides, carbonates, sulfates, etc); nevertheless, many of these were known and used as pigments from very early times. Already in the third millennium B.C. bright colored minerals became articles of trade and were transported to regions far from their places of origin.

Artificial pigments are also of very ancient origin. Carbon black, in the form of soot, charcoal, and even charred bones, was readily available about any hearth, from very earliest times. At a later stage in history the use of naturally available pigments was supplemented by the manufacture of "lakes." A "lake" is formed by treating a water-soluble dye (e.g., indigo or purple) with certain chemicals that render it insoluble and cause it to precipitate on a solid base or substrate. This can then be powdered and used in the same way as a pigment. Lakes were prepared in ancient times from many dyes, using powdered chalk or kaolin clay as substrates. They were prized for brightness but tended to fade rapidly in sunlight.

The most important properties of a pigment are chemical inertness and stability, to ensure compatibility with other materials and to prevent fading. The simple metal oxide pigments are generally the most stable, particularly to light, air, and moisture. The carbonates, sulfates, and phosphates of metals are also very stable in this respect.

Our knowledge of the pigments used in antiquity is derived from various sources: (1) ancient texts, (2) the classical writers, especially Pliny the Elder, Dioscorides, Theophrastus, and Vitruvius, (3) ancient recipe books and technical schedules, and (4) the scientific examination of materials.

In ancient texts there is often a confusion of nomenclature that can be very puzzling. An ancient recipe for "azure blue," for example, when made up, turns out to be vermilion (Thompson 1936). To give another example: the name "minium" (today applied to red oxide of lead, Pb_3O_4) was used at one time to mean cinnabar (mecuric sulfide, HgS).

INDENTIFICATION OF PIGMENTS

As a rule, pigments of mineral origin can be easily identified. Old and faded organic lakes, however, are more difficult to recognize. Lumps of pigments, architectural decorations, and ceramic and wall paintings may serve as sources of information. Very often only a small amount of material is available for analysis. This necessitates the use of micro- or semimicro-chemical techniques. The procedures employed for identification include, in addition to classical chemical methods, spectroscopic analysis, ultra-violet and infra-red photography, X-ray diffraction and microprobe techniques. When objects are too precious to be submitted to chemical examination, the nondestructive methods, such as X-ray and gamma-ray radiography and backscattering of beta-particles have been proved very useful. (Kennedy, Tolmie et al. 1970; May and Porter 1975).

CLASSIFICATION OF PIGMENTS

Pigments may be classified according to three criteria:

1. Origin (natural or synthetic).
2. Composition (organic or inorganic).
3. Color.

The color criterion is the most convenient for the purposes of this book and is used from here on. In the following pages, like pigments will be treated in groups, following the order: white, black, yellow, brown, red, green, blue, and violet.

WHITE PIGMENTS

Two natural white pigments were commonly used in antiquity: calcium carbonate (chalk, whiting) and calcium sulfate (gypsum). A light clay (kaolin) also seems to have been used occasionally, and historical records indicate that a synthetic pigment, white lead, was also known from early times. The powdered forms of calcium carbonate and calcium sulfate are white but do not possess the opacity or "hiding power" of heavy metal whites such as white lead.

Calcium carbonate (CaCO₃) was available as limestone, marble, chalk, and the shells of mollusks. The pigments were produced by wet or dry grinding minerals selected for the purity of their color. The use of natural chalk as a pigment goes back to classical times. Levigated chalk was also used as a mild abrasive for polishing gold and silver (Gettens, Fitzhugh, et al. 1974).

Calcium sulfate is widely distributed in nature as gypsum ($CaSO_4 \cdot 2H_2O$). The rock was selected for its color and was then crushed, pulverized, and screened.

Clays are hydrated silicates of aluminum. There are many forms and types of clay varying one from the other in chemical composition, crystalline structure, and properties (see Chapter 8). The type of clay most widely used as a white pigment in ancient times was kaolin or china clay, $[Al_2Si_2O_5(OH)_4]_2$. This material was almost always found in nature mixed with unwanted contaminants from which it had to be separated before use.

Bone white was obtained by calcining (burning) bones. Almost any sort of bone would do for this purpose. Particularly recommended were the wings of birds and the legs of domestic animals. The bones were dried and then put into a fire till they turned white. The product was then cooled and ground. White bone is composed chiefly (85-90 percent) of tricalcium phosphate ($Ca_3(PO_4)_2$); calcium carbonate and minor constituents made up the rest. Bone white is a grayish white and slightly grittish powder.

White lead is basic lead carbonate, (2PbCO₃·Pb(OH)₂). Synthetic white lead is the ultimate product of the corrosion of lead with vinegar. Films pigmented with white lead weather in such a manner that the paint film keeps in a relative good condition for long periods of time.

White lead was mentioned in almost all source materials and catalog lists of pigments from ancient times up into the nineteenth century. Although the basic carbonate of lead occurs naturally as the rare mineral *hydrocerussite,* the pigment was artifically produced since early historical times. Indeed, white lead is one of the oldest of all synthetically produced pigments (Gettens, Kuhn, et al. 1967).

BLACK PIGMENTS

The black pigments of antiquity fall into two chemical categories: (1) elemental carbon and (2) pyrolusite. The elemental carbon pigments were the first man-made, artificial pigments. They are subdivided according to the source of the carbon into: (a) lampblacks, (b) vegetable blacks, and (c) animal blacks.

Lampblack was made by collecting the smoke of oil lamps.

Vegetable blacks were made by charring various kinds of vegetable matter, usually wood, in ovens or closed pits. The resulting charcoal was washed to remove soluble matter and then ground to powder. Between 50 and 95 percent of charcoal black consists of elemental carbon, and the remainder is mineral material.

Animal blacks were made by the destructive calcination of bones or ivory in the absence of air.

Pyrolusite is a very abundant, naturally occurring black ore. It is manganese dioxide (MnO_2). The raw mineral was ground and sieved to the desired powder grain size.

YELLOW PIGMENTS

Two different types of yellow pigments were known to the ancients: (1) yellow ochers and siennas and (2) orpiment.

Yellow ochers and *siennas* are hydrated iron oxides obtained from limonitic ores. Although the purity and composition of limonites varies, they consist mainly of hydrated ferric oxide. The formula $Fe_2O_3·H_2O$ approximately representsd their composition, although the proportion of ferric oxide to water varies from place to place. After mining, the ores were purified and pulverized to a suitable fineness. There is no sharp compositional difference between ochers and siennas in either color or iron content. Usually, however, pigments of lighter color and lower iron-oxide content are classified as ochers and those of darker color and higher iron oxide content, as siennas.

Orpiment is a soft, highly poisonous lemon-yellow compound, arsenious sulfide (As_2S_3). The natural mineral was used as a pigment, although it had other uses, such as the removal of hair from animal hides during the tanning process. Orpiment has been identified in Egyptian paintings at Tell-Amarna. It is mentioned by Vitruvius and by Pliny as a pigment that cannot be used in fresh plaster.

RED PIGMENTS

The principal red pigments consisted of iron ores. Minerals of other elements (mercury, arsenic) were also used, as well as a synthetic material—red lead.

Red Ochers

Deposits of natural oxide of iron, hematite (Fe_2O_3) are common, but only a few of these yield material of interest as red pigments. Red ochers have excellent obliteration and staining powers. When mixed with white, some red ochers yield tints with a yellow undertone.

Vermilion is bright red mercuric sulfide (HgS), identical in composition with the mineral cinnabar. Two kinds of mercuric sulfide pigment were known in the past: the natural mineral finely ground and the synthetic material made by a dry process. The mineral may sometimes be liver-brown or even black. The synthetic preparation of vermilion from sulfur and mercury seems to have been invented by the Chinese and to have been introduced into Europe in the eighth century. Table 10.1 lists some notable ancient occurrences of vermilion (Gettens, Geller, et al. 1972).

TABLE 10.1   Vermilion: Notable Ancient Occurrences

Origin	Date (Century C.E.)
Egyptian painting from Fayum	1st-3rd
Roman wall painting from the Palatin	1st
Roman and Pompeiian wall paintings	1st
Dry pigment found at Pompeii	1st

Red lead, also called *minium,* was well known in ancient times. It is composed mainly of tri-lead tetroxide ($Pb_3^1 ♀_{44}$) and smaller amounts of lead monoxide, also called *litharge* (PbO). The pigment is bright red and has good hiding power. Red lead occurs naturally, but it was also made by the oxidation of lead, and especially of litharge, at a high temperature (ca. 500°C). The heating was continued until a sufficient amount of lead monoxide was oxidized and converted into the tetroxide, and the desired color was obtained. The kind of raw litharge used and the final content of red lead determined the quality of the pigment. Red lead was not used in Egypt until Greco-Roman times. It has been identified in wall paintings in China

and Central Asia and was a favorite pigment with Persian and Byzantine illumina-
tors.

Madder Lake

Madder is a natural dyestuff from the root of the herbaceous perennial *Rubia*
tinctorum. Its principal coloring agents are alizarin and purpurin (see below, Red
Dyes). The making of madder lakes is a difficult process, and even today old,
established firms are unwilling to publish their own recipes. In preparing modern
madder lakes the root extract is mixed in solution with alum and precipitated with
an alkali. The pink color from tombs of the Greco-Roman period was identified as
consisting of madder root extract on a base of gypsum (Russell 1893).

GREEN PIGMENTS

Copper ores in various forms were the main sources of green pigments. Occasionally, natural green pigments based on other metals were used.

Malachite, a basic carbonate of copper $[CuCO_3 \cdot Cu(OH)_2 \cdot]$ having a bright
green color, is widely distributed in nature. It occurs in Sinai and was used by the
Egyptians as a pigment and as an ornamental material from predynastic times. It
was also widely used in Western Chinese paintings of the ninth century. To
prepare the pigment for use, selected lumps of malachite were crushed and ground
to a fine powder. Malachite has been identified in illuminated manuscripts from
the seventh century onward (Laurie 1914) and in the form of bulk pigment in
Pompeii (Augusti 1967).

Chrysocolla is a hydrated silicate of copper $(CuSiO_3 \cdot 2H_2O)$ of variable composition. The naturally occurring mineral ranges in color from bright green to
bright blue.

Verdigris is an artificially made basic acetate of copper $[CuO \cdot 2Cu(C_2H_3O_2)_2]$. It
was made in wine-growing areas by treating either copper or copper minerals with
vinegar for several weeks while allowing free access of air. Verdigris is mentioned
in Greek and Roman literary sources, but although copper has been found in
numerous green and grayish green paint samples from Rome and Pompeii, it is not
yet established whether they contained verdigris or malachite (Kuhn 1970).

Glauconite, a constituent of ''green sand,'' has a dull green color. It is a hydrous
silicate of potassium and iron and is of variable composition.

BLUE PIGMENTS

Azurite. This is a native basic carbonate of copper $[2CuCO_3 \cdot Cu(OH)_2]$ and
occurs as a bright blue mineral. Early known as *cuprum luzureum,* azurite was
powdered and used as a pigment often named *mountain blue.*

Lapiz lazuli.   This is a rock composed of a mixture of minerals that include the blue felsparthoid lazurite. It was highly esteemed for its deep blue color. The rock was fashioned into polished ornaments or powdered and used as a pigment.

Glaucophane.   A sodium-magnesium-aluminum hydroxide-silicate of formula $Na_2Mg_2Al_3Si_8O_{22}(OH)_2$. Some of the magnesium may, occasionally, be replaced by iron. The Greeks used glaucophane as a pigment as early as the Bronze Age (seventeenth century B.C.) (Filipakis, Perdikatsis et al., 1976).

Egyptian Blue

It is remarkable that one of the synthetic pigments most difficult to prepare satisfactorily should have been made more than 4500 years ago. Egyptian blue is an artificial blue frit used in ancient Egypt as a pigment. The use of the term "Egyptian blue" has led to some confusion with "Egyptian faience," which is usually blue or greenish, and that was made from much the same materials. Faience and its manufacture, however merit separate study, and the reader is referred to the appropriate discussion in Chapter 8.

According to Vitruvius, Egyptian blue, which he calls *coeruleum*, was made by heating together a mixture of sand, copper filings, and soda (*nitri flore*) in a furnace. In recent times Egyptian blue has been made by heating fine sand, copper carbonate, and calcium carbonate. Table 10.2 lists the components of a mixture that, when heated to 850°C for about an hour, yields Egyptian blue. The temperature is very critical; below 820°C much of the silica is still uncombined, whereas above 900°C the product consists of quartz and bottle green glass. At 850°C however, Egyptian blue is formed without the reactants reaching true fusion. Incipient fusion causes the component particles to coalesce into a solid mass. When cold, the product can be crushed in a mortar and used as a blue pigment.

TABLE 10.2   Composition of Egyptian Blue

Component	Percent
Silica (SiO_2)	64.5
Basic copper carbonate	
$[CuCO_3 \cdot Cu(OH)_2]$	15.4
Calcium carbonate ($CaCO_3$)	12.9
Fusion mixture	
Sodium chloride (NaCl)	1.5
Sodium sulfate (Na_2SO_4)	2.3
Sodium sesquicarbonate	
($NaCO_3 \cdot NaHCO_3 \cdot 2H_2O$)	3.4

It is only recently that the exact chemical composition and structure of Egyptian blue have been established. The frit is a definite chemical compound having a fixed composition: copper calcium tetrasilicate, $CuCaSi_4O_{10}$ or $CuO \cdot CaO \cdot 4SiO_2$ (Fauque 1888; Laurie, McLintock, et al. 1914; Schippa and Torraca 1957; Nicolini and Santini 1958; Ludi and Giovanoli 1967). The compound has a definite crystal structure which was determined by X-ray diffraction (Pabst 1959).

EGYPTIAN PIGMENTS

In a fourth-dynasty tomb at Medum, dating from about 2600 B.C., there are inscriptions incised and undercut in the stone and filled in with paste colors. On examination, the pigments were found to be:

White	:	gypsum
Black	:	lampblack
Brown	:	hematite
Red	:	hematite
Green	:	malachite

Azurite and Egyptian blue were found in other decorated surfaces. A study of the inner surfaces of the Bersheh Sarcophargus, from the Middle Kingdom, revealed the elements listed in Table 10.3, which also lists the compounds present as pigments. Pigments found at Luxor (Thebes) are also listed in the table (Saleh, Iskander et al. 1974).

TABLE 10.3   Egyptian Pigments

Color	Element Found	Pigment
White	Calcium	Calcium carbonate
Yellow	Arsenic	Arsenic trisulphide
Pink	Calcium, iron trace	Mixture of iron oxide and calcium carbonate
Red	Calcium, iron	Iron oxide and calcium carbonate
Green	Calcium, silicon, copper, boron	Copper metaborate, $Cu(BO_2)_2$
Blue	Calcium, silicon, copper	Egyptian blue

Note: The calcium found in most of the specimens comes from a base coat of calcium carbonate.

All the pigments mentioned thus far were in general use through the succeeding dynasties. During the eighteenth dynasty (1580-1320 B.C.) a fine golden-yellow pigment made from orpiment came into use. White lead was used during the later periods of ancient Egypt. *Huntite,* a calcium-magnesium carbonate, $CaCO_3·3MgCO_3$, was identified as a white pigment in decorated pottery from Koshtamna, Nubia, and dated about 1600 B.C. (Riederer 1974).

A most surprising result of the analysis of Egyptian decorated terracotta was the identification of *cobalt blue,* a cobalt aluminate of formula $CoO·Al_2O_3$ that was found on terracotta vases from the El Amarna period (ca. 1370 B.C.) (Riederer 1974). In Egypt there are no cobalt deposits mentioned in the literature. It is likely that the pigment was imported from other countries. Persian cobalt deposits, for example, were known and mined extensively to get a blue glaze for ceramics and tiles.

GREEK PIGMENTS

Only a few pigments were in use in ancient Greece; most of them were naturally occurring mineral ores, although one or two were prepared by chemical processes. Some were undoubtedly imported products. *Realgar,* for example, is not known to occur in Greece. The nearest source was in Western Asia Minor and possibly Transcaucasia. There is also no evidence that blue frit—Egyptian blue—was made in Greece; in all probability it was imported from Egypt. Table 10.4 summarizes the present knowledge concerning pigments used by the Greeks in classical times (Filipakis, Perdikatsis et al. 1976). Mixtures of these pigments were also used; iron ochers, for example, were diluted with chalk to produce pale tints (Caley, 1946; Higgins, 1970; Profi, Weier, et al. (1974).

TABLE 10.4   Ancient Greek Pigments

Color	Pigment
White	Chalk, gypsum, white lead
Black	Carbon, bitumen, pyrolusite
Gray	Mixture of carbon and white clay
Yellow	Yellow ocher, orpiment
Orange	Red and yellow ocher
Pink	Madder lake
Red	Red ocher (hematite), cinnabar, realgar
Brown	Mixture of red ocher and Egyptian blue
Green	Malachite, verdigris, chrysocolla
Blue	Azurite, Egyptian blue, glaucophane

Not all the pigments listed in Table 10.4 have actually been identified by chemical analysis. Verdigris, for example, has not been identified in finds from modern excavations. Theophrastus, however, describes its preparation, and there is no doubt that it was made and used in his day. The chemical instability of this artificial basic copper acetate may be the cause of its disappearance. It is not improbable that some pigments identified today as basic copper carbonate were, in fact, originally verdigris.

Lumps of rose madder lake were found at Corinth (Farnsworth 1951). Chemical and spectrographic examination revealed that aluminum was the principal constituent of the pink-colored lumps. This indicates that alumina was used in their preparation. A reflectance curve (see Figure 10.1) left no doubt that madder is

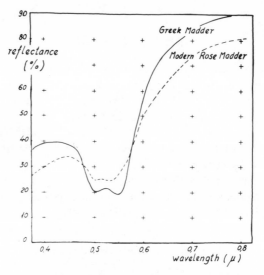

Fig. 10.1.   Spectral-reflectance curves of ancient and modern madder lakes.

present in the sample. The material of the pink lumps is, therefore, a madder-alumina lake. The use of alumina is not surprising since alums were well known to the ancients. The Greek lake, incidentally, is very similar in composition to modern madder lakes.

ROMAN PIGMENTS

In 1815 Sir Humphrey Davy, a distinguished chemist of the time, visited Italy and collected archaeological material from ancient Roman ruins, including the Baths of Titus. Among the material were specimens of bulk pigments as well as color fragments from walls. On his return to England he carried out one of the earliest scientific examinations of ancient pigments (Davy 1815). His findings are listed in Table 10.5. It is curious that white lead does not appear in the list, since historial records indicate that it was known in Rome from the earliest times.

TABLE 10.5 Ancient Roman Pigments

Color	Pigment
White	White chalk, clay
Black	Carbon black
Yellow	Ocher, Yellow oxide of lead, orpiment
Red	Red lead, red ocher, vermilion
Brown	Ocher
Green	Copper green, glauconite
Blue	Egyptian blue

A study of stocks of dry pigments found in color merchants' shops at Pompeii and Herculaneum made available a great deal of valuable data (Augusti 1967). The pigments were well preserved in jars or bottles, protected from light and air and probably closer to their original condition than are pigments found on wall paintings. It is interesting to notice that only a single example of vermilion was found among numerous samples of red pigments. Egyptian blue was the only blue pigment identified.

JAPANESE PIGMENTS

No Japanese pictorial art in color has been found dating from before the third century C. E. There is some early pottery, decorated in red with ocher or cinnabar. Funerary chambers during the fifth and sixth centuries were decorated with paintings. Paintings covering entire walls were also found in a tomb in the Fkuoka region; the pigments used are listed in Table 10.6. In mural paintings from the seventh century, near Nara, the surface of the walls was coated with fine white clay on which figures were painted with mineral pigments (also listed in Table 10.6). The introduction of Buddhism, in the middle of the sixth century, brought new materials and new techniques on which Japanese painting has been based ever since. Artificial pigments, such as red lead and litharge, and bright-colored minerals like azurite and malachite came into use. Red ocher is purer in the mural paintings near Nara than in the earlier funerary chamber from the Fkuoka region. (Yamashaki 1967, 1970).

TABLE 10.6   Ancient Japanese Pigments

	Sample (Site)			
Color	Decorated Tomb, Fkouka Region (5th-6th Cent. C.E.)	Mural Paintings at Nara (7th Cent. C.E.)	Shosoin Repository (8th Cent. C.E.)	Polyhrome Sculptures (Prior to 15th Cent. C.E.)
White	White clay (kaolin)	White clay (kaolin)	White clay (kaolin), white lead	White clay, white lead
Red	Red ocher	Red ocher, vermilion, red lead	Red ocher, vermilion, red lead	Red ocher, vermilion, red lead
Yellow	Yellow ocher	Yellow ocher, litharage	Yellow ocher	Yellow ocher litharge, gamboge
Green	Powdered rock of green color	Malachite	Malachite	Malachite
Blue	—	Azurite	Azurite	Azurite mixture of yellow ocher and indigo
Violet	—	—	Mixture of azurite and vermilion	—
Black	Carbon black mineral containing iron and manganese	Chinese ink	Chinese ink	Chinese ink
Gold	—	—	Powdered gold	Powdered gold
Silver	—	—	Powdered silver	Powdered silver

A large number of relics from the eighth century have been preserved at the Shosoin repository of Imperial Treasures, near Nara. Pigments used in paintings and ornaments now at the repository and pigments used for polychrome sculptures from periods prior to the fifteenth century are also listed in Table 10.6.

Red lead in powder form was found in the Shosoin repository. It was contained in paper packages and was marked with original labels, written at the time of packaging. Three grades were found, "superior," "medium," and "low." The composition of the red-lead samples is given in Table 10.7. Compared with

TABLE 10.7   Japanese Red Lead Samples (Composition in Percent)

Sample	Minium (Pb_3O_4)	Litharge (PbO)	Lead Metal (Pb)
Ancient (found at Shosoin, 8th Cent. C. E.)			
Superior	26.2	70.9	0.8
Medium	24.0	73.0	0.6
Low	6.2	92.2	0.6
Modern	78.1	21.1	—

modern red lead, the Japanese samples are very poor in lead tetroxide. The spectral reflectance curves of the ancient samples, shown in Fig. 10.2, are poorer than those of the modern red pigment (Yamasaki, 1967).

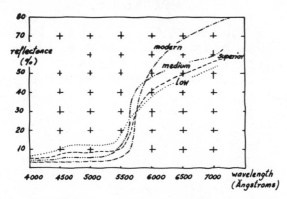

Fig. 10.2.   Spectral-reflectance curves of red-lead samples from Shosoin, Japan.

DYES

Dyes are intensely colored organic substances used to impart a more or less permanent color to textile fabrics and, sometimes, to leather, wood, bone, ivory, and other materials. Not all colored organic substances are dyes. Only organic molecules of considerably complexity are useful as dyes.

The color of dyes arises from the selective absorption of specific wavelengths of radiation within the visible spectrum. The absorbed radiation brings about *elec-*

tronic transitions within the dye molecule: the energy differences involved in these transitions determine the color of the dye. Light reflected from, or transmitted by, the dye produces at the eye the effect of color, and the color perceived is complementary to that absorbed by the dye.

Organic materials such as wool, linen, wood, and leather are readily "wettable," that is, capable of absorbing water. If such a material is soaked in water in which a dye has been dissolved, part of the dye leaves the solution and become attached to the material, imparting to it its own color. The fixation of the dye on the dyed material or "substrate" depends on the nature of the bond formed between them. Some dyes are retained by chemical bonds, and others by physical forces, and the method of retention varies with the chemical nature of the substrate. The main problems of dye technology have been (and still are) concerned with methods by which dyes may be fixed to fibers and by which fastness of color (to washing and light), may be achieved.

From a dyeing standpoint, the ancient coloring matters may be classified into three groups: (1) direct or substantive dyes (2) mordant dyes, and (3) vat dyes.

DIRECT DYES

Some dyes can be applied directly to a fiber of wool or silk, since the protein materials from which these fibers are made up contain chemical groups which confer on them an affinity for the dye. Such dyes are called *direct dyes* (see Fig. 10.3). With cellulose fibers, such as cotton and linen, the binding forces between dye and fiber are of a much weaker nature and dye fastness may not be achieved directly.

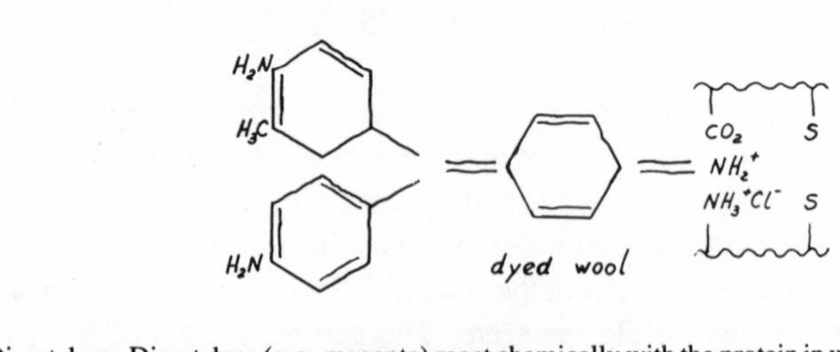

Fig. 10.3. Direct dyes. Direct dyes (e.g. magenta) react chemically with the protein in wool or silk to form a firm bond without the intervention of other chemicals.

MORDANT DYES

Dyes can be fixed to fibers with the aid of insoluble metal complexes or *lakes* formed by interaction of the dyes with certain metallic salts. This process is known as *mordanting* and the metallic salts used are called *mordants*. Mordants were employed at a very early stage in the history of dyeing. Alum, natron, soluble salts of iron (e.g., the sulfate) and of tin were among the commonest ancient mordants. They were applied to the fibers before, during, or after the application of the dye, with which they formed chemical compounds that adhered strongly to the fibers. In many cases not only the fixation on the fiber, but also the color development of the dye, could take place only in the presence of mordants.

In mordant dyeing, cloth was either penetrated with a solution of the metallic salt or was dyed in a bath containing the salt. Saffron, kermes, and madder are examples of *mordant dyes,* that is, dyes that could not be fixed on a fiber on their own. The ancient process of dyeing with madder is said to have consisted of a sequence of mordanting and dyeing operations involving at least seven different steps and requiring days for its completion (see Fig. 10.4).

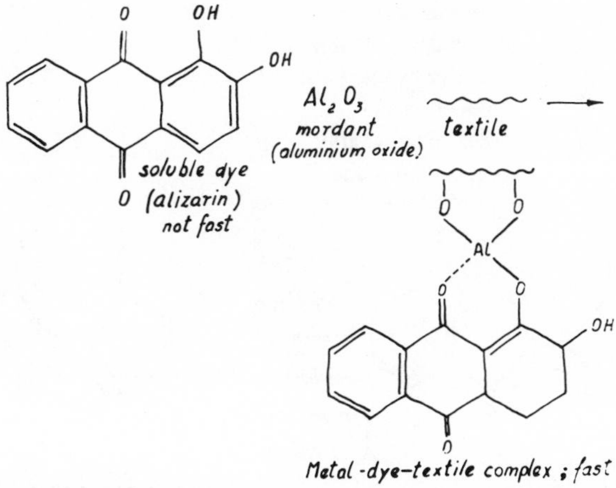

Fig. 10.4. Mordant-dyes textile bonding. The fixation of mordant dyes (e.g. madder) to textiles, involves the formation of a metal-dye-textile chemical compound.

VAT DYES

Some dyes are insoluble in water and cannot be used in a really satisfactory way unless subjected to a long fermenting process. The age-old practice of preparation of the dye liquor of these dyes was carried out in tubs or vats, from where the terms *vat dyes* and *vat-dyeing* seem to have originated. In this process an insoluble organic pigment was converted, by putrefying organic materials in the presence of mild alkalies, into a reduction product—called *leuko compound*. The leuko compound was soluble in alkali and penetrated the fibrous materials to be dyed. After dyeing and in the presence of air the leuko compound readily decomposed, reverting to the originally colored insoluble dye that was then *fast*. Indigo, which has been known for thousands of years, is such a vat dye (see Fig. 10.5). A classification of dyes based on usage is given in Table 10.8.

TABLE 10.8   Dyes: Classification Based on Usage

Class	Method of Application	Major Substrates
Direct	Applied from neutral or slightly alkaline bath containing a salt (e.g., common salt (NaCl) or Glauber salt, $Na_2SO_4\text{-}10H_2O$)	Cotton
Mordant	Applied in conjunction with salts of aluminum, tin, iron, etc.	Wool, silk, linen
Vat	Applied by solubilizing the dye by chemical reduction and reoxidizing it after deposition of the fiber	Cotton, wool

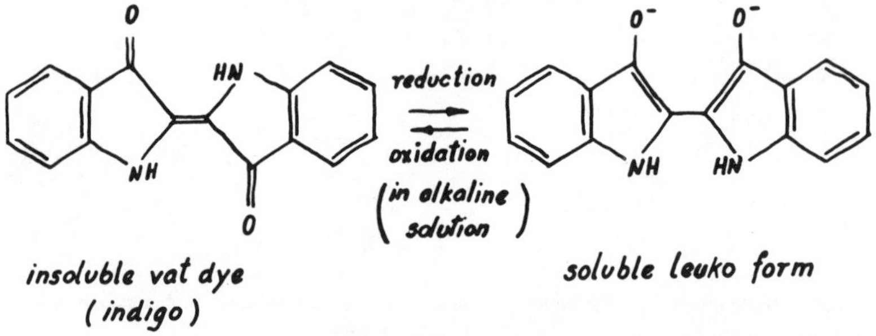

insoluble vat dye
(indigo)

soluble leuko form

Fig. 10.5. Vat dyes. Vat dyes (e.g. indigo) are insoluble in water and cannot be fixed directly to fibers. In alkaline dye baths, however, they can be reduced to soluble "leuko" forms, which can be applied directly to the fiber and then oxidized.

The end result of the dyeing process is that a color is imparted to the substrate by the dyestuff being fixed on it. The color obtained need not necessarily be that of the dye itself, since it may be changed or developed by the mordant. Logwood, for example—a dyestuff used for centuries by Mexican Indians and later introduced to

Europe by the Spaniards—can yield a variety of colors, depending on the metal salt that is used as mordant, listed later in this chapter.

Colored fibers obtained after dyeing are generally resistant to light, washing and other color-removing operations. If the color resists washing and light well, it is said to be a *fast color*. If it fades quickly or is easily removed, it is called a *fugitive dye*.

IDENTIFICATION OF ANCIENT DYES

When dissolved in suitable solvents, dyes give solutions of characteristic color well defined by their spectral absorption. The technique most often used for the identification of ancient dyes is based, therefore, on the examination of the spectral absorption of solutions in appropriate solvents. The spectrometric absorption curve obtained from an unknown dye is compared with those of known materials. Figure 10.6 shows the infrared spectral absorption curves of an ancient

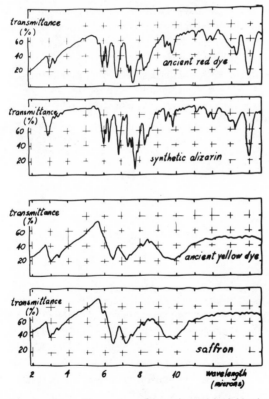

Fig. 10.6. Infrared spectra of dyes. The spectrum of the ancient red dye is identical to that of alizarin; the ancient dye is thus identified as being madder. In a similar manner the ancient yellow dye is shown to be saffron.

dye compared with that of a standard material used for identification (Saltzman, Keay, et al. 1963; Abrahams and Edelstein 1967). The major disadvantage of this technique is that it requires the destruction of some of the sample, of which the amount available for examination may vary from a relatively large quantity down to a single dyed thread. In many cases good spectral curves can only be obtained by using relatively large amounts of sample. The amount to be used for identification, therefore, is a function of the importance to the archaeologist of the identification proper.

Chromatographic methods of analysis have also been used to identify ancient dyes (Masschelein-Kleiner and Heylen 1968). Thin-layer-chromatography was found to be the most sensitive method for extremely small samples (Fedorovich 1965; Takagi 1964).

ANCIENT DYES

Ancient man's knowledge of dyes was limited to those from natural sources. Trial and error—and no doubt chance—taught man to use coloring materials found in plants such as extracts from barks, and roots of trees or juices from flowers, berries, and fruits. Animal sources—insects, mollusks, and eggs—also provided some highly prized coloring materials. The dyes obtained from natural sources were rather impure, and reproducibility of results was almost unobtainable. Nevertheless, the dyeing methods of ancient civilizations remained in use, almost unaltered, until the discovery of synthetic dyes, in the middle of the nineteenth century.

RED DYES

Quite a variety of red dyes were known in antiquity, of both vegetable and animal origin. Of the many naturally available, only a few ever attained technological importance. One of the most important, and in its heyday probably the most important of all, was *madder*.

Madder

The art of dyeing with madder appears to have originated in the East, passed to the ancient Egyptians and Persians and thence to the Greeks and Romans (Fieser 1930; Mell 1932; Schaefer 1941; Brown 1943a; Wiskerke 1952). Known under the name *rubia*, madder was extracted from perennial plants of the order *Rubiacea*, of which there are about 35 species. The most popular species is *Rubia tinctorum* found naturally in Palestine and Egypt, abundant in Asia and Europe and cultivated in the ancient world for dyeing purposes since remote antiquity.

The coloring material of madder consists mainly of *alizarin,* an organic compound having the formula:

After alizarin, the most important coloring matter in madder is purpurin, of formula:

Purpurin is not found in appreciable amounts in the roots of *Rubia tinctorum.* It is, however, the principal coloring matter in the roots of other Rubiacea plants that are the South American equivalent of madder and that were used for dyeing textiles in Peru more than 2000 years ago (Fester 1953).

Madder dye is concentrated in the roots of the plant; old roots are richer in coloring matter than young ones. After removal from the soil the roots were washed with water, dried, and chopped. The dye was then extracted with water, separated from the solution, and finally dried.

Dyeing with madder yielded the most brilliant and most permanent red color that was known: the so called "Turkey red." In conjunction with salts of different metals, used as mordants, a variety of fast colors were obtained, as follows:

Mordant Metal Salt	Color
Aluminum	Red and pink
Aluminum + iron	Brown
Calcium	Blue-red
Iron	Purple-black
Tin	Orange-yellow

Madder was popular in the Middle East and has been identified on cloth found in Egyptian Tombs (Lucas 1962) and on woolen fabrics found in the Judean Desert

(Abrahams and Edelstein 1967). It was also used as a paint on Greek terracotta statuettes (Farnsworth 1951). No evidence for its use in Mesopotamia has yet been uncovered (Levey 1959).

Madder enjoyed popularity also for its medical properties, and the dye has been identified in human bones found at Kumran (in Palestine). It appears that people of the Community of the Dead Sea Scrolls employed a decoction of madder as a beverage: this was apparently done in the belief that it had magic properties (Steckoll, Goffer, et al. 1971).

Kermes

Kermes was probably the earliest used red dye of which there are records (Heise 1895; Dimroth 1911; Born 1924, 1936b; Michell 1955). The coloring matter, *kermesic acid,* has the formula:

Kermes was obtained from an insect, *Kermococcus vermilia* Planch, (formerly *Coccus ilicis*), which is widely distributed in the Middle East, Armenia, North Africa, and Spain. The female insects were collected prior to hatching eggs and killed by exposure to the vapors of vinegar or by immersion in it. After drying and crushing, a carmine-red dye, soluble in water, was obtained.

In antiquity the use of kermes reached a high degree of perfection. Wool, mordanted with alum or tartar, dyed a rich crimson or scarlet. Kermes was the most important dye known to the ancient Babylonians (Levey 1959). It appears to have been used by the Phoenicians, and it is mentioned in the Old Testament. Egyptian fabrics were occasionally dyed with kermes (Lucas 1962). In the Hellenistic period kermes was used to dye wool, silk, and leather.

Cochineal

Cochineal was known for many centuries to the pre-Columbian inhabitants of America. Its native country is Mexico, and there is evidence that it was also used for dyeing in Peru during the Inca period (Fester 1953; Saltzmann, Keay, et al. 1963; Kashiwagi 1976). Cochineal remained unknown to the rest of the world until the Spanish conquest of Mexico in 1518, although it has been suggested that "true cochineal" was also known in the "ancient world," where it was produced from insects native to the Ararat valley (Kurdian 1941).

Cochineal was derived from the dry bodies of *cocus* insects found on various *Cacti*, especially in that of the species *Nopalea coccinillifera* S., which was cultivated in ancient Mexico for the sake of the dye. The female insects were collected and killed by exposing them to boiling water. The dye was also extracted with boiling water and was then dried. The coloring matter of the dye is carminic acid (Dimroth and Scheurer 1913) having the formula:

Lac dye

Lac dye is contained in the abdomen of *Coccus laccae*, an insect native to India and Southeast Asia. The red dye is no longer used commercially, although a by-product of its manufacture—*gum lac*—is still of value.

The active agent of the dye is *laccaic acid* (Schmidt 1887):

The dyestuff was extracted from the body of killed insects with a solution of sodium carbonate in water and was later precipitated from the solution with lime or alum.

Lac dye yielded similar but somewhat faster dyeing than cochineal. Carmine red shades were produced on wool with aluminum mordants, whereas a beautiful scarlet was obtained with tin mordants. The dyestuff was used for many centuries in Southeast Asia.

Henna

The extract of the plant, *Lawsonia alba Lam.* or *Lawsonia inermis L.* is a red dye—*henna*. The coloring matter is lawsone, having the formula (Tomassi 1920; Cox 1938):

Henna dyes wool and silk and orange shade. The dye was used less as a textile dyestuff than as a dye for hair and as a stain for nails, feet, hands, and other parts of the body. It is found in Palestine, North Africa, Persia, India, and other countries with warm climates, where it is still widely used.

YELLOW DYES

A number of yellow dyes were known in antiquity. Weld and saffron seem to have been the most widely used, but Persian berries and safflower have also been identified in ancient fibers (Augusti 1950). Safflower, for example, has been found in fabrics from the twelfth dynasty in Egypt (Hubner 1909).

Weld

Weld is said to be the oldest European dyestuff known. It is the extract of the dried herbaceous plant *Roseda uteola* (Mell 1932). The coloring matter is disseminated throughout the entire plant but is greatly concentrated in the upper branches and seeds. The active dye is *luteolin* (Perkin 1896; Perkin and Horsfall 1900):

Weld yielded the purest and fastest shades of yellow. In conjunction with different mordants, it gave a variety of colors, as listed follows:

Mordant Metal Salt	Color
Aluminum	Lemon yellow
Tin	Lemon yellow
Iron	Olive green

Safflower

A yellow dye was derived from the safflower, *Carthamus tinctorius*, a plant native to the East Indies. The yellowish-red flowers were washed with water to remove the yellow dye. The structural formula of the dye has not yet been elucidated. Yellow bindings of Egyptian mummies of the twelfth dynasty (ca. 2500 B.C.) were dyed with safflower.

The flowers were also dried to obtain an additional red dye. The red coloring matter is *Carthamin,* of formula:

$$CH=CH-COH$$

It was extracted after washing the dried flowers with acidulated water to remove the yellow coloring. The red dye was once widely used in the East for the manufacture of lakes and cosmetics.

Saffron

Saffron is a reddish-brown or golden-yellow odoriferous substance consisting of the dried *stigma* of *Crocus sativus* L. The active coloring matter is the compound *crocetin,* of formula (Karrer and Miki 1929):

Saffron is used today mainly as a food colorant. In the past it was also used as a dye. Tin-mordanted fabrics dye yellow with saffron, whereas with aluminum salts a dull orange color is obtained.

Saffron is indigenous to Southeast Asia. It was cultivated in the Mediterranean region and in parts of Europe, having been identified in textiles from the Bar Kochbah finds in the Judean Desert (Abrahams and Edelstein 1967).

GREEN DYES

In antiquity, green colors were obtained by using mixtures of other dyes, usually blue and yellow. Greens originating in Egypt were found on examination to be mixtures of indigo blue with a yellow dye (Lucas 1962).

BLUE DYES

Only one good blue coloring substance was known in the ancient world: *indigo*. This vat dyestuff can be prepared from either of two different plants: *indigo* and *woad*. In both the coloring matter is the same substance, indigo, of formula (Emmerling and Eugler 1870; Von Baeyer 1900):

In indigo the dye is about 30 times more concentrated than in woad; nevertheless, woad was more often used in classical times as a source of dye.

Indigo dyes also contain a red tinted coloring substance called *indigo red* or *indirubin,* of formula:

Traces of indirubin are probably present in all samples of the natural blue dye. Some varieties, such as indigo from Java, are especially rich in indirubin.

Coloring Matter in Java Indigo (Composition in Percent)

Sample	Total Coloring Matter	Indigo	Indirubin
1	73	69	4
2	71	66	5

Indigo has been identified in blue woolen fabrics from upper Egypt dating from 200-700 C.E. (Schunk 1892) in cloth found in the Judean Desert in Israel (Abrahams and Edelstein 1967) and in Incan textiles from Peru (Saltzmann, Keay, et al. 1963).

Indigo

Indigo has been known to man for more than 4000 years. Some historians believe it to be the oldest known dye (Vetterli 1950). It is produced from the leaves of various species of *Indigofera*, a plant native to Asia, Europe, and America. In such widely separated countries as Peru, Egypt, India, and Brazil, ancient men independently discovered the preparation of the blue dyestuff from the different species of *Indigofera* (Kashiwagi 1976).

The preparation of the dye involved the fermentation of the indigo plants soaked in water and subsequent formation of the dye by atmospheric oxidation of the aqueous solution obtained. The dyestuff was finally removed by filtration and dried (Rawson 1899).

Woad

The ancient Celts tattooed their bodies with designs colored with a dye obtained from the *woad* plant, *Isatis tinctoria*. The coloring matter, indigo, was obtained by grinding the dry plant to a paste and then allowing the moistened paste to ferment. *Indican*, a colorless compound present in the plant decomposed, and the product of the decomposition was oxidized by air to indigo.

TYRIAN PURPLE

Tyrian purple was undoubtedly the most renowned and highly prized of all ancient dyes. It stood as a symbol of wealth and distinction; cloth dyed with Tyrian purple was so expensive that only priests and kings could afford it (Cady 1937). Accurate information has long been available about the importance of purple in the ancient world, but only in recent times have its true appearance and the way it was produced become known (Heinisch 1957).

The purple-dye industry flourished in the ancient town of Tyre in the eastern Mediterranean. The dyestuff was produced from the *purple snail*, in reality several species of mollusks of the genus *Murex*. Each of the species yielded a special kind of purple. In Tyre, where the most prized purple dye was produced, *Murex brandaris* was used. *Murex trunculus* yielded an amethyst purple dye that was produced in Sidon. It has been suggested that purple was produced in America from another shellfish, *Purpura patula* Pansa (Saltzman, Keay, et al 1963).

The dystuff is obtained from a secretion produced by the mollusks. The secretion is colorless when fresh, and on exposure to sunlight it first changes to yellow, then to green, and ultimately to purple. In the course of a careful investigation of purple dyes, only 1.4 g was of the dye was obtained from 12,000 mollusks! The coloring matter was identified as being as *6-6''-dibromo indigo* (Friedlander 1909):

$$
\begin{array}{c}
\text{O} \qquad \text{H} \\
\text{C} \qquad \text{N} \qquad \qquad \text{Br} \\
\diagdown \text{C=C} \diagup \\
\text{Br} \qquad \text{N} \qquad \text{C} \\
\text{H} \qquad \text{O}
\end{array}
$$

Because of the great demand for purple and its high price, many substitutes were used to dilute it or to imitate its color. Thus kermes, *orseille*—a lichen growing on rocks near the sea—and *chalcantron*—a mixture of indigo and other dyes—were often used as substitutes. Excellent tests have been developed for identifying true Tyrian purple in ancient textiles and for recognizing substitutes (Driessen 1944).

It has often been claimed that ancient purple possessed fastness and brilliance of color. In actual fact, purple-dyed fabrics had a reddish shade tending toward violet. The color that was so gorgeous in the opinion of the ancients actually fell far short of the beauty of the shades produced by modern dyes. Scientific research on Tyrian purple and on modern dyes have laid to rest a very long-accepted fallacy.

INK

Inks are fluids used for writing, drawing, and marking. They contain coloring matter (either a pigment or a dye) and a binder, both evenly dispersed throughout a carrying agent or "vehicle". In ancient times the binder was usually vegetable gum or animal glue, whereas the vehicle was almost invariably water.

Ancient ink was generally black; however, red, blue, and green inks have also been found. It is generally assumed that red ink was made with iron oxide and blue and green inks, with copper compounds; however, there is no record of such inks having been subjected to chemical analysis.

Two types of black ink, differentiated by coloring matter, were used in antiquity: *carbon ink* and *iron-gall ink*. Carbon ink consisted of soot or lampblack, ground with a solution of glue or gum. An ancient recipe has recently been used to make carbon ink: gum arabic was dissolved in boiling water, and pulverized wood charcoal was added. This was boiled and the hot solution then filtered (Steckoll 1968).

Iron-gall inks were made from an infusion of crushed oak-nut galls, copperas, and gum arabic dissolved in water. *Oak-nut galls*, also called *Aleppo galls*, contain about 60-70 percent tannin and varying amounts of gallic acid. *Copperas*, or green vitriol, is a salt of iron (ferrous sulfate, $FeSO_4$). Soluble ferrous salts form a dark-colored solution with extracts of tannin. The proportion of galls to copperas necessary for ink making was determined empirically, but about 3 parts by weight of galls to 1 part of copperas could be considered a balanced formulation. Other sources of tannin were sometimes used in the manufacture of inks: sumac (leaves of *Rhus-coriana*) or spent tanning solutions (liquors previously used for tanning skins and hides), for example. However, because of the effectiveness of their combined contents of tannin and gallic acid, galls have been generally found the most suitable raw material for the purpose.

Chinese and Egyptian inks, used as long ago as the twenty-fifth century B.C., were carbon inks. Quotations from Dioscorides, Vitruvius, and Pliny support the view that the inks of antiquity were basicaly carbon inks. Inkstands containing dried ink have been found in ancient Egyptian tombs. The black ink is generally described as being made of carbon (Lucas 1962). Residues present in inkwells found at Qumran (in Palestine), contained no iron, suggesting that carbon ink was used (Steckoll 1968). Writing examined on broken Egyptian earthenware vessels was certainly made with carbon ink. In the case of the Lachish letter found in Israel and dating from the eightheenth century B.C., however, it has been suggested that ink made from an iron salt and carbon mixture was used (Lucas 1922).

It has been widely supposed that the use of iron-gall ink was introduced in the seventh century C.E. However, the ancient world was well acquainted with both oak galls and copper and with the black coloration produced on mixing them. References to mineral ink are found in Philo of Byzantium in the second century B.C. and Rabbi Meir, in the first century C.E. Carbon inks were used in papyri until a comparatively late period. Iron ink, on the other hand, was used in all but a few of the earliest manuscripts on vellum, well before the seventh century C.E. One of the oldest of these is a manuscript containing part of Demosthenes's *De Falsa Legatione,* dating from the fifth century B.C.

Out of a group of 12 vellum documents from the third and fourth centuries C.E., only three were found to be written in carbon ink; the rest were in iron-gall ink. Among the latter is the *Codex Sinaiticus,* which affords a good example of gelatinous brown iron ink (Lewis 1938).

Old carbon ink sometimes appears brown. A suggestion put forward to explain the color change is that the ink initially contained very little free carbon but contained compounds, possibly of a tarry nature, which have turned brown. The brown color masks any small amounts of carbon present. In other instances the carbon of the ink may have rubbed off the paper, leaving a brown stain due to residual ingredients other than carbon.

REFERENCES

Pigments

Augusti, S., I Colori Pompeiani, in   , De Luca, ed., *Studie Documentazione,* Ministero della Pubblica Instruzione, Direzione Generale della Antichita e Belle Arti, De Luca Ed., Roma, 1967.

Caley, E. R., *J. Chem. Ed.*, 314 (1946).

Chase, W. T., in R. H. Brill ed., *Science in Archaeology*, MIT Press, Cambridge, Mass., 1971, p. 80.

Davy, H., *Phil. Transact.*, (January 14 1815).

Farnsworth, M., *J. Chem. Ed.*, 72 (1951).

Fauque, F., *Comptes Rendus*, **108**, 325 (1888).

Filipakis, S. E., Perdikatsis, B., and Paradellis, T., *Studies in Conservation*, **1**(2), 143 (1976).

Gettens, R. J., Geller, R. L., and Chase, W. T., *Studies in Conservation*, **17**, 45 (1972).

Gettens, R. J., Kuhn, H., and Chase, W. T., *Studies in Conservation*, **4**, 125 (1967).

Gettens, R. J., and Fitzhugh, E. W., *Studies in Conservation*, **19**, 2 (1974).

Gettens, R. J., Fitzhugh, E. W., and Feller, R. L., *Studies in Conservation*, **19**, 157 (1974).

Giovanoli, R., *Chimia*, **22**, (4), 184 (1968).

Higgins, R. A., *Studies in Conservation*, **15**, 272 (1970).

Kennedy, P. L., Tolmie, R. W., Haulan, J., and Grant, J. M., *X-Ray Fluorescence Spectra of Artists Pigments*, A.E.C.C., Ottawa, 1970.

Kuhn, H., *Studies in Conservation*, **15**, 12 (1970).

Laurie, A. P., McLintock, W. F. P., and Miles F. D., *Proc. Roy. Soc.*, **A89**, 4118 (1913-1914).

Laurie, A. P., *The Pigment and Mediums of the Old Masters*, London, 1914.

Laurie, A. P., *Greek and Roman and Methods of Painting*, Repr. of 1910 edn, Longwood, 1978.

Laurie, A. P., *The Materials of the Printers Craft*, Repr. of 1910 edn, Longwood, 1978.

Ludi, A., and Giovanoli, R., *Naturwissenschaften*, **54**, 88 (1967).

May, R. W., and Porter, J., *J. Forens, Sci. Soc.*, **15**(2), 137 (1975).

Nicolini, L., and Santini, M., *Boll. Inst. Cent. Restauro*, **34**, 59 (1958).

Pabst, A., *Acta Crystallogr.* **12**, 733 (1959).

Profi, S., Weier, L., and Filippakis, *Studies in Conservation*, **19**, 105 (1974).

Riederer, J., *Archaeometry*, **16**, 102 (1974).

Russel, W. J., *Nature*, **69**, 374 (1893).

Saleh, S. A., Iskander, Z., El-Masry, A. A., and Helmi, F. M., in *Recent Advances in Science and Technology of Materials*, Vol 3, 141, Plenum, N.Y., 1974.

Schippa, G., and Torraca, *Boll. Inst. Cent. Restauro*, **31**, 97 (1957).

Thomson, D. V., *The Materials of Medieval Painting*, Allen and Unwin, London, 1936.

Yamasaki, K., in M. Levey, ed., *Archaeological Chemistry, A Symposium*, Pennsylvania U. P., Philadelphia, p. 347, (1967).

Yamasaki, K., *Studies in Conservation*, **15**, 278 (1970).

Dyes

Abrahams, D. H., and Edelstein, S. M., in M. Levey, ed., *Archaeological Chemistry, A Symposium,* Univ. of Pensylvania U.P., Philadelphia, 1967, p. 15.

Augusti, S., *Ind. Venice,* **4,**109 (1950).

Born, W., *Farbe und Lack,* **411,** 492 (1924).

Born, W., *Ciba Rundschau,* **4,** 110 (1936a).

Born, W., *Ciba Rundschau,* **7,** 218 (1936b).

Brown, B., *Text. Col.,* **65,** 93 (1943a).

Brown, B., *Text. Col.,* **65,** 427 (1943b).

Cady, W. M., *Am. Dyestuff, Rep.,* **26,** 539 (1937).

Cox, B., *Analyst,* **63,** 397 (1938).

Dimroth, O., *Ber. Deutsch. Chem. Ges.,* **43,** 1387 (1911).

Dimroth, O., and Scheurer, W., *Ann. Chem.,* **399,** 43 (1913).

Driessen, L. A., *Mell. Textilber.,* 2 (1944).

Emmerling A., and Engler, C., *Ber. Deutsch. Chem. Ges.,* **3,** 885 (1870).

Farnsworth, M., *J. Chem. Ed.,* **28,** 72 (1951).

Federovich, E. F., *Sovetskaia Arkheologiya,* **4,** 124 (1965).

Fester, G. A., *Isis,* **44,** 13 (1953).

Fieser, L., *J. Chem. Ed.,* **7,** 269 (1930).

Friedlander, P., *Ber. Deutsch. Chem. Ges.,* **42,** 765 (1909).

Friedlander, P., *Ber. Deutsch. Chem. Ges.,* **55,** 1955 (1922).

Heinisch, H. F., *Fibre Eng. Chem.,* **18,** 203 (1957).

Heise, R., *Arb. Gesunth Arnt., Berlin,* 513 (1895).

Hubner, J., *J. Soc. Dyers Col.* **25,** 223 (1909).

Karrer, L., and Miki, B., *Helv. Chim. Acta.,* **12,** 985 (1929).

Kashiwagi, K. M., *Bull. Chem. Soc. Japan,* **49,** 1236 (1976).

Kurdian, H., *Am. Orient. Soc.,* **61,** 105 (1941).

Levey, M., *Chemistry and Chemical Technology in Ancient Mesopotamia,* Amsterdam (1959).

Lucas, A., *Ancient Egyptian Materials and Industries,* Arnold, London, p. 152, 1962.

Masschelein-Kleiner, L., and Heylen, J. B., *Studies in Conservation,* **13,** 87 (1968).

Mell, C. D., *Text. Col.* 97 (1932).

Michell, H., *Class. Rev. N. SV.,* 246 (1955).

Perkin, A. G., *J. Chem. Soc.,* **69,** 800 (1896).

Perkin, A. G., and Horsfall, M., *J. Chem. Soc.,* **77,** 1314 (1900).

Rawson, C., *J. Soc. Dyers Col.,* **15,** 166 (1899).

Saltzman, M., Keay, A. M., and Christensen, V., *Dyestuffs,* **44,** 241 (1963).

Schaefer, G., *Ciba Review,* **4,** 1407, 1417 (1941).

Schmidt, R. E., *Ber. Deutsch. Chem. Ges.,* **20,** 1285 (1887).

Schunck, E., *Mem. Proc. Manchester Lit, Phil. Soc ,* **5,** 158 (1892).

Steckoll, S. H., Goffer, Z., Haas, N., and Nathan, H., *Nature,* **231,** 469 (1971).

Takagi, Y., *Osaka Gakugei Univ. Mem. B.,* **13,** 305 (1964).

Tomassi, A., *Gazetta,* **50i,** 263 (1920).

Vetterli, W. A., *Ciba Rundschau,* **93,** 3416 (1950).

Von Baeyer, A., *Ber. Deutsch. Chem. Ges.,* **3,** 885 (1900).

Wiskerke, C., *Econ. Hist. Jaarboek,* **25,** 1 (1952).

Ink

Lewis, A., in H. Torczyner, ed., *Lachish I, The Lachish Letters,* Oxford U.P. London, 1938.

Lucas, A., *The Analyst,* **47,** 9 (1922).

Steckoll, S. H., *Nature,* **220,** 91 (1968).

11

METALS

By the end of the Neolithic period man had acquired many of the skills and materials needed for a "civilized" way of life. He was also acquainted with "native" metals—those found in nature in conspicuous malleable lumps—which he used for making tools. The quantities of native metals that could be gathered from the surface of the earth and from the beds of streams were, however, not sufficient to maintain a regular supply of artifacts and tools.

Most common metals don't occur in the native state. They are found in the form of *metal ores:* minerals, often of striking color and high densities. At some point in time, probably shortly after 5000 B.C., men first recognized the connection between these minerals and metals: it was found that metals could be produced by heating ores in the proper kind of fire. Smelting was thus discovered. The moment of discovery is impossible, in retrospect, to pinpoint with any degree of certainty. Its impact in the story of man becomes discernable only well after the beginning of the Age of Metals.

METALS AND NONMETALS

The chemical elements can be divided into *metals* and *nonmetals*. The distinction is often not fully understood. Although the division of not always sharp, about 75 percent of the elements are metals.

Metals have a characteristic luster called *metallic luster*. They are *malleable* (i.e., can be hammered into sheets) and are *ductile,* (i.e., can be drawn into wires). They are good conductors of heat and electricity. The widespread use of the more common metals is due to these properties. Of particular importance is the fact that these and other properties can be varied in a controlled manner by *alloying,* that is, combining metals with each other and with other elements to form stable compounds. Strength, workability, and resistance to corrosion are among the properties of metals that can be improved by alloying. *Alloys* possess the characteristic physical properties of metals but are composed of two or more elements, at least one of which is a metal. Most of the materials known as metals in everyday life are actually alloys (see Tables 11.1 and 11.2 and Fig. 11.1).

TABLE 11.1   Alloys: Effect of Alloying on Physical Properties of Metals

Property	Effect of Alloying	Significance
Melting temperature	Lowering	Easier to work and cast
Thermal conductivity	Lowering	Property applied to make better cooking utensils
Strength	Rising	Stronger materials
Ductility	Lowering	The gain in strength more than offset the loss in ductility

TABLE 11.2   Common Alloys

Name	Composition (percent)	Use
Bronze	Copper (80-95); tin (20-5)	Many artifacts
Coin-bronze	Copper (92-95); tin (5-4); zinc (3-1)	Coins
Brass	Copper (50-90); zinc (50-10)	Many artifacts
Bell metal	Copper (75-80); tin (25-20)	Bells
Coin-silver	Silver (75-95); copper (25-5)	Coins
Electrum	Silver (less than 25); Gold (more than 75)	Coins, jewelry
Pewter	Tin (75-95); lead (25-5)	Table ware
Solder	Lead (50-70); tin (50-30)	Metal soldering
Steel	Iron (95-99.9); carbon (5-0.1)	Weapons; tools

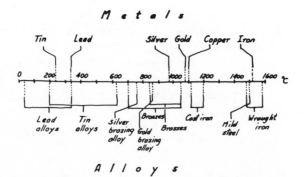

Fig. 11.1. The melting points of metals and alloys known in antiquity.

198

SOURCES OF METALS

There is great variation in the natural abundance of metals. The estimated amounts of the various metals existing in the earth's crust have been computed; Table 2.9 gives the average values for them.

Metals generally have a fairly high degree of chemical reactivity. Consequently, they are almost always found in nature in a combined state, as *minerals*, (or *ores*), from which they must be extracted by metallurgical processes. A few, however, are sometimes found free in nature in a more or less pure condition and are then said to occur *native*.

Native Metals

Native metals do not require metallurgical processing for their production. Only gold, some copper, and small quantities of iron are found native in profitable quantities.

The obvious luster of gold nuggets could hardly have been ignored by primitive man, and it is highly probable that native gold was the first metal known to him. Copper was also relatively abundant in the native state in ancient times. Even today, deposits of native copper can still be found in North America. Since copper is attacked by weathering agents, native copper appears as dark-green- or reddish-brown-coated nuggets(Coghlan 1951). Iron occurs native in the form of grains of microscopic size that would require a complex process for recovery. But large masses of iron have been brought to the earth by meteorites from outer space and are known as "meteoritic iron." Although silver occurs sometimes in large nuggests, it is only found in such a form in veins too deep in the earth's crust to have been accessible to the ancients.

For many centuries, while still relying mainly on stone as a raw material for artifacts, men used the native metals for making small objects. Native metals can usually be characterized by their chemical composition and typical microstructure. Quartz and iron oxide found in gold are a good indication of native metal since these impurities would have been removed in the course of smelting. Native copper usually has an extremely high purity that makes it distinguishable from the smelted metal. Meteoritic iron is characterized by the presence of phosphorus, and especially by a high nickel content, usually in the range of 5-10 percent.

Artifacts made from native metals have been identified at many archaeological sites. The oldest known artificially shaped metal objects made from native metal are copper beads found in northern Iraq, dating from the beginning of the ninth millennium B.C. (Smith 1965). Many other objects made of native copper have been identified all over the world. Artifacts dating from the sixth and fifth centuries B.C. were found in the Middle East (Wertime 1964). Iron beads from

Egypt, dating back to 3500 B.C. or earlier, must have been made from meteoritic iron (Wainwright 1936).

Metal Ores

Most metals occur in nature in the form of mineral compounds (oxides, carbonates, sulfides) called *ores,* which lack the characteristic malleability of metals. Geological processes result in the concentration of metals into local mineral deposits, called ore deposits, from which they can be extracted. Table 11.3 lists some of the most widely distributed metal ores.

TABLE 11.3   Metal Ores: Widely Distributed Minerals

Metal Ores	Formula
Copper ores	
Malachite	$CuCO_3 \cdot Cu(OH)_2$
Azurite	$2CuCO_3 \cdot Cu(OH)_2$
Cuprite	Cu_2O
Atacamite	$CuCl_2 \cdot 3Cu(OH)_2$
Chalcopyrite	$CuFeS_2$
Chalcocite	CuS_2
Iron ores	
Magnetite	Fe_3O_4
Hematite	Fe_2O_3
Limonite	$Fe_2O_3 \cdot 3H_2O$
Siderite	$FeCO_3$
Pyrite	FeS_2
Silver ore	
Argentite	Ag_2S
Lead ore	
Galena	PbS
Tin ore	
Cassiterite	SnO_2

All the metals known since antiquity occur in large, easily recognizable deposits and can be extracted from their ores by relatively simple processes. This is why ancient mining centers were confined to a few geographical regions. One of the main copper-mining centers in antiquity was the island of Cyprus, for which the metal may have been named. Tin was derived, for the greater part, from Britain, which was referred to as the "Islands of Tin." Much of the silver used in ancient Greece came from the mines at Laurion in Attica. Ancient gold was supplied in large quantities by the mines of Nubia. It has been calculated that the annual yield of these mines at the time of Ramses II (1300-1230 B.C.) exceeded $450,000,000!!!

METALLURGY

Ores are mined and then treated by various mechanical and chemical processes to extract the metals and convert them into the metallic form. Chemical or *extractive metallurgy* is concerned with the processes of extraction, dressing, and chemical treatment of ores that eventually produce refined metals. Other branches of metallurgy are *physical metallurgy,* which deals with the study of the metallic state and the treatment of metals to give them valuable properties, and *mechanical metallurgy,* which is concerned with the reactions of metals to loads and stresses during forming. However important to the archaeologist, the two latter branches fall outside the scope of this book and are not dealt with here.

Extractive Metallury

The process of extracting or mining of metals includes: (1) *dressing,* the breaking, pulverizing, and concentrating of the ores, (2) *reduction* of the ores to metals, and finally (3) *refining,* the purification of the metals for use (Fig. 11.2).

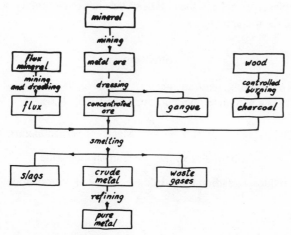

Fig. 11.2 Flow sheet of the metallurgical processes involved in the extraction of metals from minerals.

Dressing involves the elimination of specific impurities from ores to increase the concentration of the desired component and to prepare it for the reduction stage: an ore containing 5 percent copper mineral might be separated by dressing into a concentrate with 60 percent copper mineral and "trailings" with less than 0.5 percent metal. Most ores in ancient times were given some type of benefication treatment: this might involve hand picking, pounding, screening, and often washing in running water to remove the less dense nonmetallic materials (Tylecote 1976).

Crude metals were obtained in antiquity by *pyrometallurgy*, the reduction of ores to metals at high temperatures. The process is usually called *smelting;* the particular smelting process used depends on the metal to be obtained.

Fuel

The two most important fuels of antiquity were wood and charcoal. Coal and peat were apparently quite unknown to most ancient peoples, although they were found in abundance in such districts as the Ruhr in Germany and in Great Britain (Neuberger 1969).

In pyrometallurgical processes the quality of the fuel determines the quality of the product. Not only does the fuel used determine the temperature of the process, but impurities in the fuel affect the characteristic of the metal produced. There are examples of ancient metallurgical processes where wood was used as a fuel, although charcoal was usually preferred (Forbes 1950). Charcoal was used from earliest times and has continued to be used in metallurgy through most of history. It was finally replaced by coke in the middle of the eighteenth century, a change that revolutionized the metal industry.

Smelting

In the process of smelting, naturally occurring metal oxide are reduced with carbon to the metallic state at high temperatures (Grothe 1975). Thus:

$$\text{Metal oxide} + \text{carbon} \xrightarrow{\text{smelt}} \text{Metal} + \text{carbon monoxide}$$

$$\text{Carbon} + \text{oxygen} \rightarrow \text{Carbon monoxide}$$

$$\text{Metal oxide} + \text{carbon monoxide} \rightarrow \textit{Metal} + \text{carbon dioxide} \uparrow$$

Summarizing:

$$\text{Metal oxide} + \text{carbon} + \text{oxygen} \rightarrow \textit{Metal} + \text{carbon monoxide} \uparrow$$

The smelting of copper oxide, for example, proceeds according to the equation:

$$CuO + C = Cu + CO$$

$$2C + O_2 = 2CO$$

$$CuO + 2CO = Cu + CO_2 \uparrow$$

The general characteristics of the smelting process are:

1. Metal oxides are reduced to the metallic state.
2. The entire melting furnace charge become molten and the liquid metal and slag form separate layers.
3. The metal produced is usually quite impure because other elements present in the ore are also reduced and alloy with the principal metal.

Sulfide and carbonate minerals are roasted or calcined prior to smelting. *Roasting* is an oxidizing process intended mainly to convert sulfide minerals into oxides:

$$\text{Metal sulfide} + \text{atmospheric oxygen} \xrightarrow{\text{roast}} \text{Metal oxide} + \text{sulfur dioxide} \atop \text{(gas)}$$

By roasting galena (lead sulfide, PbS) lead oxide is obtained:

$$2PbS + 3O_2 \xrightarrow{\text{roast}} 2PbO + 2SO_2 \uparrow$$

The lead oxide is then smelted to produce metallic lead:

$$PbO + C \xrightarrow{\text{heat}} Pb + CO \uparrow$$

In the *calcination* process, ore concentrates are heated to drive off volatile components. Calcination of carbonate minerals causes the evolution of carbon dioxide and the formation of the metal oxide, which can then be smelted:

$$\text{Metal carbonate} \xrightarrow{\text{calcine}} \text{Metal oxide} + \text{carbon dioxide} \uparrow \atop \text{(gas)}$$

Thus the pyrometallurgical reduction of siderite (an iron ore composed mainly of iron carbonate, $FeCO_3$) proceeds in the following way:

$$FeCO_3 \xrightarrow{\text{calcine}} FeO + CO_2 \uparrow$$

$$FeO + C \rightarrow Fe + CO$$

$$FeO + CO \rightarrow Fe + CO_2 \uparrow$$

$$\overline{}$$

$$2FeO + C \rightarrow 2Fe + CO_2 \uparrow$$

Metal ores usually contain silica and siliceous materials that melt at temperatures far in excess of those attainable during smelting. A *flux* is, therefore, usually added to the ore. During smelting the flux combines with the otherwise infusible siliceous materials to form a fusible mass called *slag*. Slags are liquid at the working temperature of the furnace and for the most part are immiscible with molten metals. They float on the surface of the molten metal and can be drawn off, solidifying afterwards into glass-like waste products. Materials such as lime, magnesia, and manganese oxide are used as fluxes.

Refining

The last step involved in the pyrometallurgy of metal ores is *refining*. There are two principal classes of impurities in smelted metals: (1) impurities whose presence is deleterious to the properties of the metal and that must be removed to produce an acceptable quality of product and (2) impurities that are not necessarily harmful but have sufficient value to make their recovery profitable (e.g., gold and silver).

Pyrometallurgical refining processes vary over a wide range, but two general rules apply to all of them: (1) the crude metal must be molten before refining; and (2) the refining reactions are usually controlled oxidation reactions whereby impurities are either volatilized or converted into oxides that form slags.

Precious metals were refined or recovered from baser metals by *cupellation*. The impure metals were heated by a blast of hot air in shallow, porous cups of bone ash called *cupels*. The base metals, such as lead, tin, and copper, were oxidized by the hot gas, and the oxides were absorbed by the porous cupels. The unoxidizable noble metals-silver or gold-were left behind in the bottom of the cupel, like drops of water on an oiled surface (see Fig. 11.2).

METAL AND ALLOYS OF ANTIQUITY

The Greek poet Hesiod (eighth century B.C.) tells of four successive ages in the legendary prehistory of mankind: the Golden, Silver, Bronze, and Iron Ages. Lucretius, the Roman historian (first century B.C.), wrote: "The earliest weapons were the hands, nails and teeth. Then came stones and clubs. These were followed by iron and bronze, but bronze came first." Today it is generally accepted that the first metals known to man were those occurring in the native state: gold, copper, and meteoritic iron. There is, of course, no way of knowing which of these was used first.

The discovery of the smelting process brought about the increasing use of ores as sources of metals. By the end of the third millennium B.C. nearly all the metals producible from common minerals by reduction with carbon had been smelted and put to use, including even iron, although its use did not become widespread until about a thousand years later.

The Old Testament mentions six metals: gold, silver, copper, iron, tin, and lead. The ancient Hindus also used these six (Ray 1904), which were listed by Arabian chemists of the eighth century and by European alchemists of the thirteenth. From the repeated references to these six metals in ancient literary sources it has been inferred that the discovery of other metals came at a much later date. Nevertheless other metals were known to the ancients, although their manufacture was isolated and occasional. Among these were mercury, antimony, platinum, and zinc.

Gold

Gold was widely known through the world even before the dawn of history. Gold jewelry and ornaments have been found in Stone age tombs. Egyptian goldsmiths of the earliest dynasties were highly skilled artisans. In Mesopotamia gold ornaments from the royal graves at Ur, about 4600 years old, show a wide range of metallurgical techniques. In Palestine, at the time of Kings David and Solomon (tenth century B.C.) gold was available in large quantities. The province of Cocle in Panama has been found to be very rich in pre-Columbian gold artifacts (Lothrop 1934).

Gold occurs always as a native metal, either in alluvial deposits or in veins in quartz. Alluvial deposits were usually worked by *placer mining* methods, the simplest of which involves washing the sand and gravel from stream beds in a pan or craddle. Veins in quartz were worked by smelting and cupellation or by amalgamating.

A detailed description of the extraction of gold by the ancient Egyptians, who applied pyrometallurgical processes, has been handed down by Diodorus (first century B.C.). Diodorus described how the metal was extracted from quartz veins in Nubia, the "Land of Gold." The raw material was ground down to powder to facilitate dressing operations and to separate the gold from the stone, it was then smelted together with lead. A second smelting stage produced pure gold as a residue in the smelting pot (probably through oxidation and sublimation of the impurities and excess lead).

Vitruvius (first century B.C.) and Pliny and Elder (first century C.E.) mentioned the use of the *amalgamation* process to recover gold: the gold was dissolved in mercury forming an amalgam that was periodically removed and replaced with fresh mercury. The mercury was then displaced from the amalgam by heating and was probably recovered by distillation. The residue consisted of pure gold.

Silver

Silver rarely occurs uncombined in nature. For this reason it came into use later than gold. In Egypt, between the twentieth and fiftheenth centuries B.C. it was scarcer and perhaps more costly than gold (Mishara and Meyers 1974). The Babylonians of the same period were, however, skilled silversmiths. A silver vase from Mesopotamia dating from about 2800 B.C. was worked from an ingot of the metal. The ingot was hammered into a sheet with frequent annealings (repeated heating), raised by local hammering on a stake into the final shape, and finally inscribed with lines impressed by a chasing tool (Smith 1965).

Copper

Beads made from native copper are the earliest known metal artifacts. Shortly after the discovery of smelting, copper was being extracted from ores in the Middle East (see text that follows). The oldest copper mines known to have been used by the Egyptians are situated in the Sinai Peninsula and in Wadi Arabah in southern Israel. The metal was extracted from ores consisting mainly of green copper carbonate (malachite) and copper silicate (chrysocolla). Slags found at the afore-mentioned sites are of very varied composition, showing that the process of extraction was not always carried out in a uniform way.

The smelting of copper was well established in many different regions before the end of the third millennium B.C. By then not only carbonate ores were being used, but also ores containing copper sulfide. Extensive information has been left on the roasting process used by the Greeks to transform the sulfides into oxides. This was carried out after the model of lime burning: kilns were made of the material itself. After fuelling, the kilns were filled with the ore, the fuel was then ignited and the pyrites roasted. The end of the process was recognised by the ore having become red. Very different smelting furnaces and techniques seem to have been used in different parts of the world (Craddock 1976, 1978; Epler 1978).

Iron

Iron is the most important and widely used of all metals. Its preeminence in human technology is the result of several factors:

1. Iron is one of the most plentiful of metals in the upper crust of the earth. Large deposits of its ores are numerous and widely distributed.
2. Iron ores are comparatively easy to reduce to metal.
3. As a result of items (1) and (2), iron is the cheapest of metals to produce.
4. Iron forms, in combination with carbon, a remarkable series of useful alloys called *steels*.

Iron, like copper, was first used in the form of native "meteoritic" metal. The first reduction of iron ore to iron was probably the result of an error by a bronze smith. The spongy mass that probably was obtained was iron blooms (Bouman 1978). An iron-smelting industry existed in India probably as early as 2500 B.C. and certainly by 1500 B.C. In the Rewah Province of central India great heaps of cinders and slag, covering many square kilometers, testify to a flourishing industry. Iron smelting was also developed by the Hittites between 1500 and 1200 B.C. Most of the product then, as later, was wrought iron. By 1200 B.C. steel was being made with some degree of control (Maddin, Muhly, and Wheeler 1977). In China cast iron was being made by 500 B.C. In Europe its special properties were recognised only 2000 years later!!!

Tin

Long before metallic tin was known, bronze (an alloy of copper and tin) was in common use. In Egypt, Mesopotamia, and the Indus Valley, alloys of copper and tin were being produced in 3000 B.C. At the end of the third millennium B.C. the Cretans were adding tin to copper to lower the melting point. The earliest known Egyptian artifacts made of unalloyed tin are a ring and a pilgrim bottle from tombs of the eighteenth dynasty (1580-1350 B.C.) (Lucas 1962). Pure tin, which had evidently been prepared for use in castings, was found at Machu Pichu in Peru (Bingham 1930).

If much pure tin was made in antiquity it has perished, perhaps for two reasons: it has either been oxidized to a mixture of stannous and stannic oxides, or it has been transformed, by allotropic modification, to powdery gray tin.

The only important ore of tin is cassiterite (tin oxide, SnO_2). This was presumably the earliest source of the metal, although no direct accounts are available on the method by which tin was smelted. From the remains of furnaces it appears that smelting was a simple process of reduction carried out by heating the ore over a wood fire. The metallic tin thus obtained was simply smelted out.

Lead

Lead was an extremely important material in ancient times. Lead ores are widely distributed in nature and are easily smelted. The metal was known to the early Egyptians, Indians, and Jews. The first Pharaohs who gained victories in Asia received part of their tribute paid by the conquered peoples in the form of lead. In the time of Zecharia (fifth century B.C.) lead weights were in use. The Romans used the metal extensively for water pipes, writing tablets, anchors, and coins (Krysko and Lehrheuer 1976).

Mercury

A small vessel full of mercury was discovered by H. Schliemann, the excavator of Mycenae, in a grave at Kurna estimated to date from the sixteenth or fifteenth century B.C. (Von Lippman 1919). The element must have been known for a long time by then. Mercury was probably known to the Phoenicians in the eighth century B.C. Aristo, writing in the fourth century B.C., mention and describes it. The mode of the preparation is as follows. Native cinnabar (mercuric sulfide, HgS) was rubbed with vinegar in a copper mortar with a copper pestle yielding the liquid metal (Hill 1774).

Mercury played an important role in the *gilding* of other metals (see glossary). It was used to extract gold by forming an amalgam with it. Gold from discarded Roman garments that had been interwoven with gold thread was recovered by reducing the garments to ash in a crucible and treating the ash with mercury, which took up all the gold by amalgamation.

Antimony

Antimony was probably chiefly known in its natural compounds. Finely divided stibnite (antimony sulfide, Sb_2S_3), was sometimes used by Egyptian women for painting their eyes and by glassmakers to produce certain types of glass. The metal was also used, though only very occasionally. A fragment of a most unusual vase from excavations at Telloh was found to consist of pure antimony containing only traces of iron (Berthelot 1897). It is thus possible that the metal was known to the Chaldeans in 4000 B.C. The Greeks and Romans, like the Chaldeans, probably also knew how to obtain antimony. However, they had no adequate means of distinguishing between metals and applied the term "lead" indiscriminately to any metal that was soft, dark gray, and easily fusible.

Vessels made of antimony have been reported from very old burial grounds near Tiflis in Georgia, (in the Soviet Union). Other instances of the use of metallic antimony in antiquity have been recorded (Wirchow 1884; Cambi and Cremascoli 1957; Cambi 1958).

Zinc

Zinc was probably isolated accidentally in the course of smelting other metals. Most of the time ancient metallurgists were probably losing the volatile zinc in the form of vapor because their smelting facilities were not designed for condensing it. Still a few isolated metallic zinc finds have reported: a statuette containing 87.5 percent zinc was found in a prehistoric site at Dordosh, Transylvania, and zinc artifacts have been found at several other ancient sites (Mosso 1906).

Platinum

Platinum occurs sometimes as grains and nuggets in the alluvial sands of rivers. But the metal has such a high melting point (1774°C) that it cannot be melted with primitive equipment. It has been suggested that a little gold was mixed with grains of platinum to bond them together as the gold melted: the sintered mass thus obtained might then have been worked by alternate heating and hammering (Bergsoe 1936, 1937). Pre-Columbian Indians produced white alloys of gold and platinum from which they made many small artifacts.

An Egyptian casket found at Thebes and dating from the seventh century B.C. was proved to have been made of an alloy of platinum, indium, and gold (Berthelot 1901). This is the only known case of the use of platinum in ancient Egypt. A few other ancient objects of platinum have been found (Rivet and Arsandaux 1946).

Alloys

Soon after the discovery of metals it became clear to ancient men that the properties of metals could be improved by alloying or mixing these with other elements. Hardness can be enhanced by producing an alloy in which some atoms of the main component are replaced by others of different size. The melting points of many metals can be lowered by alloying. Other properties that can be similarly improved are strength, workability, and resistance to corrosion. Table 11.2 lists several of the more common alloys known in antiquity.

STUDY OF ANCIENT METALS

The scientific study of ancient metal objects is of obvious importance from the historical standpoint. The conclusion and inferences drawn from these studies often help in the solution of practical problems of the archeologist. It may be possible, for instance, to determine whether ancient metallurgical operations were carried out in a random fashion or under controlled conditions. Chemical analysis provides evidence as to whether metals from different archeological periods show differences in their elemental composition.

The question probably most often asked by the archeologist interested in metals is whether the composition of metal objects may constitute a guide to the geographic provenance of the metal. Modern scientific techniques may provide at least some of the answers.

Of all ancient metals, the most intensively investigaed so far has been copper and its alloys, of which many thousands of analyses have been published (Gettens 1969; Leoni 1974). Relatively little research as yet been done on gold, silver, lead, and the ferrous metals. Most of what little work has been done on gold and silver has been in connection with coins. Information on lead is also curiously scarce, even though cuprous lead was produced by the Romans, and, because of the high resistance of this metal to corrosion, ancient lead finds in good condition are plentiful.

ANCIENT METALLURGY

The analysis of ores, metals, and slags provide information on early techniques of metal extraction. Comparative analysis of metal objects produced for various utilitarian purposes shows how well the early metal workers understood the materials they worked with and how skillfully they utilized the different properties of different metals and alloys to best advantage.

The extraction of the common metals from their ores does not necessarily call for the elaborate equipment and complicated processes in use today. Lumps of copper ore, or of iron oxide, which may by chance have formed part of the ring of stones around a domestic fire and become embedded in its embers, would be reduced to metal. It is quite reasonable to suppose that some prehistoric campfire became, quite accidentally, the first metallurgical furnace. All that is needed to convert such a campfire into a smelting furnace is a small hole in the ground to receive the molten metal. A furnace of this type (see Fig. 11.3), was constructed

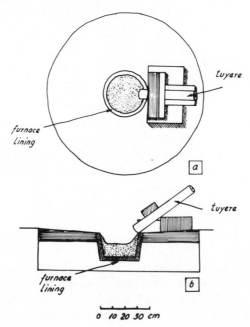

Fig. 11.3. Primitive furnace built and used experimentally: (*a*) plan.   (*b*) cross-section.

and used experimentaly under the conditions presumably available to prehistoric man. Using charcoal as fuel, a mixture of green carbonate copper-ore and tin-stone was smelted, and a copper-tin alloy containing 22 percent of tin was obtained (Gowland 1912).

Smelted copper from the Bronze Age is similar in composition to modern blister copper but, unlike the latter, usually contains only traces of sulfur. The percentage of copper in characteristic specimens is in the range 97-99 percent. Table 11.4 lists copper analyses of specimens from early civilizations.

TABLE 11.4   Composition of Ancient Copper

Origin of Sample	Composition (in Percent)				References
	Copper	Iron	Lead	Sulfur	
Early Indus Valley civilization	96.7	0.03	0.02	0.98	
Early Mesopotamian (pre-3500 B.C.)	98.8	0.98	—	—	
Roman coins (first cent. C.E.)	99.6	0.04	Traces	—	
Central European prehistoric object	95.8	—	None	—	Otto and Witter 1952
Pre-Columbian, Argentina	90.7	—	None	—	Fester 1962

Copper Smelting in Wadi Arabah

Copper-working sites undisturbed since the time of their use in the Chalcolithic period (fourth millennium B.C.), early Iron Age (twelfth century B.C.), and Roman and Byzantine times were discovered in Wadi Arabah in southern Israel (Tylecote, Lupu, et al. 1967). Two smelting techniques appear to have been used at the sites investigated. The first, characteristic of the Chalcolithic period, seems to have yielded a mixture of slag and copper in the furnace. The two were apparently separated by breaking the slags and then melting the copper in crucibles.

The second technique, used in the early Iron Age, is characterized by the use of highly sophisticated furnaces such as the one illustrated in Fig. 11.4 preliminary dressing, the copper ores (malachite, azurite, and chalcocite), along with the fluxes (chiefly iron oxide from fossilized trees, dolomite, limestone, and manganese ore), were crushed to fine granules and then mixed with charcoal to

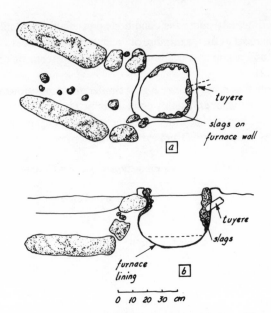

Fig. 11.4. Early Iron Age furnace from Wadi Arabah: (a) plan. (b) cross-section.

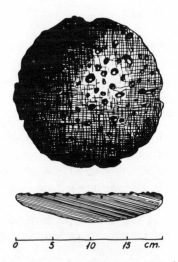

Fig. 11.5. Plan and cross-section of a Iron Age copper ingot from Wadi Arabah.

212

form the smelting charge. This was fed through the open top of the smelting furnace onto the fire, which was fed with wood from acacia trees. The composition and structure of the slags makes it possible to estimate the temperature inside the furnace as having been in the range 1200-1350°C. This was achieved with the aid of a single tuyere (see glossary) and bellows. The metal was reduced to metallic copper while the remaining material turned to slag. The heavy copper sank through the slag into the cavity in the furnace's bottom. Further quantities of charcoal and smelting charge were successively added until the furnace was full. At this stage the operation of the bellows was stopped and the slag was tapped into a circular tapping pit. The bun-shaped copper ingot remained in the furnace and was removed after solidifying. A typical ingot consisted of 90 percent copper and weighed up to 15 Kg (see Table 11.5 and Fig. 11.5).

**TABLE 11.5   Composition (in Percent) of
copper from Wady Arabah**

Sample No.	Copper	Iron	Lead	Sulfur
1	89	9.7	0.8	0.05
2	92	5.0	0.7	—
3	89	7.2	1.2	—
4	90	9.0	0.2	—

ARSENICAL BRONZES

Arsenical copper—frequently of high arsenic content—was widely used by ancient metallurgists. Finds of arsenical coppers have been reported, among others, in Argentina (Fester 1962), the Dead Sea region in Israel (Kay 1964), and the Cyclades Islands in the Aegean Sea (Renfrew 1967). Some of the results of analysis of the finds are presented in Table 11.6.

For many cultures the use of arsenical copper is characteristic of a transitional stage of development: it came to replace pure copper and was eventually, in turn, supplanted by bronze. Copper-arsenic alloys, though inferior to copper-tin alloys, are superior to pure copper in many ways: they melt more easily, are harder, and produce sounder castings.

The question arises as to whether alloys of high arsenic content were made intentionally or whether they were merely the result of the chance use of arsenic-containing raw materials. The answer to this question can be found inferentially, if all the pertinent facts are reviewed.

The arsenic content of native copper is always low, and the latter can be ruled out as a source of the alloyed metal. Copper ores from Sinai, Southern Israel, and

TABLE 11.6   Ancient Copper-Arsenic Alloys)

Sample		Composition (Percent)					
Origin	Description	Copper	Arsenic	Antimony	Silver	Lead	Reference
Dead Sea area							
	Chalcolitic artifacts	96.0	3.5	0.2	0.1	0.3	
		87.0	11.9	0.6	0.2	0.04	Kay 1964
		87.0	12.0	0.7	0.3	Trace	
Pre-Columbian Argentina							
	Armband	90.7	3.8	2.9	—	None	Fester 1962
	Breastplate	96.9	1.2	0.5	—	None	

Cyprus, known to derive from deposits exploited by the ancients, contain either zinc or tin and have a very low arsenic content. Arsenic is associated with copper in the sulfoarsenate ores Enargite ($3CU_2S \cdot As_2S_5$) and Tennantite ($CU_3S \cdot AsS_3$). Arsenopyrite (FeAsS), sometimes associated with copper sulfide ores, can lead to substantial arsenic content of the metallic copper.

Arsenic is a useful deoxidizing agent of copper, producing castings suitable for subsequent forging. With arsenic concentration of up to 8 percent, copper alloys are readily worked hot or cold and are as hard as tin bronze. The presence of arsenic is also beneficial in maintaining the workability of copper that contains residual copper oxide. It is clear, therefore, that the intentional inclusion of arsenic would have improved the casting and working properties of copper. The only way to obtain a high arsenic concentration deliberately would be by selecting copper minerals particularly rich in arsenic such as those mentioned in the preceding paragraph or by adding arsenic minerals (orpiment, As_2S_3; or realgar, AsS) to the smelting mix.

The antimony and silver contents of some of the arsenical copper objects analyzed are directly proportional to the arsenic content, as shown in the scatter diagrams in Figure 11.6. Arsenic, antimony, and silver are associated with copper only in the sulfoarsenate ores, and not in other copper ores where arsenic is absent. The presence of these elements is thus indicative that sulfoarsenate ores must have been deliberately added during the manufacturing process.

In all the Dead Sea finds the absence of zinc and tin is surprising. There is also a difference in composition between ornaments and tools, respectively (Table 11.6), and this difference is reflected in the finished articles. All the ornaments, which are made of arsenical copper, are finished off to close tolerances, whereas the tools, made of unalloyed metals, have been left rough. This may be either because the tools are older than the ornaments or because they were made by different craftsment. Still, the addition of arsenical ores probably represents a definite phase of technological development in the metallurgy of copper.

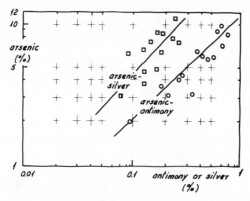

Fig. 11.6. Arsenic:antimony and arsenic:silver ratios in the composition of copper ornaments found in the Dead Sea area.

It is possible that early Bronze Age metallurgists recognized that the use of arsenic-rich copper ores or the incorporation of arsenic-rich ores into the copper-smelting mixture would result in copper of particular properties. Then they would purposely select such ores with the intention of producing a superior class of material. The high toxicity of arsenic may have been what led to the gradual discontinuation of its use in copper metallurgy, even though it is now known that toxicity had little influence on the choice of technological processes in early times; people did not always know the causes of their ailments.

An interesting sidelight on the foregoing is the fact that copper-arsenical ores were not of common occurrence in the old world, and the only known deposits of importance were in Armenia (Grigoryan 1962). This would suggest that supply routes along which the raw materials could be brought (e.g., from Armenia to the Dead Sea) were already established as far back as Chalcolithic and early Bronze Age times.

METAL-WORKING TECHNIQUES

Granulation and Filigree

Granulation and filigree were remarkable technical acievements of ancient gold-smiths. Very small gold granules (granulation) or very fine gold threads (filigree) were arranged in patterns and joined to ornaments made of the precious metal or its alloys. The very delicate decorations produced often gave a most pleasing effect.

Granulation and filigree were made by the Sumerians as early as 2000 B.C. The technique was common in Egypt and became known in Greece and Rome; however, at some time during the Roman Empire it was lost and was only rediscovered in the earlier part of this century.

Making granulation and filigree involves joining metals, a process usually done either by soldering or by brazing. In *soldering,* the metals are joined by filler alloys (solders) of low melting point. In *brazing* the filler is an alloy whose melting point is only sightly lower than that of the metals or alloys being joined. In granulation and filigree a large number of joins are made over very small areas; it is not possible to hold either soldered or brazed joints in position while others are being worked. The usual methods of joining metals cannot be used, and it is necessary to employ a special technique. In the archeological literature this technique is referred to as "colloidal hard soldering." A better appellation is *diffusion joining.*

A copper compound, usually copper hydroxide, is mixed with glue and water. The mixture, applied with a fine brush, serves to stick the gold granules or wires on to the work. The assembly is then placed in a charcoal fire. The subsequent rise in temperature brings about a series of reactions that finally result in a perfect metal joint. At about 100°C copper hydroxide is converted to black copper oxide:

$$Cu(OH)_2 \rightarrow CuO + H_2O \uparrow$$

At 600°C the glue becomes carbonized:

$$Glue \rightarrow C + CO \uparrow + H_2O \uparrow$$

At about 850°C the carbon (from the glue) reduces the copper oxide to fine particles of metallic copper, and carbon monoxide is evolved:

$$CuO + C \rightarrow Cu + CO \uparrow$$

The carbon monoxide (from the reduction of copper, as well as from the charcoal fire) prevents access of oxygen and thus oxidation of the metals being heated. A process of alloying of the fine particles of copper with the gold (or gold alloy) of the work commences at 900°C. The alloy formed *in situ* runs freely, wets the granules or wires, and the work can be removed from the fire. On cooling the joins are completed.

METAL ANALYSIS OF ARCHAEOLOGICAL ASSEMBLAGES

Native metals and those refined from ores are known as *primary* or *virgin* metals. Metals can also be derived from the remelting of scrap, which may have come from a varity of primary metal sources. Metals derived from scrap are called *secondary* metals. The distinction between primary and secondary metals is important when considering the use of their composition as a source of information about their age or provenance.

Analytical results obtained from assemblages of artifacts made of primary metals can be relied on to yield meaningful information about provenance, trade routes, and other subjects of archeological importance. With secondary metals this is hardly possible because of the probable heterogeneity of their origin. There is also the problem of alloys; different craftsmen may have used particular recipes, whereas different types of objects call for the use of different alloys. Bronze weapons, for example, generally have a higher tin content than does statuary. The metals for these alloys most probably came from different sources. Thus attempts to obtain meaningful results from the analysis of metals that may have come from secondary sources must be regarded as a wasted effort (Hall 1965).

Time of Origin

Archeological objects from different periods differ in average composition, whether in major or minor elements, or both. For this and other reasons that should by now be clear, in the study of ancient metallurgy it is customary to analyze large numbers of samples. Overall patterns, rather than the results of individual analyses, are important.

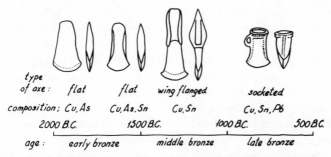

Fig. 11.7.   Style, composition, and chronology of northern British axes.

Figure 11.7 is an illustration in point (MacKerrel 1972). A group of axes from the British Bronze Age were arranged in assumed chronological order by close comparison, using typological criteria. The simplest were considered the earliest, and the more complex forms were assumed to derive from and gradually succeed them. Chemical analysis revealed a parallel increase in the complexity of the alloy used. More complex shapes succeeded versions simpler both in appearance and in alloy. The earliest arsenical copper flat axes from the beginning of the Bronze Age (ca. 2000 B.C.) were superseded first by alloys of this metal with tin, then by purer tin-bronze, and finally by leaded tin-bronze. The latter transition is concluded to have occurred at the change from the middle to the late Bronze Age. The alloy change is, in fact, the technological reason for the advent of the late Bronze Age; the new sophisticted leaded bronze provided a range of complex artifacts that was previously unattainable.

Other examples of the use of the analysis of metals are provided by variations in the composition of coins, discussed later in this chapter.

PROVENANCE OF COPPER AND COPPER ALLOYS

Archeologists often inquire whether the chemical composition of artifacts is a guide to provenance. When attempting to ascribe provenance to metal objects, two lines of approach can be followed: (1) seeking a relationship between ore and metal and (2) classification of analytical results on a statistical basis.

In principle, the linking of metal artifacts to geological deposits should be ideally served by chemical analysis at trace element level, if it is assumed that some of the minor elements present in an ore are carried over during smelting into the metal produced. Several studies along this line were undertaken (Otto and Witter 1952; Neuninger and Pittioni 1958; Pittioni 1959; Coghlan 1960). In some of the investigations copper ores were analyzed for impurities, and on the assumption that these impurities would alloy in significant quantities with the copper metal, it was sought to associate them with those present in copper artifacts. It transpires, in practice, that there are several difficulties in the way of this approach. Impurities are not always present in metals in concentrations that bear any relation to parent-ore composition (Tylecote, Ghaznavi et al. 1977). Two factors can cause considerable variability: (1) the mixed melting of metals from primary and secondary sources and (2) the variation of impurities as a function of ore depth and smelting conditions.

A different approach is based on the study of what happens to ore impurities during smelting (Friedman, Smith, et al. 1966; Fields, Milstead, et al. 1971). Copper ores, as well as native copper, were considered in this study. Representative ore samples were smelted under simulated ancient conditions, and the incorporation of ore impurities into the metal was measured. In all, the composition of 350 ore samples and 100 artifacts was analyzed. Some elements (silver, lead, iron, etc.) remained, in large part, in the finished metal. It was found that trace element levels in native copper are different from those occurring in metals extracted from ores, and that there are differences in the composition of artifacts from different geographical regions. The authors assumed that there was no mixing of ores or of copper produced from different ores. It may, indeed, be supposed with some reason that early artifacts were in fact made from metal freshly smelted from the ores. But the production of artifacts in later periods undoubtedly involved the remelting of scrap metal. An additional problem, when studying the relationship between ore and metal, is that early artifacts were made from metals originating in ore veins and mines that are now exhausted, whereas modern miners are working ore deposits not available to early man.

Useful information may be obtained by comparing the analyses of archeological objects with those of minerals in museum collections (Coghlan, Butler, et al. 1963). However, even here it is possible to find very wide variations in minor element composition. Hence not much can be said at present about establishing associations between "ore impurities" and "metal impurities," although it is sometimes possible to say with certainty that a given artifact cannot be derived from a particular ore, and that it could possibly have come from another (Tylecote 1970).

Some workers have used the analysis of metal artifacts alone and, to determine provenance, have classified the results on a statistical basis. Thus have been classified the analytical results of an enormous number of bronzes, mainly of European origin, (Junghaus, Sangmeister, et al. 1960). No less than 22,000 early Bronze Age objects of middle Europe were analyzed by emission spectroscopy and the contents of eleven elements determined, although only five (silver, nickel, arsenic, antimony, and bismuth) were eventually considered significant. The results were subjected to a statistical analysis based on two premises. The first is that if a large number of objects are found at a certain location and have a characteristic composition, they represent a local group, related perhaps to a nearby ore source. The other reason—perhaps more difficult to accept—is that if a group of artifacts in another location has the same composition as the first, the latter constitutes an import from the first location. On this basis, not considering geological and metallurgical problems, and ignoring archaeological factors, a very elaborate system of classification of the objects into analytical groups was evolved. There are, however, some objections to the premises adopted. After all, ore sources in different locations may have statistically similar compositions. Thus the question remains open as to whether the information obtained from such widespread sources can be accepted with confidence.

PROVENANCE OF NONCUPROUS METALS

Very little work has been done thus far concerning the provenance of metals other than copper. Some of the more important studies are mentioned in the following paragraphs (Muhly 1973).

Gold

Five hundred prehistoric gold objects, mostly from the Dublin Museum (Ireland), were analyzed by emission spectroscopy. Some interesting deductions were then made about the possible sources of gold used by early Irish goldsmiths, although the information was not sufficient to draw clear-cut conclusions about provenance (Hartman 1964, 1965).

Silver

Analytical data in combination with other characteristics can be used, with reasonable success, to establish groups of silver objects of common geographic provenance and to provide information on their production, use, and distribution (Meyers van Zelst et al. 1973; 1976; MacKerrel and Stevenson 1972; Gordus 1973).

Ferrous Metals

Little is known about the trace-element composition of iron produced in antiquity. A study of ancient Polish ferrous objects used chemical, spectrographic and metallurgical data (Piaskowsky 1961, 1964). It seems that chemical composition and metallographic structure make it possible to identify artifacts made in certain areas. But no single feature is sufficient for characterization; all the evidence available must be considered when attempting to assign provenance.

Lead

Galena ores of lead, occurring in different mining regions of the world, often have different lead isotope ratios, because they were either deposited during different geological periods, derived from different source rocks, or both. Table 11.7 shows that lead samples from different mining sites in ancient times can be distinguished from one another on the basis of their isotope ratios.

TABLE 11.7   Lead Isotope Ratios in Lead Ores (Galena, PbS) from Ancient Mining Regions

	Origin		
Isotope ratio	Laurion (Greece)	Great Rutland Cavern (Derbyshire, England)	Rio Tinto (Spain)
Pb-207:Pb-206	0.8307	0.8465	0.8598
Pb-208:Pb-206	2.0599	2.0814	2.1020
Pb-204:Pb-206	0.0529	0.0541	0.0548

By comparing isotope ratios determined in samples of lead from archeological objects to the isotope ratios in galena ores from different mining regions, it is

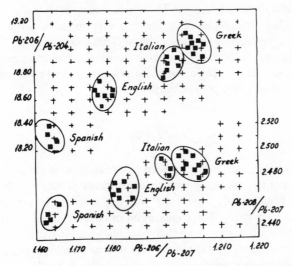

Fig. 11.8.   Lead isotopes: isotope ratios of lead in objects excavated in Spain, England, Greece and Italy. To draw these graphs, the lead from these sources was asumed to have come from known mining regions. The loops set off four major groups of samples.

possible to determine whether lead could have been smelted from ore minerals from these regions. Figure 11.8 illustrates the range of isotope ratios obtained from over one hundred specimens. (Brill and Wampler 1967; Brill 1968). Most of the archeological objects analyzed fell into one of four compositional groups: Laurion, England, Spain, and specimens from the Middle East and Italy. There are, however, a number of limitations imposed on the interpretation of the results. Although a particular set of isotope ratios may be charateristic of the ore of a given mining region, they are rarely uniquely so. Mines located in geologically similar environments may often yield identical isotope ratios even though the regions themselves are widely separated. Another complication arises from the mixing of lead in ancient time, since there is no doubt that then, as now, lead objects were often salvaged and melted down together for reuse. If leads from different sources are mixed the isotope ratios determined are no longer a useful source of information.

The high accuracy with which lead isotope ratios have been determined raises the question as to what actually constitutes a compositional group (Perlman, Asaro, et al. 1972). An asemblage of objects may be defined as a "group" when their composition are reasonably similar and distinct from others outside the group. When the dispersion of analytical results within a group is not due to experimental error, the question arises as to what causes it. In lead from leaded bronze from Laurion, for example, there is a general variation in composition that correlates with stylistic chronology. This might mean that all the lead came from Laurion ores, but from different layers that were mined at different periods; however, it might also be evidence for different mining sites with similar geological histories. Such findings need not be considered discouraging; they may eventually throw light on many of the complexities of ancient trade patterns about which little or nothing is at present known.

COINS

Of all the ancient artifacts that have been passed down to us, coins are the most numerous. Historians have used numismatic evidence in their research, but their studies have been mainly concerned with the examination of inscriptions, denominations, and types. The composition of coins is nevertheless of no less importance than their design: quantitative analysis can yield evidence relating to the sources of the metals used in minting, location of the mint of a group of coins, or technological trends in minting.

The reluctance of collectors to allow permanent damage to coins was, until a few years ago, the reason for the small amount of analytical work done on coins. This was understandable since chemical analysis meant the destruction or disfacement of at least a portion of the coin. In recent years, however, several nondestructive methods of analysis, among them X-ray fluorescence and neutron activation, have been successfully applied to obtain information about ancient coins and the men involved in their production.

Purity of Athenian Silver Coins

Athenian silver coins had a reputation for high purity. This reputation was based mainly on the appearance of the coins. It is known that the chief source of Athenian silver was Laurion in Attica. The silver was extracted from argentiferous galena (a mineral consisting primarily of lead sulfide and containing some silver) that occurred in beds in the limestone and schist of the district.

To test whether their reputation for high purtiy was justified, a number of Athenian silver coins from the sixth to the second centuries B.C. were examined for gold and copper contents, by neutron activation analysis (Aitken, Emeleus et al. 1962). The results are shown in Table 11.8.

In the Athenian coins of the middle-sixth century the copper content varies up to 3 percent and the gold content, up to 0.5 percent; in the coins of the fifth century the copper content in no case exceeds 1, nor the gold content 0.1 percent. In the second century B.C. the gold content of the coins is 0.1-0.5 percent, and the copper content shows greater variation than in the sixth century. Thus in the fifth century, at least, the reputation for purity of Athenian silver was justified. The

TABLE 11.8   Gold and Copper in Athenian Silver Coins

Percentage of Number Analyzed Having

Origin	Gold < 0.1 Percent, Copper < 1 Percent	Gold < 0.1 Percent	Copper < 1 Percent	Gold < 0.1 Percent, Copper 1-5 Percent
Athens (6th cent. B.C.)	31	47	60	27
Athens (5th cent. B.C.)	100	100	100	0
Athens (2nd cent. B.C.)	23	58	28	37

results obtained for the second century coins indicate that other sources of silver were being exploited at this time.

Macedonian Debasement

Neutron-activation analysis was used to determine differences in composition in Macedonian silver coins (Aitken, Emeleus, et al. 1962). The coins date from the reigns of three kings—Alexander I, Perdiccas II, and Archelaus I—covering a period of 100 years (498-399 B.C.). All the coins show a horse on the obverse but are divided into two groups, with rider and without; coins of both types were produced concurrently during the entire period (see Fig. 11.9). Although the two

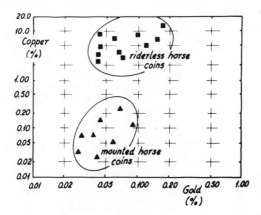

Fig. 11.9.   Relative concentration of gold and copper in Macedonian coins from the fifth century B.C.

groups have a similar gold content, they are sharply differentiated in copper content; that of the group with rider lies in the range 0.03-0.3 percent, compared with 5-24 percent for the riderless group.

The continuity of differentiation of these characteristics during the successive reigns is remarkable. The possibility of deliberate debasement has been suggested, leading to the hypothesis that the debased series was perhaps intended only for internal use.

TABLE 11.9   Roman Bronze Coins Content of Silver and Antimony, Relative to Copper

Emperor	Date (C.E.)	Silver: Copper Ratio (Arbitrary Units)	Antimony: Copper Ratio (Arbitrary Units)
Mint: Cyzicus			
Aurelian	270-275	120	—
Maximinus II	308-314	32	7
Maximinus II		27	5
Constantius II	337-361	4	5
Theodosius	379-395	7	14
Theodosius		3	7
Arcadius	383-408	4	7
Mint: Constantinople			
Constantine I	307-337	52	—
Constantine I		35	12
Theodosius I	379-395	4	10
Arcadius	395-408	10	9
Honorius	393-408	3	11
Marcianus	450-457	2	28
Leo I	457-474	2	32
Anastasius	491-518	1	32
Mint: Antioch			
Aurelian	270-275	130	—
Probus	276-282	120	3
Diocletian	284-305	104	7
Licinius	307-324	80	1
Maximinus II	308-314	43	5
Valentinian II	372-392	14	5
Eudocia	421-440	4	12

Debasement of Roman Coins

Roman bronze coins from about 260 C.E. onward contain a percentage of added silver. An investigation of the silver content of Roman coins struck in the eastern part of the Empire was made by neutron-activation analysis (Zuber 1965, 1966). Over a period of 250 years (270—512 C.E.) the silver content of the coins varies by a factor of 250; it decreases progressively throughout the period, as can be seen from the data in Table 11.9.

This trend has been confirmed by further research carried out by using X-ray fluorescence analysis of coins of the same period (Carter 1966). Table 11.10 illustrates the reduction in silver content with the passage of time.

TABLE 11.10   Composition of Roman Bronze Coins

Emperor	Date	Silver:Copper Ratio (Arbitrary Units)
Probus	276-282	5.4
Diocletian	284-304	2.4
	307-314	3.3
Delmatius	335-337	1.7
Constantius	346-350	1.2
	383	1.0

It is possible that silver was added to copper to improve the mechanical properties of the coins. The rulers of later years were perhaps no longer interested in this, or they may have tried to achieve the same effect by adding ores containing antimony. Tables 11.9 and 11.10 show an increase in the relative amounts of antimony with a decrease in silver. However, it has been maintained by some historians that silver and copper were alloyed to give the bronze coin an intrinsic value. If this is the case, the results of the analysis would show a tremendous devaluation during the period under survey (Barrandon, Callu et al. 1977).

Gold as an Impurity in Silver Coins

The amount of gold present in silver coins bears a relationship to the amount of gold in the original silver ores. It is possible to group silver coins according to their gold:silver ratios and from the results to suggest possible sites of origin of the silver ore. By determining the amount of gold present in silver coins from different mint cities, it may be possible to assess the movement of silver through a region or kingdom.

A study of gold content (determined by neutron activation) of several thousand Sassanian coins (Persia, third to seventh centuries) indicated that silver from a number of different silver sources was used for coining. It was found that monograms appearing on coins could be related, on compositional grounds, to particular mint cities (Gordus and Gordus 1974).

Minting in the Islamic Empire

History can be unfolded by determining the changes in the silver content of coins issued by various mints. Data on Islamic coins (Umayyad Dirhems) from the period 700-740 C.E. (Gordus 1971), as given in Fig. 11.10, show an increase in silver content from about 90 percent to 95 percent for all mints except Al Andalus, the present-day city of Cordoba in Spain.

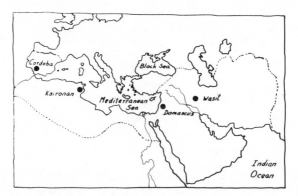

Fig. 11.10. The Islamic empire.

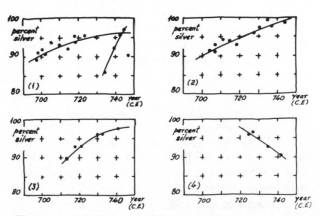

Fig. 11.11. Silver content of Islamic coins (Umayyad dirhems) struck at various mints during the years 700-740 C.E.

It has been suggested that the decreasing silver content of coinage from Spain is probable related to the geographic displacement of Al Andalus from the administrative centre of the Islamic Empire (see Fig. 11.11).

GLOSSARY—ANCIENT METALS, METALLURGY, AND METAL WORKING

Amalgam—An alloy of mercury and one or more other metals (e.g., silver amalgam, gold amalgam).

Annealing—Heating a metal, keeping it at a given temperature, and then cooling at a suitable rate. The object of annealing is to reduce hardness, facilitate cold work, and obtain desirable mechanical and physical properties.

Argentarium—A lead-tin alloy containing 50 percent tin, used by the Romans for tinning (qv).

Austenite—A form of iron, with or without minor alloying elements, existing normally at high temperatures (above 750°C).

Billion—Silver alloy containing less than 50 percent of this metal.

Blister copper—An impure product in the refining of copper. The name is derived from large blisters, which form on cast surfaces as a result of the liberation of gases.

Bloomery iron—Iron produced in a solid condition (i.e., without fusion) by the direct reduction of iron ores. Iron sponge, of no practical value until it had been forged, was produced within a mixture of iron ore and fuel (usually charcoal) piled on a hearth (see Figs. 11.12 and 11.13).

Brass—An alloy consisting of copper (50-90 percent) and zinc (50-10 percent) to which small amounts of other elements may be added. In antiquity brass was made by heating a mixture of metallic copper, charcoal, and a zinc ore called *calamine* (zinc carbonate) in a reducing atmosphere.

Brazing—The joining of two metal surfaces by allowing a fused nonferrous metal (e.g., bronze) to flow into the space between them to form a thin layer on solidifying (Cowen and Maryon 1935; Friend and Thorneycroft 1935).

Bronze—An alloy consisting of copper (80-95 percent) and tin (20-5 percent) to which small proportions of other elements, such as zinc, may be added.

Bullion—Crude lead or copper containing appreciable amounts of precious metals (silver of gold).

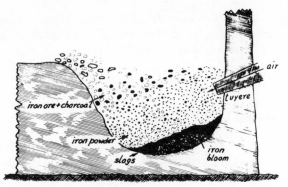

Fig. 11.12. Medieval bloomery hearth (cross section). Primitive bloomery furnaces often consisted of only a pit sunk into the ground. The furnace was fed with iron ore and fuel. Metallic iron collected at the bottom of the furnace in the form of a spongy "bloom" and was removed when cold. The furnace shown has a high wall and tuyere for blowing in air; such furnaces were used in parts of Europe up till a hundred years ago.

Fig. 11.13. Crude bloom from Mesopotamia. The lumps weighed 5-25 kilograms. They were covered with slag that was usually knocked off when cold. Holes were drilled through the blooms so they could be strung up for transportation.

Calcination—The heating of ores or concentrates to drive off volatile components such as carbon dioxide, organic matter, chemically combined water, and so on. The difference between calcination and roasting (qv.) is that roasting involves a chemical reaction between the ores and the surrounding gases, whereas in calcination the surrounding hot gas serves only to provide the necessary heat.

Carbon steel—Steel containing up to 2 percent carbon. Also called "ordinary steel."

Carburizing—Introducing carbon into the surface of a low-carbon steel by holding metal piece in contact with a suitable carbonaceous material. In antiquity the latter wax usually charcoal.

Casting—The process of pouring molten metal into a prepared mold, where it is allowed to solidify and form an object of desired shape. Four basic techniques have been used over the centuries: (1) open mold casting, (2) hollow casting, (3) false core casting, and (4) lost was casting (see Figs. 11.14 and 11.15).

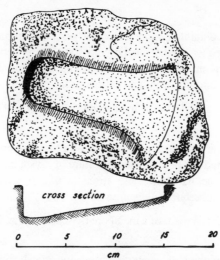

Fig. 11.14.   Open stone mold used for casting flat copper axes in Chalcolithic times.

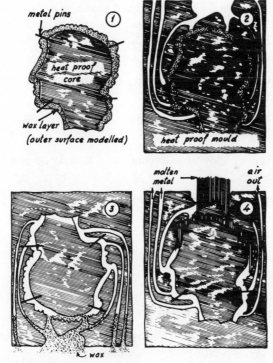

Fig. 11.15.   Lost-wax ("Cire-Perdue") casting: 1. clay core with wax layer (outer surface of wax is modeled);   2. modeled figure is surrounded by a heat-proof mold;   3. reversed mold is heated to remove the wax;   4. molten bronze is poured into runners (metal pins are later removed).

229

Cast iron—Iron containing more than 1.9 percent carbon. Cast iron is not malleable in either the hot or the cold state and is very brittle. The word "cast" is often omitted when referring to the three forms in which cast iron exists: white iron, gray iron, and nodular iron.

Cementite—A compound of iron and carbon, iron carbide (Fe_3C); it is very hard and brittle.

Cold working—The structural deformation of a metal—bending, piercing, and forging—carried out below its temperature of crystallization. Cold working brings about the hardening and strengthening of metals, which can be eliminated by subsequent *annealing* (q.v.).

Concentration—Enrichment of a metal ore by separation and removal of *gangue* (q.v.).

Cupellation—A method of separating precious metals from lead-silver-gold alloys by the oxidation of the lead base to yield lead oxide.

Damascening—A method of making sword blades with delicate alteration of higher and lower carbon areas. Crystallization effects are revealed by etching and are used for the production of decorative effects (Belaiew 1918; Smith 1957).

Dezincification—The corrosion of copper-zinc alloys, resulting in the loss of zinc and the formation of spongy copper.

Ductility—The degree to which a metal is able to undergo plastic deformation without fracturing.

Electrum—A natural alloy of gold containing a moderate proportion (ca. 20 percent) of silver.

Ferrite—A soft and ductile form of iron almost devoid of carbon. It contains not more than 0.008 percent of this element at room temperature, but may contain other alloying elements.

Fine silver—Silver with a fineness of 999 (i.e., containing a minimum of 99.9 percent silver).

Flux—A material added to metallurgical furnace charges to react with the non-metallic components of ores to form slags.

Forge welding—Welding of hot metal by striking or pressing. When carried out by hammering, it is called *hammer welding*.

Forging—The working of metal by hammering. Three basic operations are used in forging: (1) *drawing* out of the metal so that its length is increased while it cross section is reduced, (2) *squeezing* the metal between special so that the metal flows and takes up the shape of the die cavity, and (3) *upsetting* the metal, in which case the cross section is increased while the length is reduced.

Foundry—A workshop where metal castings are produced.

Gangue—The valueless minerals found in ores.

Guilding—See *Plating*.

Granulation—The production of minute metal globules used to decorate ancient goldsmith-jeweler's work.

Hammering—Beating a metal sheet into a desired shape.

Hardening—Increasing the hardness of metals by suitable treatment, usually heating followed by cooling.

Heat treatment—The controlled heating and cooling of metals to alter their structure and thus their mechanical properties. Heat treatment consists of three main steps. ´1) heating the object to a predetermined temperature, (2) maintaining the ovject at the given temperature until the metal structure becomes uniform throughout, and (3) cooling at a predetermined rate to cause the formation or retention of desirable structural properties within the metal.

Hot forming—The plastic deformation of a metal (bending, forging, etc.) at a temperature above its temperature of crystallization.

Karat—A way of defining the fineness of gold, as parts by weight of gold in 24 parts of alloy. Fine gold is 24 karat.

Karat	Percent Gold
24	100.0
18	75.0
15	62.5
12	50.0
9	37.5

Melleability—The capacity of metals and alloys to undergo plastic deformation under compression (e.g., by hammering) without rupture.

Martentsite—A hard and brittle product formed in steels that are quench-cooled above a certain temperature.

Metal leaf—Thin plates or leaves of malleable metals, obtained by hammering. Metal leaf is used to cover articles made of wood or baser metals. Gold leaf 0.001 mm thick, and silver leaf 0.0025 mm thick were in use as early as 2000 B.C.

Metallography—The study of constitution and structure of metals and alloys by optical means only (unaided eye, magnifying glass, microscope, electron microscope).

Mineral dressing—The concentration of raw ores by separation of ore minerals from those of the gangue. Also called *ore preparation, benefication,* or *milling.*

Noble metals—Another name for precious metals, such as gold silver and platinum.

Nugget—A lump of metal such as gold, silver, or copper found free in nature.

Orichalcum—A copper alloy containing zinc, used by the Romans in the production of coins (Caley 1964).

Oxidation—A chemical reaction in which there is increase in positive valency.

Pearlite—A mixture of ferrite and cementite, occurring in cast iron and steel.

Pewter—An alloy made in antiquity from tin (75-95 percent) and lead (25-5 percent). Nowadays very little lead is used in the production of pewter, since the lead (which is poisonous) may be dissolved out of certain grades of pewter, by liquids contained in vessels made from the latter.

Plating—The forming of an adherent layer of metal on an object made from another metal. A number of plating techniques were used in antiquity: (1) dipping in molten metal (Bernareggi 1965), (2) hammering metal leaf on to a base of a less noble metal, and (3) amalgamation, in which the base metal articles was coated with an amalgam of the plating metal; the mercury was then driven off by heating (Aitchison 1960). When gold was used for plating the process was called *gilding.*

Quenching—Rapid cooling, usually by submerging a heated object in a cold liquid.

Reduction—A chemical reaction in which there is an increase in negative valency.

Roasting—An oxidizing process, usually the oxidation of sulfide minerals to facilitate smelting (g.v.) (see Calcination).

Slag—The waste product of smelting resulting from the combination of non-metallic components with flux (q.v.).

Smelting—The reduction of ores to crude metals at high temperatures.

Solder—An alloy of lead (50-70 percent and tin (50-30 percent) used for joining metals.

Sorbite—A fine dispersion of cementite (q.v.) in ferrite (q.v.) formed during the tempering of steel.

Speculum—A white alloy of copper and tin, containing about 40 percent of tin. Speculum was used by the Romans for making mirrors.

Steel—An iron alloy containing carbon and often other alloying elements. There are several ways of classifying the various types of steels; that given as follows is based on end use:

Type of Steel	Percent Carbon	Uses
Dead mild	0.05-0.15	Nails, bars
Mild	0.10-0.30	Buckles, rings
Medium carbon	0.30-0.50	Buckles, rings
High carbon	0.60-0.80	Axes, swords, spearheads
Tool steels	0.90-1.40	Axes, swords, spearheads

Tempering—The softening of hard and brittle steel by heating it to a given temperature—usually 150-650°C—and then cooling at a desired rate.

Tertiarium—A lead-tin alloy containing 33 percent tin, used by the Romans for tinning (q.v.).

Tinning—Coating metal with a very thin layer of molten metal, usually a lead-tin alloy. The Romans used two types of tinning alloys: a better grade known as *argentarium* (q.v.) and an inferior grade, *tertiarium* (q.v.).

Tin pest—A change occurring in tin at very low ambient temperatures, whereby the metal changes its crystalline state and falls into a coarse gray powder. Tin pest is often confused with tin corrosion.

Tumbago—A gold alloy used by pre-Columbian Indians (copper 50 percent, gold 33 percent, silver 12 percent).

Tuyere—An opening in the shell and lining of a smelting furnace, through which air may be forced in.

Welding—The joining of two or more metal surfaces by applying heat, pressure, or both, with or without a filler metal.

Wrought iron—Relatively pure iron in which stringers of siliceous slags are embedded. Probably the earliest type of ferrous metal produced by man.

REFERENCES

Aitchison, L., *A History of Metals*, MacDonald and Evans, London, 1960.

Aitken, M. J., Emeleus, V. M., Hall, E. T., and Kray, C. M., in *Conference on Radioisotopes in the Physical Sciences and Industry*, Vol. 2, IAEA, Vienna, 1962, p. 263.

Belaiew, N., *J. Iron Steel Inst.*, **97,** 417 (1918).

Barrandon, J. N., Callu, J. P., and Brenot, C., *Archaeometry*, **19,** 173 (1977).

Bergsoe, P., *Nature*, **137,** 29 (1936).

Bergsoe, P., *Nature*, **139,** 490 (1937).

Bernareggi, E., *Revista Italiana di Numismatica*, **13,** 67 (1965).

Berthelot, M., *Ann. Chim. Phys.*, **12,** 433 (1897).

Berthelot, M., *Comptes Rendus*, **132,** 729 (1901a).

Berthelot, M., *Ann. Chim. Phys.*, **23,** (7), 5 (1901b).

Bingham, H., *Machu Pichu, a Citadel of the Incas*, Yale U. P., New Haven, Conn., (1930).

Bouman, R. W., *Proceedings, Iron and Steel Institute Meeting*, **5,** 17, 1978.

Brill, R. H., in *Proceedings, 8th International Congress on Glass*, 1968, p. 62.

Brill, R. H., and Wampler, J. M., *Am. J. Archaeol.*, **71,**63 (1967).

Caley, E. R., *Analysis of Ancient Metals*, Pergamon, New York, 1964.

Cambi, L., *Rend. ist. Lombardo sci.*, Pt. I, *Classe sci mat. e nat.* **92A,** 167 (1958).

Cambi, L., and Cremascoli, F., *Rend. ist. Lombardo sci.*, Pt. I, *Classe sci mat. e nat.* **91,** 371 (1957).

Carroll, D. L., *Am. J. Archaeology*, **78,** 33 (1974).

Carter, G. F., *Anal. Chem.*, 1264 (1964a).

Carter, G. F., *Archaeometry*, **7,** 106 (1964b).

Carter, G. F., *Chemistry*, **39,** 14 (1966).

Carter, G. F., *Science*, **151,** 196 (1966).

Coghlan, H. H., *Man*, **51,** 90 (1951).

Coghlan, H. H., *Viking Fund Publications in Anthropology*, 28 (1960).

Coghlan, H. H., Butler, J. R., and Parker, G., *Royal Anthropol. Inst. Occasional Paper*, No. 17 (1963).

Cowen, J. D., and Maryon, H., *Archaeology*, **12,** 280 (1935).

Craddock, P. T., *J. Archaeol. Sci.*, **3,** 93 (1976).

Craddock, P. T., *J. Archaeol. Sci.*, **5,** 1 (1978).

Epler, D. C., *The Metallurgist and Materials Technologist*, 303 (1978).

Fester, G. A., *Chymia*, Vol. 8, Pennsylvania U. P., Philadelphia, 1962, p. 21.

Fields, P. R., Milstead, J., Henrickson, E., and Ramette, R., in R. Brill, ed., *Science in Archaeology*, MIT Press, Cambridge, Mass., 1971 p. 131.

Forbes, R. J., *Metallurgy in Antiquity*, E. J. Brill, Leiden, 1950.

Friedman, A. M., Conway, M., Kaster, M., Milstead, J., Metta, D., Fields, P. R., and Olsen, E., *Science*, **152**, 3728 (1900).

Friedman, I., Smith, R. L., and Long, W. D., *Bull. Geol. Soc. Am.*, **77**, 323 (1966).

Friend, J. N., and Thorneycroft, W. E., *J. Inst. Metals*, **12**, 280 (1935).

Gettens, R. J., *The Freer chinese Bronzes*, Vol. 2, Oriental Studies No. 7, Smithsonian Institution, Washington D.C., 1969.

Gordus, A. A., in R. H. Brill, ed., *Science in Archaeology*, MIT Press, Cambridge, Mass., 1971, p. 145.

Gordus, A. A., in *Methods of Chemical and Metallurgical Investigation of Ancient Coinage*, Royal Numism. Soc. Spec. Publ., 8, 127, 1972.

Gordus, A. A., and Gordus, J. P., in C. W. Beck, ed., *Archaeological Chemistry, a Symposium*, American Chemical Society, Washington, D.C., 1974, p. 124.

Gowland, W., *J. Inst. Metals*, **7**, 23 (1912).

Grigoryan, G. O., *Geochemistry U.S.S.R.* (Engl. transl.), **4**, 388 (1962).

Grothe, H., *Erzmetall.*, **28**, 165, 1975).

Hall, E. T., in *Applications of Science to the Examination of Works of Art, Proceedings of Seminar*, Museum of Fine Arts, Boston, Mass., 1965.

Hill, J., *Theophrastus' History of Stones*, London, 1774, p. 227.

Hunt, C. B., *Science*, **120**, 183 (1954).

Junghaus, S., Sangmeister, E., and Schroder, M., *Metallanalysen Kupferzeitlicher un Fruhbronzezeitlicher Bodenfunde aus Europa*, Verlag G. Mann, Berlin, 1960.

Kay, C. A., *Science*, **146**, 1578 (1964).

Krysko, W. W., and Lehrheuer, R., *J. Hit. Metall. Soc.*, **10**(2), 53 (1976).

Laudermilk, J. D., *Am. J. Sci.*, **21**, 51 (1931).

Leoni, m., *La Fonderia Italilana*, **7/8**, 233 (1974).

Lothrop, S. K., *Am. J. Archaeol.*, **38**, 207 (1934).

Lucas, A., *Ancient Egyptian Materials and Industries*, Arnold, London, 1962.

MacKerrel, H., *Ed. Chem.*, **9**, 54 (1972).

MacKerrel, H., and Stevenson, R. B., in *Methods of Chemical and Metallurgical Investigation of Ancient Coinage*, Royal Numsim Soc. Spec. Publ. 8, 195, London, 1972.

Madden, R., Muhy, J. D., and Wheeler, T. S., *Sci. Am.*, October, 122 (1977).

Meyers, P., van Zelst, L., and Sayre, E. V., *J. Radioanal. Chem.*, **16**, 67 (1973).

Meyers P., van Zelst, L., and Sayre, E. V., Brookhaven Natl. Lab. Report Number 21513, 1976.

Mischara, J., and Myers, P., in *Recent Advances in Science and Technology of Materials*, A. Bishary, Ed., Plenum, New York, 1974.

Morton, G. R., and Wingrove, J., *J. Iron Steel Inst.*, 1556 (1969).

Mosso, A., *Atti reale Accad. Lincei*, **12**, 49 (1906).

Muhly, J. D., *American Scientist*, **61**, 404 (1973).

Neuberger, A., *The Technical Arts and Sciences of the Ancients*, Methuen, London, 1969.

Neuninger, H., and Pittioni, P., Jahressch Salzburger Mus. Carolino Augusteum, 1958.

Otto, H., and Witter, W., *Hanbuch der altesten Vorgeschichtlichen Metallurgie in Mitteleuropa,* Verlag J. A. Barth, Leipzig, 1952.

Perlman, I., Asaro, F., and Michel, H. V., *Ann. Rev. Nucl. Sci.,* **22.** 383 (1972).

Piaskowski, J., *J. Iron Steel Inst.,* **198,** 263 (1961).

Piaskowski, J., *Archaeologia Polanica,* **4,** 124 (1964).

Pittioni, P., *Archaeologia Austriaca,* 26 (1959).

Ray, P. C., *History of Hindu Chemistry,* Calcutta, 1904, p. 25.

Renfrew, C., *Am. J. Archaeol.,* **71,** 1 (1967).

Rivet, P., and Arsandaux, H., *La Metallurgie en Amerique Precolumbienne,* Paris, 1946.

Roberts, P. M., *Gold Bulletin,* **6,** 112 (1975).

Smith, C. S., *Endeavour,* 199 (1957).

Smith, C. S., *Science,* **148,** 908 (1965).

Tylecote, R. F., *Antiquity,* **44,** 19 (1970).

Tylecote, R. F., A History of Metallurgy, The Metals Soc., London, 1976.

Tylecote R. F., Ghaznavi, H. A., and Boydell P. J., *J. Archaeol. Sci.,* **4,** 305 (1977).

Tylecote, R. F., Lupu, A., and Ruthenberg, B., *J. Inst. Metals,* **95,** 235 (1967).

Von Lippman, E. O., *Entstelung und Ausbrietung der Alchemie,* Springer, Berlin, 1919, p. 606.

Wainwright, G. A., *Antiquity,* **10,** 5 (1936).

Wertime, T. A., *Science,* **146,** 1257 (1964).

Wirchow, R., *Verhandlungen der Berliner Gesellschaft fur Anthropologie,* 1884, p. 543.

Zimmer, G. F., *J. Iron Steel Inst.,* **94,** 306 (1916).

Zuber, K., Turkish Atomic Energy Commission Report, CNAEM 21, 1965.

Zuber, K., Turkish Atomic Energy Commission Report, CNAEM 35, 1966.

Recommended Reading

Caley, E. R., *Analysis of Ancient Metals,* Pergamon, New York, 1964.

Hall, E. J., an Metcalf, N. eds., *Chemical and Metallurgical Investigations of Ancient Coinage,* Royal Numismatic Society, London, 1972.

Tylecote R. F., *Metallurgy in Archaeology,* A. Edwards, London, 1962.

DECAY AND RESTORATION OF ARCHAEOLOGICAL MATERIALS

Objects of archaeological interest are very seldom found in their original state. The natural action of environmental factors causes almost all materials to deteriorate and break down: organic materials rot, dyes fade, and pigments flake; metals corrode, and even stones crumble to dust. Only a few ancient objects survive the ravages of time and are eventually rediscovered in a more or less severe state of deterioration.

The archaeologist is interested in clean, restored objects that can be conserved for furture study and exhibition. Up to a few decades ago the methods used for the restoration and conservation of antiquities were wholly empirical. Only in relatively recent times have scientific technological principles been applied to the repair and preservation of antiquities. The specialized knowledge of the chemist can be used in the conservation of antiquities in a number of fields: (1) recognition of the symptoms and causative factors of deterioration, (2) technical examination analysis and definition of decay products and decay processes, (3) development of conservation methods based on scientific principles, and (4) search for materials useful in conservation. All of these topics are discussed to some length in the next few chapters.

12

THE CAUSES OF DECAY

Jade, regarded by the Chinese as the most precious of gemstones, is a rare mineral, sodium aluminum pyroxene, $NaAl(SiO_3)_2$. It is formed deep in the earth's crust under conditions of high presure and temperature.

In ancient times the Chinese buried jade with their dead to prevent decomposition of the body. During the Han Dynasty (202 B.C.-220 C.E.) pieces of jade were placed in each of the body orifices to ward off decay of the corpse. It is ironic that the corpse caused decay of the jade: the hard, polished surfaces of the mineral became, in time, disfigured with chalky blemishes (Gaines and Handy 1975). Even the apparently most durable and resistant material may thus undergo deterioration and decay under appropriately adverse conditions.

ENVIRONMENT AND DETERIORATION

Decay is a process of degradative adjustment of materials to the conditions prevailing in their immediate environment: existing chemical compounds are converted, in the process of deterioration, to compounds of greater stability *vis-a-vis* their surroundings. Where archaeological objects are concerned, environment can be defined as the aggregate of all the external influences on the object. Moisture, oxygen, soil, and atmospheric contamination are the main chemical influences; temperature, sunlight, and wind are among the more important physical agents, whereas microorganisms and insects—not to mention rodents and other small animals—are the main biological factors involved in the decay and the breakdown of archaeological objects. The presence of oxygen leads to oxidation; water, to chemical attack (hydration or dissolution) or physical abrasion; and insects and microorganisms, to biological decay. The extent to which materials are modified from their original form during the centuries depends on their susceptibility to breakdown in a particular environment (Scossiroli 1976).

One of the major handicaps confronting the student of deterioration and decay of antiquities is the fact that one is dealing with very slow processes that take place over very long periods of time—centuries and millennia—periods so long as to be inaccessible to laboratory experiments. Accordingly, great efforts have been made to develop *accelerated tests,* where environmental conditions are artificially intensified so as to produce faster effects than those caused by natural ones. But it is impossible to do this precisely; in accelerated tests there is always an element of uncertainty in the extrapolation of results. When accelerated tests are correlated with simple, extended exposure, the results are often less than satisfactory.

CHEMICAL AND PHYSICAL AGENTS OF DETERIORATION

It is not always possible to make a clear-cut separation between physical and chemical agents of deterioration. Chemical agents act physically, physical agents act chemically, and in some cases the same agent may act in either a physical or a chemical manner, or in both simultaneously, depending on the particular circumstances. Water, for example, may wash away particles from a given body, thus effecting a physical action without chemically altering the substances that form the particles; on the other hand, water may wash away particles and then act chemically by dissolving them and later causing decomposition of the dissolved substances.

Two or more physical and chemical agents almost always act simultaneously, often in ways that are interrelated. The effects of two separate agents may even enhance each other mutually, resulting in an overall intensification of the decay process. Thus it is not customary nowadays to classify the agents of decay into separate "physical" and "chemical" categories, but to speak in terms of an all-embracing category of *physico-chemical* agents.

The physico-chemical agents active in decay include gases, liquids, solids, and the energy available in wind, water flow, and sunlight. Among the gaseous agents are included naturally occurring substances such as oxygen, ozone, carbon dioxide, and water vapor and atmospheric contaminants such as the oxides of sulfur and hydrogen sulfide. The principal liquid agent of decay is water (as rain or soil moisture), but acids or alkalies in the soil can also be included. Solid agents of decay are smoke and dust particles in the atmosphere and salts in the soil. Apart from these, sunlight, wind, temperature changes, and frost are the main climatic factors in decay.

Oxygen

The most universal chemical agent of deterioration is oxygen, which constitutes 21 percent of the atmospheric gases. Less plentiful under ordinary conditions but far more destructive is the more unstable and reactive gas *ozone* (O_3), an allotrope of oxygen (an allotrope is a modification of an element, existing in the same physical state but having different chemical properties). The degradative reactions brought about by oxygen and ozone are quite similar, although ozone is a much more powerful oxidizing agent.

Under ordinary conditions few materials are attacked by pure oxygen in the absence of other agents; at least one other active factor is required. Pure iron, for example, is not attacked by dry, pure, oxygen; moisture is the essential con-comitant in the oxidation of iron to rust. Ozone, on the other hand, attacks many materials directly; silver, for example, though unaffected by oxygen, is rapidly tarnished when exposed to ozone.

Many other factors besides moisture facilitate the oxidative degradation of materials. The oxidation of paints, for example, is accelerated (or "catalyzed") by sunlight.

Water and Moisture

Water is the most important agent of deterioration, acting sometimes chemically, sometimes physically, and sometimes (as mentioned earlier) in both ways simultaneously. Moreover, since water is a prime factor in plant and animal nutrition, it has a great influence on deterioration by biological agents. In large quantitites—as in the form of flowing streams, rain, snow, ice, and hail—water exerts strong erosive action on most materials.

Another way in which water can bring about a physical breakdown of materials arises from the fact that it has a maximum density at precisely 4°C. When the temperature is either raised or lowered from this point, water expands. With raising or lowering of temperature, moisture confined in the pores of porous materials (certain stones, bricks, ceramics, wood, etc.) may thus effect a sort of localized explosive action that, over long periods of time, can reduce stones to gravel and timber to dust.

The strength and rigidity of wood are greatly weakened by continuous soaking. The wood becomes soft and flexible. Wood is also subject to dimensional changes caused by variations in its moisture content. When moisture is taken up or given off by wood, a moisture gradient is established, and not all the fibers swell or shrink at the same rate. The internal stresses set up in this manner often result in warping.

Glued wood joints are broken down by the actions of moisture: the adhesive may give way under the force of mechanical stresses developed in the wood through swelling or shrinking, or the adhesive itself may be dissolved or chemically decomposed.

Water, in its various forms, exerts an unfavorable influence on many other, most diverse materials. Textiles subjected to periodical wetting and drying are weakened. Parchment exposed to an excess of moisture for long periods of time breaks down completely; the constituent proteins are decomposed by the water into a decay product known as "parchment size." The adhesion of paints to their substrates is destroyed by the action of moisture, with the consequent formation of blisters, which eventually break and scale off. Water is essential to the corrosion of metals (a subject discussed in Chapter 14). Usually, the more water is present, the more serious are the detrimental effects that it has on archaeological objects and the more rapidly they occur.

Water can contribute to the breakdown of some materials by its absence as well as its presence. In fact, for most materials there is some optimum moisture content for the maintenance of useful properties. For example, papyrus that is too dry is brittle. Over-dry parchment and leather become stiff and unworkable.

Soil Contaminants

Contaminants in the soil are powerful agents of decay. Acids and salts come under this heading. In the presence of water, salts undergo *hydrolysis,* the process in which the ions of a salt and water interact to yield acid or alkaline solutions. All soils are indeed either acid or alkaline; they are rarely, if ever, neutral. Acids and alkalies readily attack many different materials, eventually destroying their original properties.

Salts such as chlorides, sulfates, phosphates, nitrates, and carbonates, all abundant in the soil, are also agents of deterioration. Their most common mode of action is in contributing to the corrosion of metals, either by direct chemical attack or by faciltating the electrolytic process of corrosion.

Salts that crystallize in the pores of cement or ceramics can give rise to internal pressures sufficient to disrupt the whole matrix and to turn it into gravel. Chemical reactions between salts and buried materials can contribute to the scaling and disintegration of cements, blistering and flaking of paints, and other types of corrosion.

Ground water dissolves atomspheric carbon dioxide, and the result is an aqueous solution of the weak carbonic acid:

$$CO_2 + H_2O = H_2CO_3$$

The weakly acid solution disolves limestone: insoluble limestone is converted into soluble calcium bicarbonate. Thus:

$$CaCO_3 + H_2CO_3 = Ca(HCO_3)_2$$

Insoluble limestone	Soluble calcium bicarbonate

The evaporation of water containing calcium bicarbonate causes carbon dioxide to be driven off and calcium carbonate to precipitate:

$$Ca(HCO_3)_2 = CaCO_3 + H_2O \uparrow + CO_2 \uparrow$$

Dissolved calcium bicarbonate	Precipitated calcium carbonate

Evaporation of water from soaked archaeological objects leaves on them a hard scale or incrustation of reprecipitated calcium carbonate. Such a scale is also commonly found on archaeological objects that have undergone cycles of soaking and drying.

Atmospheric Contaminants

The atmosphere is a reservoir of aggresive contaminants that eventually settle out and can cause serious deterioration. The most important are sulfur dioxide and hydrogen sulfide. Sulfur dioxide yields, with moisture, sulfurous acid, which is oxidized by atmospheric oxygen to sulfuric acid, one of the strongest known acids:

$$SO_2 \quad + H_2O \ = \ H_2SO_3$$
Sulfur dioxide   Sulfurous acid

$$H_2SO_3 \ + \ \frac{1}{2}O_2 \ = \ H_2SO_4$$
Sulfuric acid

Thus sulfur dioxide leads to the attack of all the metals known in antiquity (with the exception of gold), forming with them metallic salts.

Sulfur dioxide can also act as a reducing agent; as such, it causes textiles (especially cotton) to be rapidly broken down. Under its influence leather changes in texture, becomes brittle and eventually breaks down completely.

Hydrogen sulfide (H_2S) is a foul-smelling gas that is released into the atmosphere in the course of decay of animal tissue and the combustion of natural fuel. Hydrogen sulfide reacts with almost all metals to form dark-colored corrosion layers commonly found in archaeological finds.

TEMPERATURE EFFECTS ON DETERIORATION

The physical properties of all known materials are greatly modified by changes in temperature. The latter also influence the rates of chemical reactions. Heat and cold act as powerful agents of chemical and physical deterioration of antiquities.

Chemical Effects

The rate at which most chemical reactions take place is doubled with every 10°C increase in temperature. Consequently, an increase in ambient temperature intensifies the rate of reaction of many materials with the elements of environment such as water, oxygen, and soil and atomspheric impurities; the result is faster chemical degradation.

Physical Effects

When a body is heated, the heat is transferred from one part of the body to another by conduction. The rate of heat conduction in any material is dependent on the *thermal conductivity* of the particular materials (i.e., the quantity of heat transmitted in the material per unit time per cross section per unit temperature gradient). The thermal conductivity of some archaeologically important materials is given in Table 2.2.

Metals are good conductors of heat, whereas nonmetals and organic materials usually have low thermal conductivity and are used as insulators of heat in the form of clothing, building materials, and so on.

Perhaps the most important physical effect of temperature changes on materials is that of dimensional changes. The change in volume of water when heated or cooled from 4°C has already been discussed. For solid materials the *coefficient of linear expansion* (α) is defined as the change in length per unit length per unit change in temperature. Table 12.1 gives the values of α for some archaeologically important materials.

TABLE 12.1 Coefficient of Linear Expansion of Common Archaeological Materials

Material	Coefficient of Linear Expansion: ($\alpha \times 10^6$) between 0 and 80°C
Brass	10.7
rick	5.3
Bronze	10.0
Glass (soda-lime)	9.2
Iron	
Cast	5.9
Forged	6.3
Marble	10.0
Solder (lead-tin)	13.4
Wood	
Oak	
Along fibers	2.7
Across fibers	3.0
Pine	
Along fibers	5.4
Across fibers	3.4

Most materials of low thermal conductivity do not heat up or cool down uniformly, but do so by a series of temperature gradients throughout their volume. Consequently, in an object of uniform composition whose temperature is changing, different portions may be expanding or contracting at different times and at different rates. The net result is the production of internal stresses that the material may (or may not) be able to resist. The stresses set up by repeated heating and cooling may lead to distortion of shape, cracking, and breaking down of objects composed of a single material.

Structures composed of two or more materials may also break down when subjected to temperature variations; since the coefficient of thermal expansion is characteristic for every material, it is unlikely that different materials will expand or contract to the same extent. Therefore, when objects assembled from different materials are heated or cooled, some of the components are unduly compressed whereas others are forced apart; the final result is physical breakdown.

SUNLIGHT

Sunlight is often the single natural factor responsible for most of the physico-chemical degradation of archaeological materials (especially organic substances). Only a very narrow portion of the solar spectrum causes deteriorative changes. It is the shortwave ultra violet that is responsible for far more damage than either visible or infrared radiation.

Sunlight falling on organic materials brings about a series of reactions called *"photochemical degradation"*; this is essentially the destruction of the material by atomspheric oxygen, which is both initiated and subsequently accelerated by solar radiation. On absorption by the irradiated materials, the energy carried by the sun's rays brings about the displacement of atoms in their constituent molecules, making it easier for oxidative reactions to occur.

The effect of solar irradiation on materials depends on a number of interrelated factors: the chemical structure of the exposed material, the presence of reactive substances (oxygen, moisture, contaminants), and the ambient temperature. Textiles (cotton, wool, silk, etc.), more than any other archaeological materials, suffer deterioration on exposure to solar radiation. They usually disintegrate completely in quite a short time. Paints are also seriously affected by the photo-chemical activity of sunlight, and dyestuffs are usually readily bleached.

OTHER DETERIORATING AGENTS

Only a few of the most important physicochemical agents in the deterioration of archaeological materials have been discussed. The phenomenon of deterioration is actually exceedingly complex and involves a host of other agents: wind, sand, grit, and other windblown abrasives; hygroscopic dusts—alkaline or acidic—present in the atmosphere; and many others.

REFERENCES

Gaines, A. M.,and Handy, J. L., 1975, *Nature,* **253,** 433 (1975).

Scossiroli, R. E., *Elementi di Ecologia,* Aznichelli, Bologna, 1976.

DETERIORATION OF ARCHAEOLOGICAL MATERIALS

BUILDING MATERIALS: STONE, CEMENT, AND MORTAR

Good building materials are very durable if used properly. Nonetheless, they are subject to the agents of weathering and decay. Textured igneous rocks are deterimentally affected by temperature changes. Calcareous minerals are susceptible to attach by water containing dissolved carbon dioxide. They are also more or less readily damaged by soluble salts.

Atmospheric water vapor, ground moisture, or rain can all be equally damaging. Pure water, when subjected to changes in temperature, may expand sufficiently to disrupt stone or mortar, whereas repeated cycles of wetting and drying are known to cause serious damage to buildings and monuments.

In arid regions, where hot days are followed by cool nights, the alternate expansion and contraction of moisture in porous stone gives rise to internal pressures, which lead to flaking and bursting. In cold climates, frost is a major agent in the disruption of stone and cement. This is brought about by either the displacement of pore water away from the advancing frost or the expansion that water undergoes when cooled below 4°C.

Salts dissolved in water that migrates from the soil into stone can act in several different ways as deteriorative agents. Acid ions (in particular the ubiquitous sulfate) may directly attack the chemical components of stone or cement, converting them into products that scale off and ultimately disintegrate. Surface efflorescence, salt hydration, and salt crystallization can produce internal pressures strong enough to have disruptive effects.

Carbon dioxide and sulfur dioxide in the atmosphere dissolve in rain and ground water to form acidic solutions. The solutions attack stone and mortar; the chemical constituents of these are solubilized and washed out (more rapidly in carbonaceous than in siliceous stone). The end result of this is softening, pitting, and finally crumbling of the stone surface. The defacement commonly seen in outdoor statues and monuments is mostly the result of this type of process (Lal Gauri 1978).

PATINATION OF STONE AND GLASS

Patination is a general term meaning the geochemical alteration of an exposed surface. Many stone, glass, and metal artifacts—when exposed to the atmosphere, buried in the soil, or submerged in water for long periods of time—undergo chemical deterioration that results in the formation of a crust of weathering products, or *patina*.

Natural patination is a inimitable and irreversible geochemical process. It need not, however, be visible to the naked eye. The nature of the patina is dependent on the nature of the rock, glass, or metal and on the environment's chemical and climatic conditions.

The following are the most common types of stone and glass patina formations:

Hydration.   The penetration of waterinto the surface layer. The change in volume due to the absorption of water causes a strained surface [see Chapter 6; Friedman, Smith et al. (1966)].

Induration.   Leaching out of soluble silica from subsurface layers and its redeposition at the surface in the form of a hard layer.

Crust formation.   Zonal leaching of impurities and their reprecipitation at the surface.

Bleaching.   Leaching out of silica and its replacement by calcium salts.

Desert varnish formation.   Deposition of dark stains of iron and or manganese oxide (Hunt 1954; Laudermilk 1931).

Attempts to correlate the thickness of patina with the age of objects or artifacts and thus to use patina as a chronometric tool must take into account other effects besides time that can influence the thickness of patina. The texture, microstructure, and chemical composition of a material, as well as the conditions of the environment should be evaluated. Only after allowances have been made for these factors will it be possible to clarify the age dependence of patina.

POTTERY

Well-fired pottery (i.e., that fired at temperatures above 800°C) is an exceedingly stable material; it is, in fact, practically inert and indestructible. If the firing temperature is low, however (below 700°C), inferior wares are obtained that, under the influence of rain or ground water may soften and even disintegrate.

As in the case of stone, the penetration of salts into pottery can cause serious deterioration. The salts (the cumulative effect at repeated solution and crystallization and the result of cyclic variations in humidity) will tend to set up internal stresses in the material. Eventually, the surface of the pottery (where the rate of moisture evaporation is fastest) will crumble, and even the whole body may ultimately be reduced to dust.

GLASS

Glass, contrary to appearances, is to a certain extent soluble in water. Some of the chemical constituents of glass are more soluble than others. Hence certain types of glass are more susceptible to damage by water than others. The deterioration of ancient glass—usually that of the soda-lime type—is due to leaching out of the glass "modifiers" by moisture, rain, or ground water. When, acid or alkaline contaminants are present in addition to water, the solubilization process is accelerated (Bettenbourg 1976).

Modifiers (i.e., the oxides of sodium and potassium; see Chapter 9,) were used in the manufacture of ancient glasses in an higher proportion than in modern glass. In this way the melting temperature was lowered, making the glass more workable. However, the high proportion of modifiers made the glasses more soluble and thus more susceptible to deterioration. Moisture selectively dissolves the alkaline modifiers, leaving on the surface of the glass a layer of silica. Seasonal cycles of soaking and drying, repeated over long periods of time, result in the formation of successive layers of silica built up in an onion skin fashion. These are responsible for the iridescent color effects so often found on the surface of ancient glass.

In alkaline surroundings the deterioration of glass is even faster, since alkalies attack the silica that constitutes the main structural framework of glass. Prolonged subjection to alkaline conditions will weaken the glass to such an extent that any slight pressure may be enough to cause the glass object to crumble into powder.

If a large excess of modifiers is present in the glass at the time of manufacture, a part of them may remain uncombined within the body of the glass. This may cause a form of deterioration known as *sweating* or *weeping* of glass. The glass modifiers, especially potassium oxide, are strongly alkaline and very *hygroscopic;* that is, they tend to absorb water from their surroundings. In condition of high humidity, beads of potassium hydroxide solution form on the surface of the glass. These absorb atmospheric carbon dioxide to form potassium carbonate, which further attack the silica skeleton and hasten deterioration. The potassium carbonate formed also absorbs further moisture, thereby facilitating the movement of even more alkali to the surface. This form of deterioration continues as long as the glass is kept under conditions of high humidity. Unless the glass is transferred to a dry environment, it eventually decays completely.

WOOD

Wood is subject to deterioration through the action of physical, chemical, and biological agencies whether separately or all together. Because of its *anisotropy* (the quality of having different properties in different directions), wood is subject to warping caused by variations in atmospheric humidity that lead to different dimensional changes across or along the grain.

If the moisture content of wood is higher than that necessary for wood-fiber saturation (ca. 30 percent), it deteriorates. Wood in contract with the ground is exposed not only to the effect of moisture, but also to attack by bacteria, fungi, and insects. When kept off the ground, wood is exposed to such deteriorating agencies as wind, high temperature, and fire, which may cause mechanical wear, shrinking, checking, and warping.

Waterlogged Wood

Wood submerged in water or embedded in mud or peat undergoes partial degradation; whereas the cellulose fibers decompose, leading to an increase in porosity, the thick lignin structure does not decay and preserves the original shape and appearance of the object. As long as waterlogged wood is kept wet, it will keep its shape. If allowed to dry, however, the weak lignin structure will collapse beyond repair, unless it has been specially pretreated (see Chapter 15).

TEXTILES

Ancient textiles were made from natural fibers. Linen, cotton, and hemp are examples of vegetable fibers, whereas wool and silk are of animal origin. Deteriorative changes may occur as the result of any of a great number of agents, whether physicochemical (e.g., oxygen, sunlight, and moisture) or biological.

The most evident effects of physicochemical deterioration are loss of strength and change in appearance. The effect of excessive moisture is to cause swelling, which is always accompanied by softening. Desiccation, on the other hand, leads to embrittlement. Sunlight is the single factor most responsible for textile decay: it prepares the way for the chemical breakdown of the fibers by oxidation.

Wool seems to be the natural fiber most resistant to physicochemical decay. Cotton and linen come second, having similar resistances, whereas silk is the most easily degradable natural fiber.

LEATHER

Skins and hides are animal products, composed mainly of proteins and obtained mostly from mammals. In antiquity they were used almost untreated for clothing and other purposes. Hides are unstable to environmental conditions and deteriorate readily through biological action. Leather is a more complex material, obtained by drastic mechanical and chemical treatment (tanning) of animal skins. The object of the chemical treatment is to stabilize the skin and make it resistant to the agents of decay. This is achieved through the formation of a protein-tannin chemical complex.

Different leathers vary materially in their resistance to deterioration, which may occur through a variety of mechanisms. Fundamentally, however, most leather decay is caused by the *hydrolysis* of the protein-tannin complex (hydrolysis is the process in which the components of the complex react with water). In extreme cases degradation by moisture may be total, with nothing remaining in an excavation to indicate that leather was once present there.

Sulfur dioxide from the atomsphere is absorbed by leather. With age, it causes what is called *red rot*. After absorption by the leather, the sulfur dioxide is oxidized to sulfur trioxide through surface-catalyzed reactions. This leads to a gradual buildup of sulfuric acid in the material, which that has been correlated with age and extent of deterioration (Frey and Clarke 1931). Other deteriorative processes may occur due to heat or oxidative conditions. Many leathers may last for long periods in dry conditions, but eventually, even with no apparent physical or chemical demand on them, they crumble to a powdery residue. It seems that oxygen is slowly taken up by the leather and then converted to hydrogen peroxide by the catalytic effect of the leather surface and of small amounts of sulfuric acid present in the leather; traces of iron also present reduce the hydrogen peroxide to produce oxygen in a very highly reactive state that attacks the proteinic substance (Cheshire 1946).

Sunlight has relatively little effect on leather. Occasionally it may act to accelerate other forms of deterioration; leather alternately exposed to water and sunlight, for example, is known to deteriorate rapidly.

REFERENCES

Bettembourg, J. M., *Verres Refract.*, **30,** 36 (1976).

Cheshire, A., *J. Intern. Soc. Leather Trade Chem.*, **30,** 134 (1946).

Frey, R. W., and Clarke, I. D., *J. Am. Leather Chem. Assoc.*, **26,** 461 (1931).

Lal Gauri, K., *Sci. Am.*, (June) 126 (1978).

Weier, L. E., *Bull. Inst. Archaeol., London Univ.*, **11,** 131 (1973-1974).

CORROSION OF METALS

Corrosion is a general term for the deterioration of metals through chemical reaction with their environment. Oxidation and rusting are other terms having similar connotations. *Rusting,* for example, generally implies the corrosion of iron; rust is iron oxide. Another term connected with the deterioration of metals is *tarnishing.* This refers to a loss of metallic luster as a result of the formation of metallic-decay products on a polished surface.

"Lay not up yourselves treasures on earth where moth and rust doth corrupt," says the admonition in the New Testament, implying that the inevitable course of corrosion was already familiar in ancient times. An unprotected surface of copper grows dull within a few weeks; a coating is formed through chemical reaction of the metal with the components of the atmosphere (oxygen, carbon dioxide, water, and salts carried by the wind). Within a few years, even in an unpolluted atmosphere, the copper surface will take on green bloom and eventually turn green all over. Copper roofs turn green in country air within 50-100 years, but in sulfur-polluted urban air the change takes place in about half that time (Gettens 1970). The corrosion of copper or bronze buried in soil or immersed in water will usually take place even more rapidly.

Different metals react differently in various environments, as shown in Table 14.1 (Cornet 1970). However, the data given and similar data from other sources should be regarded with reservations. Soils vary widely in corrosivity, and steel exposed in soil may pit at much more rapid rates than those indicated. If steel buried is in contact with copper, for example, the rate of corrosion of the steel will be greatly increased and that of the copper greatly decreased.

CAUSES OF CORROSION

Corrosion takes place because in natural environments most metals are not inherently stable and tend to be converted into more stable compounds; the metallic ores found in nature are the most familiar cases in point. Natural phenomena always favor the formation of stable compounds. Since the products of metal corrosion are more stable than the metals themselves, it follows that corrosion is a perfectly natural thing to expect (Scully 1975). The chemical reactions taking place during corrosion can be expressed by the general equation:

TABLE 14.1   Representative Corrosion Rates of Steel and Copper in Air, Seawater, and Soil

	Corrosion Rate			
	Steel		Copper	
Environment	mdd[a]	μpy[b]	mdd	μpy
Rural atmosphere	—	—	0.1	0.06
Marine atmosphere	2.9	1.3	0.3	0.13
Seawater	25	11.4	8	3.2
Soil	5	2.3	0.7	0.3

[a]Milligrams per square decimeter per day.
[b]Microns (μ) per year ($1 = 10^{-6}$m).

$$\text{Metal} + \text{environment} \rightleftharpoons \text{Stable metal compound} + \text{energy}$$

In man-made metallurgical processes this natural sequence is turned about-face (see Fig. 14.1) through the local application of large amounts of energy, and the metals, with their invaluable properties, become available for human use. But this is only a temporary situation; nature takes its inexorable course and the process of corrosion reestablishes the natural equilibrium. All man can hope to do is to halt, or rather to slow down corrosion for longer or shorter periods.

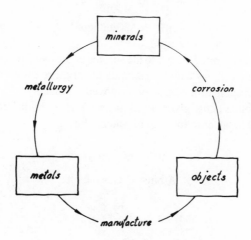

Fig. 14.1.   Corrosion: metallurgy in reverse.

CHEMICAL MECHANISM OF CORROSION

On exposure to air, almost all new metal surfaces soon become coated with a thin film of metallic oxide. If this film is continuous and stable, it may render the surface less susceptible to further corrosion. However, if the film is discontinuous or chemically unstable and the metal comes in contact with an electrolyte, corrosion may proceed even more rapidly. An *electrolyte* is an aqueous solution of any compound that undergoes ionization, that is, one dissociating into positively and negatively charged "ions" that then move freely in the solution. When a metal comes into contact with such a solution, atoms of the metal tend to take up a positive electric charge and move into the solution themselves, whereas positively charged ions in the electrolyte will be simultaneously displaced from the solution to reestablish the electrical balance. These ions will give up their charge to the metal atoms and themselves become electrically neutral. Thus:

$$\text{Metal} \; + \; \text{ions}^+ \; \rightarrow \; \text{Metal ions}^+ \; + \; \text{atoms}$$

It follows that the presence of water is essential for corrosion to take place. Water for the corrosion process is provided by moisture in the soil, or, in the atmosphere, by water vapor condensing on the surface of the metal objects.

Water itself is an electrolyte that dissociates into hydrogen ions and hydroxyl ions:

$$H_2O \; \rightleftharpoons \; H^+ \; OH^-$$

During corrosion metals react with the hydrogen ions to form metal ions and hydrogen atoms:

$$\text{Metal} \; + \; n\text{H}^+ \; = \; \text{Metal ion}^{n+} \; + \; n\text{H}$$

The hydrogen atoms formed adhere to the surface of the metal in the form of a thin, invisible film that interferes with the continuation of the reaction. If corrosion is to proceed, the hydrogen film has to be removed. Usually, removal takes place through the hydrogen being given off as gas bubbles, or by its recombining with oxygen dissolved in water to form more water. Thus:

$$2\text{H} \quad \rightarrow \quad H_2$$

$$\text{Atoms} \qquad \text{Gas bubbles}$$

or

$$2\text{H} \; + \; 1/2O_2 \; \rightarrow \; H_2O$$

The removal of hydrogen allows more metal to go into solution, and corrosion proceeds with the accumulation of metal ions in the solution. These metal ions may react with the hydroxyl ions produced by the dissociation of water to form the metal hydroxide:

$$M^{n+} + nOH^- = M(OH)_n$$

Alternatively, the metal ions may be directly oxidized by atmospheric oxygen. In either case an insoluble precipitate is formed; in the case of iron, rust:

$$Fe + 2H_2O = Fe(OH)_2 + H_2 \uparrow$$

$$4Fe(OH)_2 + 2H_2O + O_2 = 4Fe(OH)_3$$

Ferric oxide, or rust

Corrosion entails an interchange of electrically charged particles (ions) and is thus considered an *electrochemical process* (see Fig. 14.2). This calls for the

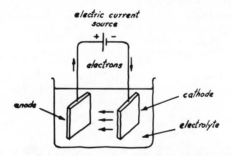

Fig. 14.2.   Electrochemical cell.

presence of an *anode,* a metallic surface from which metal ions are given off; a *cathode,* where hydrogen ions are formed; an *electrolyte,* the liquid medium through which the ions move; and a *metallic path,* to close the electric circuit (Fig. 14.3).

FACTORS AFFECTING CORROSION

The rate at which corrosion proceeds in a buried metal object depends, among other factors, on the moisture, porosity, and acidity of the burial site, the presence of soluble salts that may affect the process, and the temperature. In dry air, for example, metals are resistant to attack and even when pure water is present,

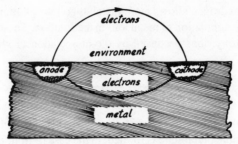

Fig. 14.3. Corrosion. An electrolytic cell can form on a single piece of metal (even of microscopic size) because of local differences in the composition of the metal. On a surface undergoing corrosion millions of such cells may be present.

corrosion is slight. However, when pollutants are present—soluble salts and acids, for example—the rate of corrosion increases measurably. The effect of humidity on corrosion may be systematized, as illustrated in Table 14.2.

TABLE 14.2   Effect of Humidity on Metal Corrosion

Relative Humidity (Percent)	Degree of Corrosion
Less than 60	None
More than 60	low but definite
Less than 80	Decided increase
More than 80	Very high

SUSCEPTIBILITY OF METALS TO CORROSION

Metals exposed to the environment can be divided into three groups according to their susceptibility to corrosion: (1) metals that resist corrosion in all natural environments (e.g., gold), (2) metals that are at first easily attacked but subsequently form a corrosion-resistant film and become resistant to further attack (e.g., copper, and bronze), and (3) metals that corrode rapidly and do not form protective corrosion products (iron is the major metal in this group).

Metals that resist chemical attack have been referred to, since ancient times, as *noble metals*. The metals known in antiquity can be arranged in a series starting from gold, the most "noble," downward to the "base" metals through silver, copper, lead, tin, and iron. The *base metals* are more easily corroded than those higher in this series. When two objects made of dissimilar metals are in contact with each other in the presence of an electrolyte, an electric current flows between the two (Fig. 14.4). Similarly, when two archaeological objects made of different

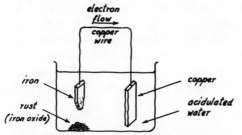

Fig. 14.4. Cathodic protection. When two metals are in electrical contact the "baser" metal is preferentially corroded, whereas the "nobler" metal remains unaltered.

metals are in contact with each other, an electrical current flows from one to the other. The direction in which the current flows is determined by the *electro-chemical potential* (a magnitude expressing the tendency of metals to oxidize), characteristic for each metal. Noble metals have a lower potential, and reactive metals, a high one. The *galvanic* series grades metals in terms of their electro-chemical potential (Table 14.3). When two metals are in contact, the one having a higher potential will be preferentially corroded, whereas the other survives by what is known as *cathodic protection*.

TABLE 14.3   Galvanic (electrochemical) Series of Metals

1.	Platinum	(Noble Metal)
2.	Gold	
3.	Silver	
4.	Copper	
5.	Bronze	
6.	Brass	
7.	Lead	
8.	Tin	
9.	Lead-tin solder	
10.	Iron	
11.	Steel	
12.	Zinc	(Base Metal)

Electrochemical decay is promoted when metals are alloyed. For instance, the susceptibility to corrosion of bronze—an alloy of tin and copper—is greater than that of the pure, unalloyed metal components and, once it sets in, the decay of bronze proceeds more quickly than that of copper. Base silver is also more intensely corroded than pure silver. In bronze-plated iron objects the bronze aggravates the rusting of the iron and is itself protected in the process.

Corrosion crusts on ancient metals sometimes have layered structures, containing two or more chemical compounds. The outer, more stable and mineralized crusts often serve as protective layers for less stable compounds lying underneath. Some corrosion crusts may also act as cementing agents for ugly and disfiguring earthy accretions.

PRODUCTS OF CORROSION

Analysis of the corrosion products found on metal antiquities reveals that they are closely related chemically to some of the minerals of the earth's crust. This is not surprising if it is remembered that corrosion is a natural phenomenon in the course of which stable compounds are formed.

Gold

Gold is not susceptible to corrosion or chemical attack under natural conditions; thus it is always found in the metallic state. Sometimes ancient gold objects are covered with what seems to be a red patina. This is not, however, a true patina, but an adventitious color developed through heating or by accidental contamination.

Silver

Silver is readily tarnished by hydrogen sulfide, which is an atmospheric contaminant: a thin layer of dark silver sulfide (Ag_2S) is formed. Occasionally silver sulfide forms thick black layers of patina. The presence of chlorides in solids where silver is buried causes the formation of stable, gray silver chloride (AgCl)— *horn silver*) which may, through the great expansion accompanying the change from metal to mineral, cause severe deformation of the original object. Silver objects retrieved from desert soil are often encrusted with silver chloride, which sometimes penetrates deep into the metal structure. Silver objects recovered from Ur, for example, were heavily encrusted with horn silver (Graham 1929).

Copper and Its Alloys

The main corroding agents of copper and its alloys are atmospheric and soil contaminants in the presence of air and moisture (Gabel, 1976). Even small amounts of chlorides and nitrates may, in the presence of moisture, bring about mineralization of the surface of copper and bronze and, eventually, the entire mass of metal. The process begins with the formation of copper chlorides and nitrates of surface metal. In the presence of oxygen and carbon dioxide, these are slowly

TABLE 14.4 Corrosion Products Found in Copper Antiquities

Corrosion Product	Formula	Color	References
Oxides			
Cuprite	Cu_2O	Red	Many
Tenorite	CuO	Gray to black	Fink and Polushkin 1936
Carbonates			
Malachite	$Cu_2(OH)_2CO_3$	Dark green	Many
Azurite	$Cu_3(OH)_2(CO_3)_2$	Bright blue to red indigo	Many
Chalconatronite	$Na_2Cu(CO_3)_2 \cdot 3H_2O$	Blue green	Gettens and Frondel 1955
Chlorides			
Atacamite	$Cu_2(OH)_3Cl$	Emerald to black-green	Otto 1959
Botallacite	$Cu_2(OH)_3Cl \cdot H_2O$	Green-blue	Frondel 1950
Nantokite	$CuCl$	Pale gray	Rosenberg 1917; Caley 1941
Sulfates			
Brochantite	$Cu_4(SO_4)(OH)_6$	Green	Gettens 1933
Antlerite	$Cu_3(SO_4)(OH)_4$	Green	Otto 1961
Connellite	$Cu_{19}(SO_4)Cl_4(OH)_{32} \cdot 3H_2O$	Bright blue	Otto 1963
Sulfides			
Chalcocite	Cu_2S	Black	Rogers 1903
Chalcopyrite	$CuFeS_2$	Brass-yellow	Clarke 1924
Covellite	CuS	Indigo blue	Daubree 1875
Digenite	Cu_9S_5	Blue to black	Perinet 1961

259

converted into green oxychlorides and oxycarbonates of copper. When bronze is corroded a certain quantity of tin compounds is also formed, chiefly tin oxide. In either case—whether copper or bronze—underlying the green crust is a layer of copper oxide, below which may be a core of metal. In extreme cases, the corrosion goes all the way through the specimen, and none of the original metal remains what is left is a hard shell of mixed copper corrosion compounds covering a soft, brittle core of metal oxide.

Many agents may be instrumental in the formation of other corrosion products. Organic acids or ammonia play an important part in the mineralization process. Electric currents in the ground may also be of importance. These and other factors are undoubtedly responsible for the great variety of minerals that have so far been identified in copper and bronze corrosion products (Gettens 1964; Borrelli, 1975).

Bronze Disease.   Of all the corrosion products of copper and its alloys, the one that probably causes most damage and presents the most difficult problem for the archaeological chemist, is the unstable compound cuprous chloride, Cu_2Cl_2.

Ancient copper and bronze can survive burial for centuries without appreciable deterioration because an equilibrium is established between the corroded metal and its surroundings. This equilibrium may be destroyed in excavation of the object. Unstable cuprous chloride may then become chemically active, combining with free oxygen to form cupric chloride:

$$2Cu_2Cl_2 \quad + \quad O \quad = \quad Cu_2O + 2CuCl_2$$

Cuprous	Cupric
chloride	chloride

The cupric chloride reacts in turn with as yet unattacked metal, as it is again reduced to cuprous chloride:

$$CuCl_2 + Cu = Cu_2Cl_2$$

The result is a pale green, powdery deposit occurring in spots on the surface of the object (Fig. 14.5). The spots continue growing as a result of cuprous chloride being transformed into basic cupric chloride, by the further action of oxygen in the presence of moisture:

$$3Cu_2Cl_2 + 3[O] + 4H_2O = CuCl_2 \cdot 3Cu(OH)_2 \cdot H_2O + 2CuCl_2$$

Basic copper chloride

Green spots are characteristic features of corroding copper and bronze and are referred to as *bronze disease*. Bronze disease completely corrodes any underlying metal unless arrested by special treatment.

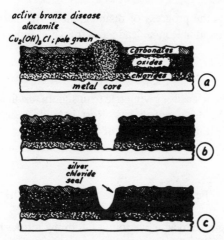

Fig. 14.5. Bronze disease. (*a*) pale green powdery spots are the result of the reaction of unstable cuprous chloride (attacamite) with oxygen and moisture in the atomosphere; (*b*) in "dry treatment," spots are first excavated and all active atacamite removed; (*c*) silver chloride is applied to form a chemical seal.

Lead

Ancient lead objects excavated from the ground are often covered by a white incrustation, which is produced by the chemical action of saline matter in the ground, on the metal. A wide variety of lead corrosion products, listed in Table 14.5 have been identified. Cerusite (lead carbonate) forms a white, protective

TABLE 14.5   Corrosion Products Found in Lead Antiquities

Corrosion Product	Formula	Color	Reference
Oxides			
Massicot	PbO	Yellow	Caley 1955
Plattnerite	PbO_2	Brown to black	Caley 1955
Carbonates			
Cerussite	$PbCO_3$	Warm gray	Many
Hydrocerussite	$Pb_3(CO_3)_2(OH)_2$	White	Many
Chlorides			
Cotunnite	$PbCl_2$	White	Lacroix1910
Phosgenite	$Pb_2(CO_3)Cl_2$	White	Lacroix1909
Sulfates			
Anglesite	$PbSO_4$	White	Perinet 1961
Sulfides			
Galena	PbS	Gray	Daubre 1875

layer and prevents the progress of disintegration. Another white lead mineral, phosgenite, was found in Roman lead pipes from hot springs in France.

Iron and Steel

Iron and steel corrode easily, giving rise to unsightly rust that causes swelling and deformation of the decaying objects. Rust is held in such disfavor that it is eliminated quickly if the iron object is of any interest.

The general chemical equation for the corrosion of iron can be written as follows:

$$Fe + 2H_2O \rightarrow Fe^{++} + 2H + 2OH^-$$

The ferrous ions pass into aqueous solution forming ferrous hydroxide, $Fe(OH)_2$, which under the influence of atmospheric oxygen is oxidized into insoluble red ferric hydroxide, $Fe(OH)_3$:

$$2Fe(OH)_2 + 1/2O_2 + H_2O = 2Fe(OH)_3$$

On drying, ferric hydroxide becomes ferric oxide (Fe_2O_3), which is common rust:

$$2Fe(OH)_3 = Fe_2O_3 + 3H_2O$$

The composition of rust is almost identical to that of the natural ores from which iron is industrially extracted.

Corrosion of Iron Submerged in the Sea. If free oxygen is present in seawater, the surface of iron submerged in the latter is gradually transformed into rust. The rusting process stops, however, when iron objects sink into the seabed, where there is little or no free oxygen. Under these conditions iron is kept fresh until brought up again and freed from covering incrustations. However, during its stay in the seabed, iron is also exposed to the penetration of salt water into the metal. Inside the metal, water and dissolved sodium chloride react with the iron in the following way (Eriksen and Thegel 1966):

$$Fe + 2NaCl + 2H_2O = FeCl_2 + 2NaOH + H_2$$

Ferrous
chloride

The ferrous chloride formed combines with water, which is present in excess, binding it to itself as ''water of crystallization'':

$$FeCl_2 + 2H_2O = FeCl_2 \cdot 2H_2O$$

Hydrated ferrous chloride

The sodium hydroxide and hydrogen formed diffuse to the surface of the iron and dissolve in the surrounding water, whereas the hydrated ferrous chloride remains within the lattice of the iron.

When a submerged iron object is removed from the seabed, the solution of ferrous chloride begins to seep outward, producing a characteristic "sweat" that, depending on the density of the iron, may continue to appear for many years. As the ferrous chloride reaches the surface of the iron, oxygen from the air oxidizes it to ferric chloride with the concurrent formation of ferric oxide:

$$6FeCl_2 + 3/2O_2 \rightarrow 4FeCl_3 + Fe_2O_3$$

Simultaneously, oxygen from the air, in the presence of water from the ferrous chloride solution, starts rusting the iron. The porous rust formed does not prevent the further penetration of air, and the rust layer continues to grow until checked by appropriate techniques of conservation.

PATINA

Under certain circumstances the corrosion products on the surface of an ancient metal object may add beauty to it. If the incrustation is nonporous, hard and chemically stable under normal conditions, it may also be protective. A corrosion layer of this kind is usually called *"patina"* (Plenderleith 1938) (Fig. 14.6). A uniform patina will preserve the details of shape and may enhance the appearance

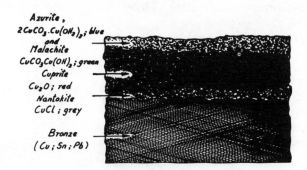

Fig. 14.6. Patination layers of ancient copper (cross section). The metal core is covered by successive layers of chlorides, oxides, and carbonates. There may be many layers, not necessarily in this order. Layers of oxides of tin or other metals, may be found in coroded bronzes.

of an object. Sometimes it may also constitute evidence of antiquity. A noble patina on a bronze or copper surface is a thin, fairly uniform green crust of basic copper carbonates with splashes of the related blue mineral azurite. Good patinas are very much admired and are often imitated and reproduced artificially by paint or by chemical means (Weil 1977; Smith 1978).

REFERENCES

Borrelli, L. V., in *ICOM Comm. for Cons.*, *4th Triennial Meeting*, Venice, 1975.

Caley, E. R., *Proc. Am. Phil. Soc.*, **84**, 689 (1941).

Caley, E. R., *Studies in Conservation*, **2**, 49 (1955).

Clarke, F. W., *U.S. Geol. Surv. Bull.*, 770 (1924).

Cornet, I., in R. Berger, ed., *Scientific Methods in Medieval Archaeology*, California U. P., Berkeley, 1970.

Daubree, A., *Compt. Rend.*, **80**, 461 (1875).

Eriksen, E., and Thegel, S., *Conservation of Iron Recovered from the Sea*, 1966.

Fink, C. G., and Polushkin, E. P., *Am. Inst. Mining Met. Eng.*, *Inst. Metals Div.*, *Tech. Publ.* No. 693, 1936.

Frondel, C., *Mineralog. Mag.*, **29**, 34 (1950).

Gabel, H., *Corrosion*, **32**(6), 253 (1976).

Gettens, R. J., *Technical Studies in the Field of the Fine Arts*, **2**, 31 (1933).

Gettens, R. J., *The Corrosion Products of Metal Antiquities*, Smithsonian Institute, Washington, D. C., 1964.

Gettens, R. J., in S. Doeringer, D. G. Milten, and A. Steinberg, eds., *Art and Technology*, MIT Press, Cambridge, Mass., 1970, p. 57.

Gettens R. J., and Frondel, C., *Studies in Conservation*, **2**, 64 (1955).

Graham, A. K., *Museum Journal (Philadelphia)*, **20**, 246 (1929).

Hunt, C. B., *Science*, **120**, 183 (1954).

Lacroix, A., *Comptes Rendus*, **151**, 276 (1910).

Lacroix, A., *Bull. Soc. Francaise de Mineralogie*, **32**, 333 (1909).

Laudermilk, S. K., *Am. J. Archaeol.*, **38**, 207 (1934).

Otto, H., *Freiberger Forschungshefte*, **B37**, 66 (1959).

Otto, H., *Naturwissenschaften*, **48**, 661 (1961).

Otto, H., *Naturwissenschaften*, **50**, 16(1963).

Perinet, G., *Rev. Men. Saint Germain-en-Laye*, **36**, 454 (1961).

Plenderleith, H. J., *Transact Oriental Ceramic Soc.*, **16**, 33 (1938).

Plenderleith, H. J., *The Conservation of Antiquities and Works of Art*, Oxford U. P., London, 1966.

Rathgen F., *The Preservation of Antiquities*, Cambridge U. P., 1905.

Rogers, A. F., *Am. Geol.*, **31**, 43 (1903).

Rosenberg, G. A., *Antiquities en Fer et en Bronze,* Copenhagen, 1917.

Scully, J. C., *The Fundamentals of Corrosion,* Pergamon, Oxford, 1975.

Smith, A. W., *Archaeometry,* **20,**2 (1978).

Weil, P. D., in N. B. S. Special Publication 479, p. 77, Washington, D. C., 1977.

Recommended Reading

Brown, B. F., Burnett, H. C., Chase, W. T., Goodway, M., Kruger, J., and Pourbaix, M. Eds., *Corrosion and Metal Artifacts. A Dialogue between Conservators and Archaeologists and Corrosion Scientists,* N.B.S. Special Publication, No. 479, U.S. Dept. of Commerce, Washington, D. C., 1977.

Weier, L. E., The deterioration of inorganic materials under the sea, *Bull. Inst. Archaeol.,* London Univ., **2,** 131 (1973).

CONSERVATION OF ANTIQUITIES

The physical remains of the arts and crafts of past ages—in the form of objects and artifacts recovered by archaeological research—are a precious part of man's cultural heritage. Very little has survived out of the vast quantity of material manufactured in days long past from clay, metal, glass, wood, and the like. Most has succumbed to the combined ravages of man and environment. It is important, therefore, to ensure that the relatively few artifacts that remain to us be protected as completely as possible from decay. The aim of *conservation practice* is to preserve antiquities unchanged for as long as possible. To do this, antiquities are restored, made structurally sound, and then kept in a controlled environment (Madsen 1976).

The natural deterioration of antiquities is the result of physical and chemical changes taking place in them. It is reasonable to expect, therefore, that the restoration and conservation of such objects should be best achieved by the application of scientific principles. Up to a few years ago, however, the treatment of antiquities was an empirical and casual matter. It was mainly the purlieu of museum craftsmen, working in mechanical workshops and using whatever materials seemed to them most appropriate for the task in hand. In the last quarter of the nineteenth century, a change in the approach to conservation got under way, and empirical procedures of conservation began to be displaced by scientific methods. The first laboratory for conservation was established in the Berlin State Museum in 1880. Since then, and especially after 1945 a new concept, that of *scientific conservation,* has emerged.

The application of scientific principles to conservation implies much more than mere repair. Specialized knowledge is applied to the scrutiny of the decay processes, diagnosis of factors responsible for deterioration, and study of the problems that may arise during treatment. New materials are used to substitute for decayed ones, matching their physical and aesthetic properties.

In later years there has been a marked increase in public concern for the conservation of antiquities. A newly developed interest in archaeology and ancient art have brought an increase in awareness of the need for effective conservation. Training centers for conservators have been established at several universities. Scientific laboratories are devoted to the examination and conservation of antiquities. The combined efforts of a large number of chemists, physicists, metal-lurgists, and biologists have lead to the acceptance of the scientific approach to conservation.

INFORMATION AND RECORDS

A basic requirement before embarking on the restoration of antiquities is that a maximum of information be available regarding the objects to be treated. This should be in the form of a systematic inventory of data regarding composition and condition of the objects, urgency of the treatment required, and other factors.

The examination of antiquities prior to conservation is usually nondestructive. It is aimed at obtaining information of the chemical composition, physical structure, and mechanical stability of the antiquity, the nature of the decay products, and the consistency of the incrustations. As more and more information becomes available as work proceeds, it should all be carefully collated and added to the existing store of data. Nothing should be dismissed as insignificant. Thus illustrations are also very important, especially photographs of known magnification and radiographs, where available.

CONSERVATION

The range of objects that may, at one time or another, be considered worth conserving is very wide. The present work is confined to a review of some of the most important chemical conservation principles and techniques available. It is in no way meant to be all-embracing or to be an archaeologist's *vade mecum* of conservation.

STONE

Two types of stone antiquities may need conservation: monuments in open places, and objects kept under cover. The heterogeneity of stone makes it impossible to arrest completely the natural process of decay of monuments in open places; all that can be done is to slow it down. Stone antiquities kept under cover present a much easier task (deWitte, Huget et al. 1977; Lal Gauri 1978; Moncrieff 1976).

Stone Buildings and Monuments

In the early days of stone conservation the modes of migration of moisture and salt within stone were not understood. Attempts to isolate the objects from weathering and moisture by totally sealing their exposed surfaces generally created more problems than they solved; the waxes and oils that were applied for surface sealing often caused accelerated flaking, efflorescence, and undesirable stains.

Modern stone-conservation techniques fall into two categories: hardening and sealing (Sleater 1977). *Hardening* or consolidation consists of impregnating stone with a material that will prevent further disintegration. Stone to be treated in this manner must be sufficiently porous to permit proper penetration of the hardening fluid into its fabric. Such compounds as silicon esters (Schmidt-Thomsen 1969), epoxy resins, and polymethyl methacrylates (Gairola 1959) have been used; the hardening they cause is so complete that no discoloration or efforescence occurs after treatment.

Sealing of stone is intended to develop a tight, impervious skin that prevents the access of moisture. Watertight sealers, as mentioned before, are no longer applied, since moisture trapped inside the stone may cause much damage. The treatment of stone surfaces with solutions of compounds that are converted to insoluble precipitates on contact with the stone has given good results in the stabilization of marble and limestone. The precipitates reduce the size of microchannels and diminish the absorption of water, whereby surface attack is retarded (Burgess and Schaffer 1952; Lewin 1968).

The chemical treatment of stone buildings and monuments has proved generally unsatisfactory in the long run. If, however, maintenance treatment can be given periodically, the application of hardeners and sealers may be of some use (Amoroso 1977; Sleater 1977).

Stone Objects Kept under Cover

Stone antiquities kept under cover do not usually require chemical treatment. They do, however, require removal of salts and protection against biological attack. It sometimes happens that the amount of salts in a stone object is so large that their removal is out of the question since the object would collapse. In such cases sealing the surface can prevent hygroscopic action, provided that the object is subsequently kept in a humidity-controlled atmosphere.

POTTERY

Glazed pottery is waterproof and is, therefore, very stable. Unglazed pottery, although chemically inert, is easily disrupted by soluble salts that tend to migrate to the surface, where evaporation is greatest, and there crystallize out. Pottery saturated with soluble salts must be thoroughly washed to prevent breakage through the cumulative effect of salt crystallization. Washing is a lengthy and difficult process that may take several months to complete. However, it can be accelerated in several ways: by frequently changing the wash water, by raising the temperature of the object and of the bath, or by regular cycling of the object between alternate baths of hot and cold water (Organ 1955; Jedrzejewska 1970).

Insoluble salts, such as calcium sulphate or calcium carbonate, cause unsightly incrustations on pottery. These are best removed by mechanical rather than chemical means. Soaking in acids may cause the disolution of incrustations as well as of salts that actually form part of the pottery itself (Olive and Pearson 1975). With unglazed painted pottery or paint-decorated pottery, there is a danger that the paint layer may be damaged by washing. Preliminary consolidation, by applying a stabilizer—such as soluble nylon—will cause adherence of the paint layer after which the object may be washed without risk.

Objects of unfired clay—such as clay tablets from Mesopotamia—or badly fired pottery are frail and can only be handled when dry since they soften and turn to mud on wetting. Such objects can be stabilized by rebaking.

Pottery from the Sea

Pottery survives well under the sea. Disintegration may result when it is recovered, and correct procedures have to be followed to prevent this from happening.

The most important conservation process is *desalination* (the removal of salt contamination), which has to be started immediately after removal from the sea. When storage previous to treatment is required, the wares have to be kept under seawater. Desalination should proceed slowly; if carried out too rapidly—by a forthright change of saltwater by deionized water, for example—high osmotic pressures may set up within the objects and cause damage to susceptible glazes.

Most pottery recovered from the sea are covered with some type of concretion. These can be easily removed by treatment with an acid, but there is danger of damaging either the body or the glaze. In pottery containing a high proportion or iron oxides (e.g., earthenware or terracotta) the acid dissolves the iron oxides and increases the tendency to exfoliation. The most satisfactory way to remove concretions from pottery recovered from the sea is by manual techniques while objects are still wet. If allowed to dry, the concretion becomes much harder, probably by reaction with atmospheric carbon dioxide to form calcium carbonate in a manner analogous to the setting of lime in mortar. Once cleaned and freed from salts, pottery can be allowed to dry. If desalination has been carried out efficiently there is little likelihood of pottery deteriorating at later stages.

GLASS

Most ancient glasses are of the soda-lime type. Many contain an excess of lime, which makes them water-soluble and easily damaged by moisture; such glasses loose their transparency easily and can never by fully restored.

Decaying glass—devitrified, flaking, or weeping glass—is a difficult material to restore. Soluble salts often cannot be washed out, and insoluble deposits must be removed mechanically or even left untouched to prevent total disintegration.

Unstable glass changes color on exposure to sunlight; it is advisable, therefore, to keep excavated glass objects in darkness and in a damp atmosphere and to color photograph them shortly after discovery.

METALS

Metal objects retrieved by the archaeologist are often in an advanced state of corrosion. The accepted procedures for cleaning and stabilizing metal antiquities are aimed at achieving either one (or both) of two objectives: (1) arresting corrosion and (2) removal of corrosion products.

Arresting Corrosion

It often occurs that the corrosion of ancient metal artifacts has proceeded to a point where very little actual metal is left. It may also happen that the products of corrosion are in themselves worth preserving, as in the case of an attractive patina. In such cases the conservator's primary objective is to stabilize the artifact, preventing further chemical change, as far as possible and then covering it with a protective coating.

Stabilization of Copper and Bronze. Stabilization of excavated bronzes may be effected by exposing them to vapours of ammonia (Thouvenin 1958, 1959). Bronze disease spots (q.v.) may be mechanically removed from a stable patina that it is desired to preserve and the resulting cavities then filled with a fine paste of silver oxide. Insoluble silver chloride thus formed seals off the underlying harmful copper chloride and thus renders it harmless (Organ 1963). Prolonged immersion in a solution of sodium sesquicarbonate stabilizes the patina of copper (Plenderleith 1966).

Stabilization of Iron. The presently accepted practice for the stabilization of iron objects involves four stages:

1. Restoring the appearance of the artifact, as far as possible, by selective rust removal, using such methods as electrolytic reduction and mechanical stripping or chemical agents,
2. desalination (removal of salt contaminants),
3. drying, and
4. providing a physical barrier against contamination (i.e., a protective coating).

Soluble salts are removed from antique iron with great difficulty, regardless of the desalination procedure used. Repeated washing in alternate baths of water at near boiling and room temperatures respectively improves penetration of the water into the pores of corroded iron; after 16 cycles of treatment no chloride was detected in the water (Wihr 1975). Negative chloride ions can be removed from the corrosion product by the combined effects of iron reduction and an applied electric field (cathodic desalination); in this case iron objects are made the cathode in an appropriate electrolytic bath.

Removal of Corrosion Products

Metal objects that have undergone surface corrosion can often be made more presentable by removing the crust of corrosion products. These are often soluble in acid and can be removed by dissolving them in the latter. However, it is almost impossible to regulate the action of acids in such a way as to avoid pitting of the underlying metal core. The usual result of acid treatment is that the corroded parts are dissolved together with some of the metal core, leaving a rough, unsightly metallic surface on which fine details are obliterated. Alkaline solutions are usually milder in their action and do not usually attack metal, dissolving only the products of corrosion. In consequence, the use of alkaline solutions is usually preferred.

Electrolytic Reduction

The corrosion of metals is usually an electrolytic process (see Chapter 14). A reversal of the natural electrolytic process of corrosion is today the most accepted method of restoration of ancient metal objects. In this method, called the *electrolytic method,* the metal compounds in the corrosion crust are reduced by the action of an electric current and converted back to metal (Fink and Eldridge 1925). The fundamental idea is to cathodically reduce the metal that has passed into the corrosion layer. Reduction is carried out in an *electrolytic cell,* (Fig. 15.1). A solution of an alkali—such as caustic soda or potash—is used as the electrolyte.

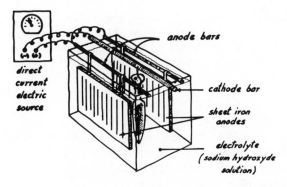

Fig. 15.1.   Apparatus for the electrolytic reduction of corroded metal objects.

The alkali neutralizes the harmful chloride and sulfate ions that are ever present in corroded metal objects. The electrolytic cell is connected to a source of direct current. The object being treated is made the cathode (negative pole) of the electrolytic cell, whereas the anode (positive pole) is of platinum or carbon. When an electric current is passed through the cell, a "potential difference" is established between the electrodes; the anode becomes deficient in electrons (i.e., acquires a positive charge), whereas the cathode having a surplus of electrons, takes on a negative charge. The positively charged ions (cations) in the solution migrate—by electrostatic attraction—to the negatively charged cathode where they acquire electrons (i.e., they are reduced). Simultaneously the negatively charged ions (anions) migrate to the positively charged anode where they give up electrons (i.e., they are oxidized). The following reactions take place:

1. The alkali in solution is ionized, forming (in the case of sodium hydroxide), sodium cations and hydroxyl anions:

$$NaOH = Na^+ + OH^-$$

The water of the solution is also ionized:

$$H_2O = H^+ + OH^-$$

2. At the anode hydroxyl ions are decomposed; then water is formed and oxygen loses electrons and is released as a gas:

$$4OH^- = 2H_2O + O_2\uparrow + 4e^-$$

Chloride and sulfate ions in the corrosion products go into solution and undergo chemical changes, electrolytic or otherwise. Thus the one important effect of the electrolytic treatment is the removal of harmful chloride and sulfate from the object under treatment.

3. At the cathode, sodium ions from the sodium hydroxide and hydrogen ions from the water, compete for electrons. Less energy is required to reduce hydrogen than to reduce sodium, so hydrogen is reduced at the cathode while the sodium ions remain in solution. The cathodic reaction, therefore, produces free hydrogen:

$$4H^+ + 4e^- = 2H_2\uparrow$$

The other important effect of the electrolysis is the evolution of hydrogen gas on the surface of cathode, i.e. of the object being treated. The evolved hydrogen reduces the metallic corrosion products to finely divided or spongy metal, which can later be removed mechanically or by other means. The hydrogen bubbles themselves help to loosen gross incrustations. Electrolytic treatement reveals most of the details, markings, and inscriptions made on ancient metal objects, if these are still preserved to any extent under the corrosion crust.

Electrochemical Reduction

The relative ability of metallic elements to undergo oxidation depends on the ease with which they lose electrons. If, for example, a polished iron nail is dipped into a solution of copper sulfate ($CuSO_4$), the copper ions (Cu^{2+}) in solution will take up electrons from the atoms of the nail. Then they will precipitate out as atoms of metallic copper (Cu^0), whereas metallic iron atoms, having lost electrons, will pass into solution in the form of ferrous ions (Fe^{2+}). On withdrawing the nail from the solution it can be seen that copper metal has deposited on the iron surface. If left long enough in the solution all the iron in the nail wil be replaced by copper in this way. This is because iron atoms lose electrons more readily than copper atoms, that is:

$$Fe^0 + Cu^{2+} \rightarrow Fe^{2+} + Cu^0$$

In this reaction the iron loses electrons (i.e., it is oxidized) while the copper gains electrons (i.e., it is reduced).

The relative tendencies of the metallic elements to be oxidized have been determined experimentally. Table 14.3 lists metals in order of their increasing "electrochemical potential", the magnitude that measures the tendency to lose electrons (i.e., to be oxidized). Elements high in the table, such as gold and silver, have low potentials. They are *noble* metals, that is, are very resistant to oxidation. Those lower down have higher potentials. They have baser properties and are more readily oxidized (corroded).

When two different metals are in contact with each other in the presence of an electrolyte, such as the solution of a salt, the baser metal is oxidized preferentially, the nobler metal surviving by *cathodic protection*. There follows the possibility of removing the corrosion deposit on a metal object by placing it in contact with a baser metal and then immersing the couple in an electrolyte solution. Under these conditions the metal ions in the corrosion products are reduced to a metallic condition, whereas the baser metal is oxidized. If M_C is a corroded metal and M_B is a baser metal, the oxidation will proceed according to the equation:

$$M_C^{2+} + M_B = M_C + M_B^{2+}$$

The overall result of this *electrochemical reduction* is that the corrosion products are eliminated: the oxidized metal ions are reduced to metal powder that can be easily brushed away; the anions—mainly chloride and sulfate—combine with the cations that result from the oxidation of the baser metals.

Other specialized methods are available for treating particular metals; however, the two outlined in the preceding paragraph form the basis of the most useful general methods for the reduction of corroded metal objects (Plenderleith 1966; Plenderleith and Torraca 1968; Stambolov 1968).

Much can be done to preserve metal antiquities by the rational application of the principles of cleaning and stabilization just mentioned. There are, however, many pitfalls and ill-judged attempts at cleaning and treatment carried out by inexperienced practitioners can do much more harm than no treatment at all.

Protection of Cleaned Metal Objects

Cleaned, reduced, and stabilized metal antiquities are in need of protection to prevent the recurrence of corrosion. Stabilizing measures are required to prevent the detrimental effect of airborne contaminants. *Air conditioning* keeps the water-vapor content of the air low. *Impregnation* with a layer of some corrosion-protecting material prevents the penetration of contaminants.

Coating of clean metal objects with a satisfactory protective layer is not always as simple as might be thought. The coated object must be thoroughly covered, yet it should not look as if it has been coated. Coating with paraffin, vaseline, an similar substances seldom provides adequate protection, unless these are applied in thick layers that spoil the object aesthetically, making it look greasy and damp.

A very thin coating of a substance providing an effectively impervious layer 20-25 μ thick may afford adequate protection yet not be asthetically disturbing (Hofmann 1965). Solutions of either natural or synthetic resins may be applied with good results; after evaporation of the solvent, a very thin, effective coating is formed (Organ 1963; Ypey 1964; Moncrieff 1966). Mixtures of resins and waxes usually combine the properties of toughness, plasticity, and adherence. Micro-crystalline wax (Plenderleith 1966), vinyl resins like chloride-acetate copolymers, and cellosic ethers such as ethyl cellulose (Burns and Bradley 1962) are very successful and make good coatings.

ORGANIC MATERIALS

Leather

Leather exposed over long periods to very dry conditions (i.e., relative humidities below 40 percent) becomes dry, hard, and brittle. Impregnation with a suitable grade of polyethylene glycol restores its flexibility, permitting unfolding and reshaping (Werner 1968). The same chemical can be successfully applied to waterlogged leather. Being hydrophilic, (i.e., having a strong affinity for water) polyethylene glycol removes the water and consolidates the leather.

Textiles

Conservation measures for archaeological textiles include cleaning, sterilization against biological attack, protection from environmental dangers, and reinforcement to consolidate weakened material (Leene 1972).

Cleaning is best performed using water soluble detergents. Household detergents contain many active additives that may damage frail, delicate fabrics and are thus not suitable for cleaning archaeological specimens. Neutral, nonionic detergents are efficient and harmless and hence most suitable for use with ancient textiles (Plenderleith 1966; Beecher 1968). Textiles which should not be wetted for any reason are best dry-cleaned with organic solvents.

Clean textiles must be protected from attack by living organisms and microorganisms to which they are vulnerable. The growth of molds is best prevented by keeping the relative humidity of the atmosphere at a low level (i.e., by air conditioning). Fumigation eliminates existing infestation of molds, moths, and beetles. Gases commonly used for fumigation include methyl bromide, hydrocyanic acid, and sulfuryl fluoride. Fumigation provides no safeguard against subsequent attack. Preventive measures include the application of insecticides and the use of organic insect repellants such as paradichlorobenzene and naphthalene.

Fastening a weak fabric to a strong backing or impregnating its fibers with a material that will set to a tough, flexible solid, are methods usually employed to consolidate textiles and to arrest further disintegration.

Bone and Ivory

The condition of bone and ivory antiquities may vary greatly, depending on the environmental conditions where they have been kept, such as the nature of the soil in which they have been buried, or prolonged exposure to heat or sunlight. A chalky (alkaline) soil will dissolve the organic matrix; acid conditions may cause fragility due to the absorption of soluble salts, and wet soil may cause bone to become soft.

Prolonged washing will remove absorbed salts. Incrustations of insoluble salts call for mechanical removal or treatment with acids. However, if applied in a uncontrolled way, the latter may totally disrupt the object. This may be avoided by resort to local cleaning, making sure that just sufficient acid is applied to soften the incrustation on a limited area. An operation of this sort should be carried out using a strong magnifying glass or microscope.

Clean bone and ivory objects can be consolidated by impregnating them with some resin applicable in aqueous solution (Plenderleith 1966; Werner 1968; Rixon 1976).

Wood

Ancient wooden objects may have suffered from biological attack and from the damaging action of a hostile environment. These result in a loss of mechanical strength. The consolidation of such antiquities involves the use of liquid consolidants for impregnation, solid ones to fill in gaps left by material that has rotted or otherwise been lost, and adhesives for joining parts of objects that have become detached.

The use of consolidants applied in solution raises the problem of shrinkage, which occurs when a liquid sets to the solid state. Shrinkage is likely to cause distortion of fragile objects. The chemical reactions involved in the setting of epoxy or polyester resins is such that no volatile matter is given off, and as a consequence, no appreciable shrinkage or distortion take place. This makes them particularly suitable for the consolidation of wooden antiquities.

Waterlogged Wood.   Wood excavated from wet soil—*waterlogged wood*— may have a water content exceeding the total mass of dry matter present. The cellulosic components of wood undergo biological degradation in wet soil, and if degradation has reached an advanced state, the wood may be very soft and have very low mechanical strength.

The preservation of waterlogged wood was a long-standing problem for the archaeologist: when removed from its site it underwent severe changes in shape and size thus developing cracks and flaking. several special techniques of consolidation have been developed (Rosenquist 1959; Seborg and Inverarity 1962; McCrawley, 1977). Soaking waterlogged wood in an aqueous solution of polyethylene glycol has proved to be a very valuable treatment: the water in the wood is replaced by the polyethylene glycol that fills up voids. Polyethylene glycol dries into a flexible consolidant, very compatible with wood. The grade of polyethylene glycol most suitable for the treatment of waterlogged wood is grade 4000, which is a hard, white, nonhygroscopic solid (Werner 1968). Spongy waterlogged wood treated in this way can be freely handled after drying, and objects that would otherwise be impossible to save can be reconstructed (Bouis 1973).

GLOSSARY—CONSERVATION

Acrylic resins.   Synthetic resins derived from acrylic acid ($CH_2 = C-C-OH$)
$$CH_3 \quad O$$
or from related compound such as acrylic ester and acrylonitrile. Acrylic resins (e.g., ''Perspex'') can be highly transparent (Gaynes 1976).

Adhesives.   Materials that can be prepared and stored in plastic form and that harden on application. They are used for binding together solid surfaces or bodies. An important property of adhesives is their propensity to shrink on setting, thus causing stresses that can weaken or distort the bonded joint. The less the shrinkage the more valuable the adhesive. A useful way of classifying adhesives is according to the various ways in which they set: (1) setting due to loss of solvent, [e.g., polyvinyl acetate (PVA)], (2) setting due to change in temperature (e.g., sealing wax), (3) setting due to change in temperature accompanied by a loss of solvent, (e.g., most animal glues), (4) setting due to chemical reactions (e.g., polyesters and epoxy resins). Adhesives of the last group shrink very little, if at all, on setting, and are especially valuable in conservation.

Anion. A particle in an electrolyte that carries a negative electric charge and deposits at the anode (q.v.) when a current flows through the electrolyte.

Anode. The positive terminal in an electrolytic cell.

Benzotriazole. An organic chemical compound used for the stabilization of copper and bronze antiquities. The formula of benzotriazole is:

$(C_6H_4NHN:N)$

The chemical mechanism of the protective reaction of this compound is not yet completely understood. There is general agreement, though, that it is strongly adsorbed on to metal surfaces, forming a protective film that is stable and chemically inert. This surface film is believed to be polymeric in structure. It is insoluble in water and in most organic solvents. Benzotriazole protects clean copper surfaces from tarnish and prevents further corrosive attack on surfaces covered with corrosion products (Madsen 1967, 1971; Greene 1971, 1972, 1975; Walker 1976).

Calgon. A proprietary name for sodium hexametaphosphate $[(NaPO_3)_6]$, which is used in water solution for the removal of calcareous deposits from metal antiquities.

Cathode. The negative terminal in an electrolytic cell.

Cation. A particle in an electrolyte that carries a positive electric charge and deposits at the cathode when a current flows through the electrolyte.

Cellulose acetate. A tough plastic material derived from acetic acid and cellulose and compounded with suitable plastisizers. Used in conservation for the repair of porcelain, pottery, and glass in the form of a solution in an appropriate solvent.

Cellulose acetate butyrate (CAB).   A material similar in properties to cellulose acetate but tougher and having better moisture resistance and dimensional stability.

Cellulose adhesives.   Water-soluble adhesives extensively used in conservation. Ethyl cellulose, methyl cellulose, and sodium carboxymethyl cellulose are useful as general purpose adhesives in the conservation of textiles.

Cellulose nitrate (nitrocellulose).   A tough plastic material similar in physical properties to cellulose acetate. It is derived from nitric acid and cellulose.

Curing.   A change in the physical properties of a plastic material (setting, hardening, etc.) brought about through chemical reaction. This is sometimes accomplished by the action of heat with or without pressure.

DDT.   Dichlorodiphenyltrichloroethane, a white crystalline compound highly toxic to many insect species.

Detergents.   Substances used for their cleansing action. Detergents are capable of dislodging and removing dirt from a surface. The oldest and best known class of detergents are the *soaps*, the sodium salts of certain organic acids, usually natural fats. Synthetic soap substitutes, or *soapless detergents*, are also much used. They are more effective cleaning agents than soap. Synthetic detergents may be grouped in three major categories, depending on whether the "active" group in the molecule is ionic (positive or negatively electrically charged) or whether it is non ionic (without an electric charge). There are several hundred kinds of synthetic detergents, sold under many trade names.

Commercial cleansing powders should never be used for conservation work; they may contain additives harmful to delicate materials. Soaps have many drawbacks for they form insoluble stains that are difficult to remove. Ionic detergents may be deleterious to fine textiles. Nonionic detergents are the safest for cleaning ancient textiles.

Dip impregnation.   Application of a protective layer by dipping an antique object into a bath of molten resin or wax.

Efflorescence.   The outgrowth of saline matter on stone and similar surfaces.

Electrolysis.   A process by which chemical reactions are carried out by means of a electric current. Ionizable compounds, for example, are split into positive and negative ions that migrate to and collect at the electrodes.

Electrolyte.   A solution which conducts electricity by the movement (migration) of ions.

Epoxy resins.   Plastic materials available as liquid and solid resins. They consist of two components: one based on ethylene oxide (CH_2–O–CH_2) and having epoxy rings ($-\overset{\text{O}}{\underset{}{\text{C}}}=\text{C}-$) that can react with a second component, a suitable hardener. Epoxy resins set to a hard solid mass without the evolution of volatile matter. Consequently, setting takes place without shrinkage, and no strains are set up in the treated object. Epoxy resins find wide use as adhesives and in coatings, castings, and encapsulations.

Fluoroplastics. A family of plastics [e.g., polytetrafluoroethylene (PTFE) or "Teflon"] and others, characterized by such properties as nonadhesiveness and thermal and chemical resistance.

Hygroscopic. A term used to describe a material that becomes wet by absorbing moisture from the atmosphere.

Humidity. (1) *Absolute humidity,* the percentage by weight of water vapor in air; (2) *relative humidity,* the water-vapor content of air usually expressed as a percentage of the saturated water-vapor content of air at a given temperature. Air is saturated when it is unable to hold any more water in the form of vapor. On cooling saturated air, the water is at once precipitated in the form of dew.

Inhibitor. A substance that diminishes the rate of a chemical reaction. Rust inhibitors, for example, are used to retard the corrosion of metal antiquities.

Microcrystalline waxes. Semisynthetic waxes produced as by-products in the refining of petroleum. They have a microcrystalline structure that confers on them a plasticity useful in the treatment of ancient skins, metals, and stones.

Monomer. A relatively simple chemical compound that can react to form a polymer (q.v.).

Nitrocellulose. See Cellulose nitrate.

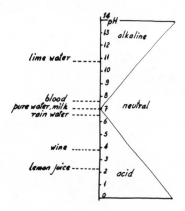

Fig. 15.2. pH of some substances.

*p*H. A term used to express the acidity or alkalinity of solutions. It is the reciprocal of the logarithm of the hydrogen ion concentration ($[H^+]$) expressed in gram-ions per liter, thus:

$$pH = \log \frac{1}{\log [H^+]}$$

A completely neutral material (e.g. pure water) has a pH of 7. The pH of some common materials is given in Fig. 15.2.

pH of Various Materials	
Distilled water	7
Strong acids	1
Strong alkalies	14
Acid soil	3-7
Chalky alkaline soil	8-10
Seawater	7.7-8.3
Cotton	6-8
Oak	5-8

Polyester resins. Plastic materials formed by the reaction between polybasic acids and polyhydroxy alcohols. A large variety of products are available, depending on the nature of the reacting components. Polyesters set without loss of volatile material and without shrinking. They are widely used as adhesives and coatings in conservation.

Polyethylene. A widely known plastic formed by the polymerization of ethylene ($CH_2=CH_2$) molecules. Used to make containers of all kinds.

Polyethylene glycol. A series of polymers of ethylene glycol $O \cdot CH_2 \cdot CH_2 \cdot O \cdot CH_2 \cdot CH_2$. The lower members of the series (those of low molecular weight) are viscous liquids, the intermediate members are pastes, and the higher members (having high molecular weights) are solids having the physical appearance of waxes, although they are readily soluble in water. Polyethylene glycol waxes are used to restore the flexibility of fragile, brittle leather. Heavier grades replace water in the conservation of waterlogged wooden objects.

Polymer. A substance composed of large molecules built up from a number of identical simple molecules that join together—polyermize—to form long chains.

Polymerization.   A chemical reaction in which two or more molecules of the same compound—monomer—combine to form a new substance—polymer—with different properties from the monomer (see Fig. 15.3). When two or more different monomers are involved in a given reaction, the process is called *copolymerization.*

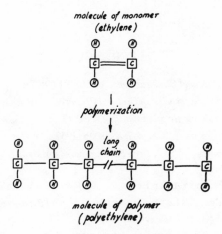

Fig. 15.3.   Polymerization.

Polyvinyl acetate (PVA).   A plastic material composed of polymers of vinyl acetate. It is a colorless solid extensively used in conservation for adhesion and coating.

Polyvinyl alcohol.   A resin produced by the partial or complete hydrolysis (reaction with water) of polyvinyl acetate. Widely used as a water-soluble adhesive and coating. It has been used for glueing fragile textiles to plastic supports.

Resin.   A class of solid or semisolid organic products (usually polymers) of natural or synthetic origin (Mills and White 1977).

Rochelle salt.   A double tartrate of potassium and sodium ($KNaC_4H_4O_6 \cdot 4H_2O$) used in the cleaning of ancient metals.

Set.   Conversion of a liquid resin or adhesive into the solid state by curing or by the evaporation of a solvent or suspending medium.

Silica gel.   Product of the decomposition of sodium silicate that has a strong affinity for moisture and is used as a drying agent.

Silicon esters. The name given to the silicates of organic products prepared by the controlled hydrolysis of ethyl silicates and used as preservatives for stone.

Soluble nylon. A modified nylon polymer that can be dissolved in alcohols and in solvent mixtures containing alcohols. Chemically, it is n-methoxymethyl nylon. It is used as an adhesive, as a size for paper, and as a consolidant for fragile surfaces.

REFERENCES

Amoroso, G. G., *Materiaux et Constructions,* **10**(56), 91 (1977).

Beecher, E. R., in *The Conservation of Cultural Property,* UNESCO, Lausanne., 1968.

Bouis, J., *Cahiers d'Archeologie Subaquatique,* 2, 1973.

Burgess, G., and Schaffer, R. J., *Chem. Ind.,* 1026 (1952).

Burns, R. M., and Bradley, W. W., *Protective Coatings for Metals,* Reinhold, New York, 1962.

de Witte, E., Huget, P., and van den Broek, P., *Studies in Conservation,* **22,** 190 (1976).

Fink, C. G., and Eldridge, C. H., *The Restoration of Ancient Bronzes and Other Alloys,* The Metropolitan Museum of Art, New York, 1925.

Gairola, T. T., *Handbook of Chemical Conservation of Museum Objects,* University of Baroda, Baroda, 1959.

Gaynes, N. I., *Metal Finishing,* **74**(10), 47 (1976).

Greene, V., Thesis, Institute of Archaeology, London University, 1971.

Greene, V., in *ICOM Committee for Conservation,*Working Group on Metals, Madrid, 1972.

Greene, V., in *Conservation in Archaeology, IIC Congress Proceedings,* Stockholm, 1975.

Hofmann, F., *Metallwissenschaft und Technik,* **10,** 1065 (1965).

Jedrzejewska, H., in *New York Conference on Conservation of Stone and Wooden Objects,* IIC, 1970, p. 19.

Leene, J. E., *Textile Conservation,* Butterworths, London, 1972.

Lewin, S., in *Conference on the Weathering of Stones,* Council of Monuments and Sites, ICOMOS, Paris, 1968, p. 41.

Madsen, H. B., *Studies in Conservation,* **12,** 163 (1967).

Madsen, H. B., *Studies in Conservation,* **16,** 120 (1971).

Madsen S. T., *Restoration and Anti Restoration,* Universitets Forlaget, Oslo, 1976.

McCawley, J. C., *J. Canadian Conserv. Inst.,* **2,** 17 (1977).

Mills, J. S., and White, R., *Studies in Conservation* **22,** 12 (1977).

Moncrieff, A., *IIC News,* **4,** 6 (1966).

Moncriff, A., *Studies in Conservation,* **21,** 179 (1976).

North, N. A., and Pearson, C., in *Conservation in Archaeology, IIC Congress Proceedings,* Stockholm, 1975.

Olive, J., and Pearson, C. in *Conservation in Archaeology, IIC Congress Proceedings,* Stockholm, 1975.

Organ, R. M., *Museums Journal,* **55,** 112 (1955).

Organ, R. M., in G. Thomson, ed., *Recent Advances in Conservation,* Butterworths, London, 1963a, pp. 104, 128.

Organ, R. M., *Studies in Conservation,* **8,** 6 (1963b).

Pelikan, J. B., *Studies in Conservation,* **11,** 109 (1966).

Plenderleith, H. J., *The Conservation of Antiquities,* Oxford U. P., London, 1966.

Plenderleith, H. J., and Toracca, G., in *The Conservation of Cultural Property,* UNESCO, Lausanne, 1968, p. 237.

Rixon, A. E., *Fossil Animal Remains: Their Preparation and Conservation,* The Athlone Press, London, 1976.

Rosenquist, A. M., *Studies in Conservation,* **4,** 13 (1959).

Schmidt-Thomsen, K., *Deutsche Kunst. und Denkmalpflege,* 11 (1969).

Schreir, L. L., *New Scientist,* 332 (1964).

Seborg, R. M., and Inverarity, R. B., *Science,* **136,** 650 (1962).

Sleater, G. A., *Stone Preservatives,* N.B.S. Tech. Note No. 941, U.S. Dept. of Commerce, Washington, D.C., 1977.

Stambolov, T., *The Corrosion and Conservation of Metallic Antiquities,* Central Research Laboratory for Objects of Art, Amsterdam, 1968.

Thouvenin, A., *Revue Archaeologique (Paris),* **2,** 180 (1958).

Thouvenin, A., *Berliner Blater fur Vor-und Fruhgeschichte,* **8,** 63 (1959).

Walker, R., *Corrosion,* **32**(10), 414 (1976).

Werner, A. E., *Museums Journal,* **64,** 5 (1964).

Werner, A. E., in *The Conservation of Cultural Property,* UNESCO, Lausanne, 1968, p. 265.

Wihr, R., in *Conservation in Archaeology, IIC Congress Proceedings,* Stockholm, 1975.

Ypey, J., *Das Preparator,* **10,** 39 (1964).

DATING

The evolution of man can be understood properly only if the time element is taken into consideration. Dating the past has been a preoccupation of historians since history began to be recorded. For the archaeologist, the dating of finds is of fundamental importance; he is continually looking for better methods of determining the time and sequence of events not described in written records.

Events may be located in time by two kinds of dates: *absolute*, or *direct* dates, which are expressed in terms of distance in years from the present time; and *relative*, or *indirect* dates, which are expressed in relation to other events. Obviously, the most desirable chronology is that obtained by absolute dating: contemporaneity between distant events can then be determined.

The archaeologist's primary chronological tool is *stratigraphy*. If disturbance of deposits has not occurred, a sequence of objects can be established, based on the superimposition of younger strata on older: and a satisfactory relative time scale can then be derived. Where superimposition is lacking, chronology can be established by *seriation*. Once the typological variations of an object have been classified they often fall into a developmental series. A mathematical approach to seriation has been evolved.

Another approach to chronology is *chemical dating*, that is, the determination of the age of ancient materials and artifacts by measurement of their chemical composition or physical properties. A particularly impressive effort has been made to develop *radiometric* method of age determination. When the relative amount of a radioactive isotope, or the ratio of radioactive parent to stable daughter isotope have been determined in a material, an "age" can be calculated. This age has a particular meaning only if the material under study has been a "closed system" with respect to the elements involved. The main problem facing the chronologist is to devise means for determining whether the material has or has not been a closed system. Most of the difficulties encountered in radiometric dating arise out of this problem.

Four main chemical dating methods can be distinguished:

1. methods making use of the relative amounts of a stable element or compound present in a material (e.g., uranium or aminoacid dating),

2.   methods based on the relative concentration of a radioactive isotope as measured by its present activity (radiocarbon dating is the best known),

3.   methods making use of the ratio of radioactive parent:stable daughter element (i.e., the ratio of radioactive elements still present to the products of radioactive decay; potassium-argon dating comes into this category), and

4.   methods based on the measurement of the damage to materials caused by radioactive decay (e.g., thermoluminescence and fission-track dating).

Most chemical methods of dating furnish absolute dates. All of them involve fewer assumptions than do archaeological dating methods. Each method, however, can only be applied to a limited number of materials.

The past decade has seen advances in the application of chemical dating methods in archaeology. These advances are largely due to improvements and innovations in analytical methods.

The use of chemical dating, and particularly of nuclear methods, has changed many long-held notions about the timetable of human development. Dates hitherto universally accepted have been radically adjusted. As a consequence of radio-carbon dating, the end of the Ice Age has been brought forward and the beginning of urbanization has been pushed backward. Potassium-argon and fission-track dating have changed assumptions about the age of man, which has been shown to be very much greater than the half million years accepted as late as 1960. The acceptance and utilization of these new techniques by archaeologists and anthropologists are rapidly leading to a new, more precise ordering of the sequence of human events than was ever possible before.

CHEMICAL DATING OF BONES AND FOSSILS

Bone consists of two structural components: inorganic mineral and organic tissue. Compositional differences exist according to species, sex, and so on, but significant quantitative facts are general and well known. The typical composition of fresh bone is given in Table 16.1. Bone mineral is made up chiefly of the elements calcium, phosphorus, hydrogen, and oxygen: the atoms are symmetrically arranged in a lattice, now generally considered to be hydroxyapatite, a calcium phosphate of formula $Ca_{10}(PO_4)_6(OH)_2$. The main organic components of bone are fat and collagen. The latter is a protein which forms the fibrous tissues of bone. Microscopic examination of bone tissue shows that the collagen fibers are organized in bundles, around which crystals of calcium phosphate are packed (see Fig. 16.1).

TABLE 16.1   Typical Elements
in Fresh Bone

Element	Percent
Calcium	20-22
Phosphorus	10
Fat	10
Water	10
Nitrogen	6
Iron	Traces
Fluorine	Traces

The interment of bones after death leads to a slow process of chemical alteration, commonly called *fossilization*. Fossilization is governed by physical, biological, and chemical factors that vary according to the nature of the surrounding soil. Physical factors include ground water flow and degree of aeration; among biological factors the most important is bacterial activity. The chemical components of the soil are determined by the parent rock, the surrounding vegetation, and the process of soil formation. There is, in consequence, no such thing as a

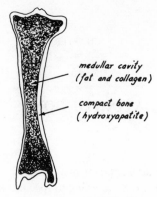

medullar cavity
(fat and collagen)

compact bone
(hydroxyapatite)

Fig. 16.1. Longitudinal cross section of a bone, illustrating the relationship between hard and soft tissues.

simple or uniform process of fossilization: *fossilization* can thus only be defined in general terms as whatever happens bone in the place of burial.

During fossilization the inorganic compounds present in bone may be either diminished—if conditions are such that hydroxyapatite is broken down—or augmented—if, for instance, the bone should lie in a very calcareous soil where concretions could deposit within its lattice. Organic matter, on the other hand, decomposes, and this decomposition is accompanied by changes in both the relative and the absolute quantities of its component elements.

For whatever physical and chemical reasons, bone appears to be a very favorable substrate for the deposition of a number of minor elements. It seems reasonable to assume, therefore, that the chemical analysis of bones could yield clues as to their age. The course of bone alteration cannot be measured against any absolute and generally valid time scale; chemical analysis does not establish whether two bones, found in widely separated areas, were formed or buried at the same time. Nevertheless, an evaluation of the parameters involved, combined with the availability of local controls, makes possible the relative dating of the bones.

Chemical methods for the dating of bones use the relative concentration of some stable compound or element as an index to the age of a material. They include: (1) fluorine dating, (2) uranium dating, (3) nitrogen dating, and (4) amino-acid dating.

Other elements and chemical groups of geochronological interest are sometimes determined: total and organic carbon, iron, phosphate, and carbonate (Glover and Phillips 1965; Sietz and Taylor 1974). It has been suggested that if yttrium is present in the environment, the yttrium content of fossil bone might serve as an indicator of age (Parker and Toots 1970). Manganese has also been proposed as an age indicator, though researchers are not all in agreement about this (Shimoda,

Endo, et al. 1964; Shimoda and Ozaki 1967; Parker and Toots 1970). Most of the work published till now, however, has been concerned with the methods discussed in the following pages.

FLUORINE DATING

The oldest of the chemical methods of dating bone is *fluorine dating*, based on the progressive increase in the fluorine content of bone with increased time elapsed since interment (Middleton 1844; Carnot 1892, 1893).

The increase in fluorine results from the contact of bones with groundwater containing fluorides. Hydroxyapatite—the main inorganic component of bone—is slowly altered through the exchange of hydroxyl (OH^-) ions for fluoride (F^-) ions:

$$Ca_{10}(PO_4)_6(OH)_2 + 2F^- \rightarrow Ca_{10}(PO)_4)_6F_2 + 2OH^-$$

The result of the exchange is the formation of fluorapatite, $Ca_{10}(PO_4)_6F_2$, a compound much more stable and less soluble than hydroxyapatite. Once fluorine has been fixed in bone, therefore, it will not easily be dissolved out.

The rate of increase of fluorine content of bones varies from site to site. Fluorides may be relatively abundant in groundwater in one locality and practically absent in another. Thus very old bones in a fluorine-poor site may acquire less fluorine than recent bones in a fluorine-rich site. Calcite deposition on bones buried in calcareous soils may interfere with the process, by *sealing* the bone and rendering it impervious to penetration by groundwater. Nevertheless, the results of fluorine analysis can given a clue as to the age of the bones. This is shown in Tables 16.2 and 16.3. It should be emphasized once again that fluorine analysis does not provide the means for estimating absolute age, but only the possibility of dating bones relative to others found in the same site. Therefore, attempts to cross-apply the results of one site with those of another should not be relied on.

TABLE 16.2 Fluorine in Fossil Bone

Bone Sample	Maximum Age (Years)	Fluorine Contents (Percent)	Fluorine: Phosphate Ratio
Recent	11,000	<0.3	0.058
Pleistocene	2,000,000	1.5	0.390
Tertiary	30,000,000	2.5	0.650
Mesozoic	130,000,000	3.5	0.907
Flourapatite	—	3.8	1

TABLE 16.3 Fluorine Analysis of Bone Samples Provenant from the Caribbean Region[a]

Sample	Age (Years)	Fluorine (Percent)
Contemporary	<10	<0.02
Montalvan	<100	0.15
El Chao, 1	700	0.06
El Chao, 2	700	0.04
Los Paredones	1000	0.15
La Betania, 1	1700	0.25
La Betania, 2	1700	0.35
Taima-Taima	14000	1.0
Muoco	16000	1.1

[a]Data From Tammers (1969).

Bones buried in sandy or gravelly soil are most suitable for relative dating by the fluorine method. One the other hand, bones provenant from volcanic soils will not yield useful dates. Bones from calcareous caves are of little use, since calcite deposits on the surface of the bone may have prevented fluorine penetration.

The fluorine method of dating is subject to many uncertainties, such as variations in the fluorine content of ground waters, or even to variations in the rate of flow of the water. In many contexts, unresolved geological and analytical problems make the method useless as a chronometric tool (McConnell 1962). Dates obtained from fluorine analysis results should therefore, be used with great caution.

URANIUM DATING

Trace amounts of uranium enter the living body via the normal intake of food and water and are incorporated into the blood. Uranium circulating in the blood stream is exchanged with calcium atoms in the bones and becomes fixed in them. In this way the skeletons of all living animals come to contain minute amounts of uranium.

Natural uranium is a mixture of several isotopes (uranium-235, uranium-238, etc.), most of which are radioactive. The concentration of uranium can thus be determined by measuring radioactivity emitted. As uranium approaches equilibrium with its daughter elements, there is and increase in radioactivity over a period of several thousand years. The content of disintegration products does not reach a maximum until equilibrium is reached, that is, until the amount of each transient daughter element newly generated in a given time is equal to the amount lost in the same time by decay. Hence if uranium were neither added nor removed from bones after burial, the absolute age of the bones could be determined by measuring the increase in their radioactivity (Bowie and Davidson 1955).

Unfortunately, however, the amount of uranium in bones increases during interment, in the same way as in the living body; calcium is replaced by uranium in bones buried in sites where the groundwater contains the latter. The longer the time during which the bone remains in contact with uranium-charged groundwater, the more uranium it will contain. As in the case of fluorine, the rate of increase is dependent on environmental conditions: soil permeability, hydrological flow, and concentration of uranium in the groundwater. Analytical results show that although bones from different sites vary in their uranium content, there is a rough correlation between uranium content and age: different bones of the same age from a single site have also the same levels of radioactivity. Thus the radiometric assay of uranium can be useful for distinguishing older from younger bones at the same site; but no more than this (Szabo, Malde, et al. 1969; Howell, Cole, et al. 1972). The concentration of uranium is most affected by fossilization; also, it seems that uranium is absorbed in bones during the decay of organic material and is lost after decay ceases. If these loses are accounted for, the concentration of uranium may be used to date skeletal material accurately.

NITROGEN DATING

Proteins are the main organic components of fresh bone. *Collagen,* a fibrous protein comprising more than 90 percent of the organic matter in bone, is resistant to decay after death and disappears very slowly. It may survive burial for thousands of years. Amino acids, the constituents of collagen, might persist almost indefinitely: collagen fibres have been found, for example, in upper Pleistocene fossil bones (Abelson 1954; Garlick 1969).

Collagen is gradually lost from buried bones after death. The rate of loss is dependent on the physical, chemical, and bacteriological conditions of the buried bone's environment. The loss is relatively rapid in the period immediately after burial. Following the initial drop, which lasts a few hundred years, the loss of collagen continues at a relatively slow, uniform, and consistent rate. Hence the relative ages of bones can be determined by comparing their respective collagen contents.

Nitrogen is a basic component of collagen. As fossil bone protein disappears, there is also a change in the quantity of nitrogen found by direct analysis (Barber 1939). Nitrogen determinations are useful as cross checks on the results obtained from fluorine and uranium analysis; Table 16.4 for example, summarizes the results of fluorine, uranium, and nitrogen analysis made on fresh, ancient, and fossilized bones.

TABLE 16.4   Fluorine, Uranium, and Nitrogen Contents of Bone

Sample	Fluorine (Percent)	Uranium (ppm U_3O_8)	Nitrogen (Percent)
Fresh bone	0.03	0	4.0
Neolithic skull	0.3	—	1.9
Epimachaerodus tooth (upper Paleolithic)	1.6	30	—
Swanscombe skull (lower Paleolithic)	1.7	27	Traces
Elephas planifrons (Pleistocene)	2.7	610	Traces

[a]Data From Oakley (1963).

Apart from collagen, fats are the main organic components of bone. After death and interment fats undergo *autolyis* (chemical breakdown caused by dead cell sera) and are rapidly lost. However, traces of fat remaining in fossil bone undergo changes that can be followed on a progressive scale of time (Gaugh 1936). The data obtained are far from being precise but are consistent with results obtained from fluorine analysis. It is possible that further research into the subject might lead to interesting conclusions.

RACEMIZATION OF AMINO ACIDS IN BONES

Many organic compounds exhibit a property known as *isomerism*. Isomers are compounds whose molecules contain the same atoms arranged in different ways. One type of isomerism in *optical isomerism*.

The carbon atom can be likened to a regular tetrahedron, with one valency bone at each apex. When four atoms or molecular groups, each different from the other, are attached to a given carbon atom, they may be located in two different ways relative to each other, as illustrated in Fig. 16.2. A carbon atom linked in this manner to four different groups has neither a plane nor a centre of symmetry and is called an *asymmetric carbon atom*. The two similar (but not identical) compounds thus formed are called *optical isomers* or *enantiomers*; each of the two optical isomers is a mirror image of the other. To distinguish between the optical isomers of a compound, the prefixes D- and L- respectively are used; thus D-lysine and L-lysine, D-arginine and L-arginine, and so on. The prefixes D-and L- are arbitrarily allocated, by international agreement.

The D- form of an optical isomer can be transformed into the L- form, and vice versa. A mixture of the D- and L- isomers of a compound is called a *racemic mixture,* and the transformation of a single isomer into a racemic mixture is called

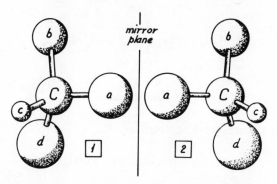

Fig. 16.2. Optical isomers. Mirror image forms of molecules of the same compound, such as (1) and (2) are called enantiomers.

racemization. Many substances racemize spontaneously at ordinary temperatures.

Amino acids, the building bricks of naturally occurring proteins, exhibit optical isomerism; they may exist in either the D- or L- form. For reasons not yet understood, the amino acids in living tissues consist almost exclusively of the L-form. After an organism dies, however, racemization begins and part of the amino acids in it changes into the D-form.

The rates of racemization of some amino acids are slow enough to be of use in geochronology. Amino acids in shells, for example, are partially racemized, with the amount of racemization increasing with the age of the shell (Hare and Mitterer 1969); in shells of Miocene age (i.e., ca. 30,000,000 years old), racemization is complete.

Isoleucine, one of several amino acids isolated from fossil bones, shows a progressive increase in the degree of racemization, which depends on the age of the fossils. This amino acid's racemization is slow, in geochronological terms: its *half-life* at 20°C is in excess of 100,000 years and can be used to date bones too old to be datable by radiocarbon (Bada 1972). The product of the racemization of isoleucine is a nonprotein amino acid, D-alloisoleucine. Both optical isomers have the formula $C_6H_{13}O_2N$ (1-amino-2 methyl-n-valeric acid); they only differ in the spacial arrangement of the component groups, one being the enantiomer of the other:

$$CH_3—CH_2—\underset{\underset{H}{|}}{\overset{\overset{CH_3}{|}}{C}}—\underset{\underset{NH_2}{|}}{CH_2}—CO_2H$$

L-Isoleucine

$$
\begin{array}{c}
\text{H} \\
| \\
\text{CH}_3\text{—CH}_2\text{—C—CH}_2\text{—CO}_2\text{H} \\
| \quad | \\
\text{CH}_3\text{NH}_2
\end{array}
$$

D-Alloisoleucine

The racemization reaction can be written as:

L-Isoleucine $\rightleftharpoons$ D-Alloisoleucine

The reaction is highly sensitive to temperature, being faster at higher temperatures. Therefore, for the racemization of L-isoleucine to be applicable to the dating of bones, the temperature history of their environment must be known fairly accurately. For example, an uncertainty of $\pm 2°C$ in the temperature history of the bones could lead to an error of ± 50 percent in the age determined.

Among the places where temperatures would be expected to vary little over long periods are ocean deeps and the insides of caves. Palaeotemperatures investigations suggest, in fact, that the latter provide essentially constant environments. If temperature changes have occurred in them, nonetheless, it is still possible to determine the actual variation from the oxygen isotopes ratio in speleothems (cave deposits) (Hendy and Wilson 1968; Duplessy, Labeyrie, et al. 1970).

Cores raised from the deep-sea floor have been dated (Hare and Mitterer 1969; Bada, Luyendyk, et al. 1970; Wehmiller and Hare 1971), and the method seems also applicable to the dating of fossil bone (Turekian and Bada 1972). A bone from the Muleta Cave in Majorca, Spain, for example, yielded a L-isoleucine-D-alloisoleucine racemization age of 28,000 years, consistent with that obtained by radiocarbon analysis.

Age determined from the racemization reaction of L-isoleucine may have an uncertainty of ± 20-30 percent. Although this is nowhere near the accuracy obtained with radiocarbon, the method can be used to obtain estimates of the age of bones too old to be dated by the latter.

L-Aspartic acid ($C_4H_7O_4N$, also called *aminosuccinic acid*) is another amino acid found in fossil bone. Its enantiomer is D-aspartic acid.

$$
\begin{array}{c}
\text{NH}_2 \\
| \\
\text{CO}_2\text{H·CH}_2\text{·C·CO}_2\text{H} \\
| \\
\text{H}
\end{array}
\qquad\qquad
\begin{array}{c}
\text{H} \\
| \\
\text{CO}_2\text{H·CH}_2\text{·C·CO}_2\text{H} \\
| \\
\text{NH}_2
\end{array}
$$

L-Aspartic acid   D-Aspartic acid

L-Aspartic acid racemices substancially faster than isoleucine. The racemization reaction can be written:

$$\text{L-Aspartic acid} \rightleftharpoons \text{D-Aspartic acid}$$

In bone, this reaction has a half-life, at 20°C, of 15, 000-20,000 years.

By determining the extent of racemization of L-aspartic acid in bones from a particular site that have previously been dated by the radiocarbon technique, it is possible to calibrate the rate of racemization at that site (obviating the tedious investigation of temperture history that would otherwise be necessary). Once this callibration has been done, the aspartic-acid racemization reaction may be used to date other samples from the area that are either too old to be dated by radiocarbon or of which insufficient amounts are available for radiocarbon dating. This approach was used to date fossil bone from Olduvay Gorge in Tanzania. Bones with aged 5,000-70,000 years were dated, and the actual ages were dependent on the temperature of the site (Bada and Protsch 1973).

TABLE 16.5   Comparison of Amino-acid Racemization and Collagen-Radiocarbon Ages of Bones

Sample	Amino-acid Racemization Age	Collagen-Radiocarbon Age
	Years B.P.[a]	
Asiab, Iran	8,600	8,700 ± 100
Szeleta Cave, Hungary	40,000	43,000 ± 1100
Muleta Cave, Mallorca	18,600	18,980 ± 200
	33,700	28,600 ± 600
Murray Spring, Arizona	10,500	11,230 ± 340

[a]Years before present.

Archaeologists have long believed that men first reached the North American continent about 20,000 years ago. However, when the extent of aspartic acid racemization of human fossils was determined (Bada, Schroeder, et. al. 1974), the results indicated that Palaeo-Indians lived in California at least 48,000 years ago. This figure has been confirmed by radiocarbon dating: on laboratory analysis of tools and bones found in Santa Rosa island, off California, no carbon-14 was found in them (Berger 1978). Since it takes over 40,000 years for an object to lose all the radiocarbon it originally contained, the site and the people who used it must have been at least that old.

By measuring the extent of racemization of bone amino acids that racemize at slower rates than aspartic acid, it should be possible to extend this dating method well beyond the time range to which radiocarbon is applicable. Moreover, because of the small amount of material needed for racemization analysis, even bone recent enough to be datable by radiocarbon but available only in small quantities could be dated (Bada and Helfman 1976).

In summarizing this chapter it can be said that the accumulation, depletion, or alteration of a number of substances in ancient bones is a definable function of time. Ideally, such substances could furnish useful dating information. In practice, the information thus obtained must be interpreted with the greatest caution.

REFERENCES

Abelson, P. H., *Carnegie Institution Yearbook, Washington, D.C.*, **53,** 97 (1954).

Bada, J. L., *Earth and Planetary Science Letters*, **15,** 223 (1972).

Bada, J. L. and Helfman, P. M., *World Archaeol.*, **7**(2), 160 (1976).

Bada, J. L., Kvenvolden, K. A., and Peterson, E., *Nature*, **245,** 308 (1973).

Bada, J. L., Luyendyk, B. P., and Maynard, J. B., *Science*, **170,** 730 (1970).

Bada, J. L. and Protsch, R., *Proc. Nat. Acad. Sci. U.S.A.*, **70,** 1331 (1973).

Bada, J. L., Schroeder, R. A., and Carter, G. F., *Science*, **184,** 791 (1974a).

Bada, J. L., Schroeder, R. A., Protsch, R., and Berger, R., *Proc. Nat. Acad. Sci. U.S.A.*, **71,** 914 (1974b).

Barber, H., *Palaeobiologica*, **7,** 217 (1939).

Berger, R., Personal communication, 1978.

Bowie, S. H. V., and Davidson, C. F., *Bull. Br. Mus. (Nat. Hist.) Geol.*, **2,** 276 (1955).

Carnot, A., *Comptes Rendus*, **115,** 337 1892).

Carnot, A., *Ann. Mines*, **3,** 155 (1893).

Duplessy, J. C., Lalou, C., and Gomes de Azevedo, A. E., *Comptes Rendus, Acad. Sci. Paris, Ser D.*, **268,** 2327 (1969).

Duplessy, J. C., Labeyrie, J., Lalou, C., and Nguyen Hun Van, *Nature*, **266,** 631 (1970).

Gaugh, I., *Osterreich Chem. Zeitung*, **39,** 79 (1936).

Garlick, J. D., in D. Brothwell and E. Hicks, eds., *Science in Archaeology*, Thames and Hudson, London, 1969, p. 503.

Glover, M. J., and Phillips, *J. Appl. Chem.*, **15,** 570 (1965).

Hare, P. E., and Mitterer, R. M., *Carnegie Institution Yearbook Washington, D. C.*, **67,** 205 (1969).

Hendy, L., and Wilson, J., *Nature*, **219,** 49 (1968).

Howell, F. C., Cole, G. H., Kleindienst, M. R., Szabo, B. J., and Oakley, M. P., *Nature,* **237,** 51 (1972).

McConnell, L., *Science,* **136,** 241 (1962).

Middleton, J., *Proc. Geol. Soc. London,* **4,** 431 (1844).

Oakley, K., in D. Brothwell and E. Hicks, eds., *Science in Archaeology,* Thames and Hudson, London, 1963, p. 24.

Parker, R. B., and Toots H., *Geol. Soc. Am. Bull.,* **81,** 925 (1970).

Rogers, H. J., Weidmann, S. M., and Parkinson, A., *Biochem. J.,* **50,** 537 (1952).

Seitz, M. G., and Taylor, R. E., *Archaeometry,* **16,** 129 (1974).

Shimoda, N., Endo, S., Inoue, M., and Ozaki, H., *Tokyo Nat. Sci. Mus. Bull.,* **7,** 225 (1964).

Shimoda, N., and Ozaki, H., *Tokyo Nat. Sci. Mus. Bull.,* **10,** 377 (1967).

Szabo, B. J., Malde, H. E., and Irwin-Williams, C., *Earth and Planetary Science Letters,* **10,** 253 (1969).

Tammers, M. M., *Comptes Rendus,* **268,** 489 (1969).

Turekian K. K., and Bada, J. L., in W. W. Bishop, ed., *Calibration Hominoid Evolution Proceedings Symposium,* Scot. Acad. Press, 1972, p. 171.

Wehmiller, J., and Hare, P. E., *Science,* **173,** 907 (1971).

RADIOCARBON DATING

Radiocarbon dating is a method of determining the absolute age of carbon-containing materials. The method, developed by Libby and his collaborators, involves the comparison of the amount of radiocarbon present in the sample under examination, with that present in contemporary living matter (Libby 1955). The determination of the amount of radiocarbon in the sample, relative to a calibrated scale based on the radioactivity of present-day atmospheric carbon, makes it possible to calculate the age of the sample.

PRINCIPLES OF RADIOCARBON DATING

Radiocarbon, or *carbon-14* (C-14), is a radioactive isotope of carbon. Other naturally occurring isotopes of carbon are carbon-12, the most abundant, and carbon-13, both nonradioactive. The relative number of radiocarbon atoms, compared with "normal" carbon-12, is extremely small: in living matter, for example, only one in 1.000.000.000 carbon atoms is a carbon-14 atom. Radiocarbon atoms can be detected only by means of the radiation that they emit during radioactive decay.

Carbon-14 undergoes nuclear transformation, through the emission of the beta-particle, to nitrogen-14; the emitted beta-particle can be detected with a radioactivity counter. The half-life of carbon-14 is 5730 years ± 40 years. During this period the radioactivity of carbon-14 is reduced by one-half.

Radiocarbon dating depends on the formation of carbon-14 from nitrogen in the upper atmosphere (Labeyrie 1976). Cosmic rays entering the atmosphere from outer space break up atoms in air and generate neutrons. These, on colliding with ordinary nitrogen atoms (nitrogen-14), react with them to form carbon-14:

$$N\text{-}14 \; + \; \text{Neutron} \; = \; C\text{-}14 \; + \; \text{Proton}$$

The rate of radiocarbon formation depends on a number of factors, such as the intensity of cosmic radiation and the intensity of the magnetic field of the earth. But it can be safely concluded that carbon-14 is formed continually at a fixed rate of about 2.4 atoms per second for each square centimeter of the earth's surface.

Carbon-14 atoms react rapidly with atmospheric oxygen to form carbon-14-dioxide:

$$C^{14} + O_2 = C^{14}O_2$$

Carbon-14-dioxide is chemically undistinguishable from "ordinary" carbon-12-dioxide, with which it mixes, being eventually distributed throughout the atmosphere. It also dissolves in seas and oceans, becoming uniformly dispersed in the hydrosphere.

Carbon is a basic component of all living tissues. Plants absorb atomspheric carbon dioxide through photosynthesis, thus incorporating radiocarbon and "ordinary" carbon into their cells in about the same proportion as in the atmosphere. Carbon assimilated by plants is in turn consumed by animals and men. Thus there is a constant influx of carbon-14 into the *"biosphere"* (the group of all living organisms).

The atmosphere, hydrosphere, and biosphere together make up a *"carbon exchange* reservoir." The reservoir contains the three isotopes: carbon-12, carbon-13, and carbon-14. All the isotopes, in different chemical combinations, circulate freely between the components of the reservoir. The time of circulation and mixing within the reservoir varies greatly. It takes only a few years for carbon to circulate between the stratosphere and the atmosphere, but very long periods between other components of the reservoir. It is much less, however, than 5730 years, the half-life of radiocarbon. Thus although there are some differences in isotopic abundance between parts of the reservoir, the mean ratio carbon-12:carbon-13:carbon-14 is constant, approximately 100:1:0.0000000001.

Carbon 14 is the only unstable carbon isotope. It decays by reverting to nitrogen-14 through beta-decay:

$$C\text{-}14 = N\text{-}14 + \beta\text{-Particle}$$

The rate of decay (half-life 5730 years) is the same whatever the material in which the carbon is present and whether in atmospheric carbon dioxide, organic tissues, or in any other form.

As a consequence of radioactive decay, there is continuous completion of carbon-14 from the exchange reservoir. But since the mean isotopic ratio (carbon-12:carbon-13:carbon-14) remains constant, there must be a sufficient amount of carbon-14 on earth to ensure that the rate of decay is just equal to the rate of formation. This amount has been calculated to be 81 tons.

The equilibrium between formation and depletion of carbon-14 is also maintained in living organisms. While an organism is alive it exchanges carbon with the surrounding atmosphere. The amount of carbon-14 per unit mass of living matter (called the *specific activity*) is the same as the carbon-14 activity in the atmosphere.

The withdrawal of any object from the carbon-exchange reservoir—in the case of living matter through death—brings to a halt the intake of radiocarbon. In other words, the concentration of carbon-14 fixed in plant and animal matter dwindles away by radioactive decay after death. The decrease proceeds at a rate independent of internal conditions and is determined only by the half-life. An archaeological specimen 5730 years old will thus contain half as much carbon-14 as contemporary living matter, one-quarter after 11,460 years, only one-eighth after 17.190 years, and so on. This is shown in Fig. 17.1.

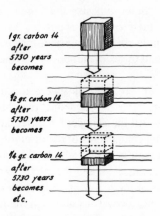

1 gr. carbon 14
after
5730 years
becomes

½ gr. carbon 14
after
5730 years
becomes

¼ gr. carbon 14
after
5730 years
becomes
etc.

Fig. 17.1.   Half-life of carbon-14.

The date obtained by radiocarbon dating is, therefore, the date of removal of the sample from the carbon-14 exchange reservoir. However, it needs to be emphasized that the date found may not be the one in which the archaeologist is interested. Materials such as wood from buildings or charcoal from hearths may have been dead for a long time before being used by ancient man; the wood may have come from a tree that was already hundreds of years old at the time it was felled. The radiocarbon method only provides, in every case, the maximum possible age. In other words, radiocarbon dates only tell when organic matter lived and died, not when it was put to use.

METHODOLOGY OF RADIOCARBON MEASUREMENT

The natural content of radiocarbon in datable samples is very low; add to this the weakness (lack of penetrating power) of the beta-radiation emitted by carbon-14, and the result is that the measurement of natural levels of radiocarbon is difficult.

Very elaborate chemical and radiochemical procedures are involved in order to obtain accurate measurements of activity.

The Sample

The size of the sample required for measurement depends on the composition and on the method of measurement. With modern radioactivity counters, 2-3 g of elemental carbon are required. Table 17.1 shows suggested amounts of different samples when using a modern proportional counter system. Usually, organic carbon samples are preferred over inorganic ones. Specimens preserved under very dry conditions furnish the best results. Less satisfactory are samples from wet caves or from sites with humid conditions.

Fig. 17.2. Chemical apparatus used to extract carbon-14 for radiocarbon dating.

TABLE 17.1   Amounts of Different Materials Needed for Radiocarbon Dating

Material	Optimum Sample (grams)	Minimium Sample (grams)
Wood	0	3
Charcoal	5	2
Bone (charred)	1,000	50
Bone (uncharred)	2,000	100
Textiles	10	3
Flesh, hair	25	5
Leaves	20	4
Soil	1,000	200

Sample preparation

The carbon in the sample has first to be extracted and converted, by chemical methods, into a gas, usually carbon dioxide. Carbon-14 activity can then be measured in the gas. More modern techniques involve the subsequent conversion of carbon dioxide into another carbon-containing gas such as acetylene, ethane, or methane, which allow a more accurate measurement of radioactivity (see Fig. 17.2).

Radioactivity measurement

To determine the amount of radiocarbon present, the gaseous sample—carefully freed from radioactive impurities—is used as filling gas of a radioactive counter, and the rate of disintegration is measured. The activity of carbon-14 is so feeble that it can easily be swamped by extraneous natural radioactivity. This is accomplished as follows: (1) the counter is made of materials that are as free as possible from radioactive impurities; (2) the counter is placed in a radioactivity-free iron shield, which may vary in thickness between 20 cm and 75 cm; and (3) a ring of counters that eliminate the counting of any particles other than those produced by the decay of carbon-14 within the sample is placed around the main counter (see Fig. 17.3) "Modern" carbon, that is, carbon in living matter, averages 13.5 nuclear disintegrations per minute per gram of contained carbon. The older the sample, the lower the rate of distintegration per unit mass. A sample 23,000 years old produces only one disintegration per minute. To minimize the error in the measurement of such a low count, the number of disintegrations is counted over a long period, usually 15-20 hr.

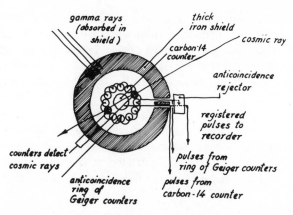

Fig. 17.3. Carbon-14 counter: the arrangement of components. Radiocarbon-bearing gas is pumped into the counter. All counts resulting from carbon-14 disintegrations are recorded. Interference by extraneous sources of radiation (cosmic rays, background radiation, etc.) is prevented by a thick shield and an "anticoincidence" ring of Geiger counters.

RESULTS

The radiocarbon age of a sample can be calculated from the specific activity of the sample relative to that of present-day living matter. The age is given by the relationship:

$$\text{Radiocarbon age } (t) = \frac{5730}{0.693} \log \frac{\text{Specific activity of modern carbon}}{\text{Specific activity of sample}}$$

Practical considerations limit the effectiveness of radiocarbon dating to a range of 200-50,000 years. Samples less than 200 years old can only be called "modern." The radioactivity of samples older than 50,000 years is so low that its measurement leads to unreliable results.

Since radioactive decay consists of a succession of discrete events (individual nuclear disintegrations) occurring at random, radiocarbon dates are necessarily subject to the statistical errors intrinsic to its measurement. Radiocarbon dates are thus expressed as a time range in the form:

$$\text{Age } \pm \text{ Error (in years)}$$

The *age calculated* is based of the half-life of carbon-14, 5730 years ± 40 years. The *error,* being inherent in the statistical nature of the data, can never be done away with completely. It is expressed in terms of a quantity known as the "standard deviation."

A *standard deviation* means that there is a 68 percent probability (i.e., a likelihood of 2:1) that the date will be somewhere within the indicated limits; the probability that the result will fall within twice the standard deviation, that is, twice the given limits is 95 percent (i.e., a likehood of 20:1); and the probability that it will fall within 3 times the limits specified is 99.7 percent (i.e., a likelihood of 300:1).

> *Example:* The half life of radiocarbon is stated to be 5730 years ± 40 years. This means that there is a likelihood of 2:1 that it is somewhere in the range 5690-5770 years, a 20:1 likelihood that it is somewhere in the range 5650-5810 years; and a 300:1 likelihood that it is in the range 5610-5850 years.

The efficiency of the radioactivity counter, variations in the background radio-activity of the counter, and contamination by extraneous radioactivity may also lead to errors of measurement in dating. These, however, are generally small in comparison with the counting error.

UNCERTAINTIES, COMPLICATIONS, AND LIMITATIONS OF RADIOCARBON DATING

When the method of radiocarbon dating was introduced in 1950, archaeologists were at first hesitant to accept the information that it provided. In the face of a rapidly accumulating mass of corroborative evidence, however, hesitation soon yielded to universal acceptance. Since its introduction the method has undergone many improvements.

Discrepancies are sometime found between the otherwise verified age of certain specimens and their radiocarbon age. These discrepancies are due to deviations from the basic assumptions on which the radiocarbon method rests. The fundamental assumptions are, essentially: (1) immutability of the decay rate of carbon-14, (2) constancy of carbon-14 formation in the upper atmosphere, and (3) uniformity of carbon-14 distribution in the exchange reservoir. These assumptions, and the factors that may cause deviations from them, are discussed in the following paragraphs.

Immutability of the Decay Rate of Carbon-14

Immutability of the rate of decay implies that the half-life of carbon-14 is constant. There is, in fact, no question of any fluctuation in the half-life; it has been measured a number of times.

The value 5730 years ± 40 years is the best presently available, even though the internationally accepted value is *5568 years ± 30 years,* based on earlier, already obsolete measurements. The decision to retain an obsolete value as the accepted half-life of carbon-14 was the outcome of a consensus of opinion of scientists engaged in radiocarbon dating (Pullman 1965). Changing the old value (5568 years) for the new and more accurate one (5730 years) would have required the revision of many thousands of published dates. It was also recognized that discrepancies between the radiocarbon chornology and other chronologies arising from the old value would not have been corrected by the change.

Constant Rate of Carbon-14 Formation

Experimental evidence indicates that the formation of carbon-14 proceeds at a constant rate, known to an accuracy of within ± 20 percent. There is no longer any question, however, that there have been changes in this rate of formation. This has been shown by comparing the age of ancient wood samples, determined by dendrochronology, with the radiocarbon age of the same samples.

The rate at which radiocarbon is formed is known to change in a cyclic fashion. Two cycles seem to be in operation: (1) a long-range one (due perhaps to geomagnetic oscillations) with a period of about 10.000 years and (2) a short-range cycle of only a few years, which is due to solar magnetic activity. Both effects combined seem to be the cause of systematic errors of several hundred years. They must be allowed for when using the radiocarbon dating method.

Uniformity of Carbon-14 Distribution in the Exchange Reservoir

A vast mass of evidence has been adduced to prove that the mixing rate of carbon-14 in the atmosphere is rapid and that the atmosphere can be considered as a homogeneous whole with respect to carbon-14 content. Contamination from extraneous sources, however, can invalidate this assumption.

Two types of contaminaion can be differentiated: (1) physicochemical contamination and (2) mechanical intrusion.

Physicochemical contamination includes fractionation of carbon isotopes (gain or loss of carbon-14) between compounds that are part of the "exchange reservoir," contamination of samples with carbon-14 produced by hydrogen-bomb tests, and dilution of the carbon-14 concentration in the atmosphere by "old" carbon released through the combustion of fossil fuel (coal and oil). The uncertainties introduced by physicochemical contamination restrict the accuracy and the effective time range of radiocarbon dating. They also make the application of the method more complex.

Mechanical intrusion means the penetration of the sample, before or after collection, by carbon of a different age than that of the sample itself. Rootlets penetrating the buried speciment, infiltration of windblown organic matter, or insertion of fibers from brushes or other instruments used to clean the sample are all examples of potential "modern carbon" contamination. Modern carbon contamination leads to the assigning of dates more recent than the true ones. "Old carbon" contamination caused by carbonate crystallization on or within the sample, petrol or oil deposited from excavating tools, and so on, is conducive to the assigning of dates earlier than the true ones.

The effect of mechanical contamination is demonstrated by the results shown in Tables 17.2 and 17.3. It becomes clear that the error will be less significant, the closer the respective ages of the sample and its contaminants.

TABLE 17.2   The Effect of Varying Degrees of Contamination by Modern Carbon on the True Age of a Sample

True Sample Age (Years)	Approximate age (years) after contamintion by Modern Carbon		
	Contamination (Percent)		
	5	20	50
5,000	4,650	3,700	2,000
10,000	9,000	6,800	3,600
20,000	16,500	10,600	5,000
30,000	21,000	12,200	5,400

TABLE 17.3   The Effect on the True Sample Age of Varying Degrees of Contamination by Carbon so Old That All Carbon-14 Radioactivity Has Decayed

True Sample Age (Years)	Apparent Age (years) after Contamination by Old Carbon		
	Contamination (Percent)		
	5	20	50
500	900	2,200	6,000
5,000	5,400	6,700	10,500
10,000	10,400	11,700	15,500
20,000	20,400	21,700	25,500

ACCURACY OF RADIOCARBON DATING

The difficulties mentioned serve to emphasize the need for close collaboration between the archaeologist and the natural scientist. Fortunately, the deviations encountered in carbon-14 dating are normally small for dates falling within the last three millennia. Curves of corrections have been published and can be used for correcting dates that fall in the period between 1000 B.C. and the present day (see Fig. 17.4); results thus obtained are concordant with historical dates (Suess 1965; Betancourt, Michael, et al. 1978; Ward and Wilson 1978).

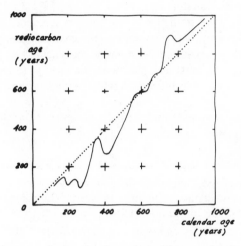

Fig. 17.4.   Calibration curve of radiocarbon dates, obtained from samples whose true calendar date is precisely known. The calibration curve is presently being extended to several millennia B.C.

The experience gained in the years since radiocarbon dating was first introduced indicates that the method is reliable in simultaneity; in other words, two contemporary samples taken from any place in the world will give the same date for any time in the past. Absolute dates, on the other hand, may occasionally be subject to errors large enough to invalidate them. This is due to the variability in time of natural phenomena (e.g., intensity of cosmic radiation) or to ecological contamination (e.g., fossil-fuel effect).

RADIOCARBON DATING USING PARTICLE ACCELERATORS

The conventional radiocarbon dating technique described so far requires waiting for the decay of radiocarbon atoms and measuring the resulting radiation. This is an ineffective way of counting if one considers that for each simple atom decaying per minute there may be 10^9-10^{12} more radiocarbon atoms present in the sample. Great improvements in the radiocarbon dating technique can be achieved by directly counting the radioactive carbon-14 atoms.

The presence of nitrogen-14 atoms in a sample of radiocarbon datable material raises a formidable problem. The mass of a nitrogen-14 atom differs only by 1 part in 100,000 from that of a carbon-14 atom. Therefore, in any direct counting technique the atoms of nitrogen-14 have first to be separated from those of carbon-14.

Using a particle accelerator as a mass spectrometer (Chapter 5), it has been possible directly to detect and count carbon-14 atoms present in samples of wood and charcoal (Nelson, Korteling, et al. 1977; Bennet, 1977; Hedges 1978).

In the experimental determination the sample is first reduced to charcoal (in an oxygen-free atmosphere). A few milligrams of the charcoal are placed in the ion source of a particle accelerator; the atoms in the sample are then ionized and accelerated (as negative ions) at a predetermined energy, to be finally counted when they emerge from the accelerator. A typical dating run may last as little as six minutes.

Counting carbon-14 atoms directly offers some great advantages in comparison to the conventional procedure of measuring radiation:

1. The *sample size* is substantially reduced (1-15 mg is sufficient).
2. The *accuracy* of the technique is increased (a 5000-year-old sample can be dated with an uncertainty of only 10 years, compared with 150 years' uncertainty with the conventional technique).
3. The *limit of the dating range* can be extended to 70,000 (possibly even 100.000) years.
4. The *time required for a determination* is reduced to a small fraction of that required for a conventional determination.

Accelerators of the type needed for radiocarbon dating are expensive. Many of them, however, are in use for other purposes and can be adapted for radioisotope dating. The cost of using an accelerator for the short time required for age determinations may prove to be little or more than that of conventional radiocarbon dating.

USES AND APPLICATIONS OF RADIOCARBON DATING

Radiocarbon dating has become a standard technique of major importance for the archaeologist. It is particularly valuable in the research of prehistory. The use of carbon-14 dating allows isolated finds of unknown stratigraphic provenance to be placed in their proper positions on the time scale. The large number of dates determined by the laboratories engaged in radiocarbon dating are published in *Radiocarbon*, a semiannual periodical issued by the *American Journal of Science*.

DATING WITH RADIOCARBON

The information accumulated during two decades of radiocarbon dating is so extensive that even the briefest of reviews is out of question here. Among outstanding chronological problems settled by radiocarbon dating, the following may be mentioned: the fixing of an age of approximately 24,000 years for charcoal from a hearth in an upper Paleolithic locality in Southern France (Rubin and Suess 1955), the dating of mining activities at ancient gold-mining sites in Saudi Arabia (Broecker, Kulp, et al. 1956), the fixing of a proper time scale for the Mayan calendar (Libby 1954), and the chronology of early man in America (Berger, 1976).

Charcoal, wood, leather, textiles, ropes, and fibers are only a few of the materials that are routinely collected at archaeological sites for dating purposes. Attempts have been made to extend the applicability of radiocarbon dating to other materials; some of these are described as follows.

Dating of Bone

Bones are among the carbon-containing materials found in greatest abundance in archaeological excavations. Nevertheless, there are few reports of radiocarbon dating of bone in the literature. Two reasons have been suggested for the dearth of bone dates: (1) the dating of bone is reputed to be difficult and tedious; and (2) discordance between determined bone dates.

Bone tissue has organic components. The inorganic component, calcium phosphate with adsorbed carbonate, is held together by a network of organic matter—collagen fibrils—composed exclusively of protein. The two fractions contain carbon and should, in theory, lend themselves to radiocarbon dating. In practice, however, the dates obtained from the two fractions do not agree at all well. Radiocarbon dates derived from the inorganic carbonate fraction are liable to inaccuracy because the carbonate in the bone is in constant chemical exchange with the surrounding medium. As mineralization, incrustation, or leaching by groundwater may cause enrichment or depletion of carbon-14, there is little value in dating the inorganic component of bone. Most of the collagen dates obtained, on the other hand, are acceptable within their archaeological context.

The basic collagen method of dating bones has been developed in the last few years. It involves the isolation of the protein by mild acid treatment of the bones. Humic acid and natural soil contaminants make it necessary to use separation procedures which isolate only amino acids native to the bone (Berger, Horney, et al. 1964; Longuin 1971; Protsch 1973; Hassan and Ortner 1977).

Some results of collagen dating of bone are given in Table 17.4. The conclusion

TABLE 17.4   Comparison of Collagen-Radiocarbon and Archaeological Dates of Bones

	Collagen Date	Archaelogical Date
Sample	Years B.P.[a]	
Calf femur	1760 ± 70	1776
Human femur	1420 ± 65	1430
Horse humerus	1935 ± 75	1950
Sacrum from cattle	3765 ± 70	3780
Human vertebra	875 ± 65	890

[a]B. P.:before present.

to be drawn from these results is that collagen seems to be as good an analytical material for radiocarbon dating as, for example, wood and charcoal.

Dating of Mortar and Plaster

Lime-based mortars and plasters have been used for building purposes since at least the third century B.C. Slaked lime, coarse sand, and water are used in their making. The mixture sets by the progressive fixation of atmospheric carbon dioxide on the lime surrounding the sand particles. In this way carbon-14 is incorporated into the material. Once the mortar has set, the carbon-14 activity decays in exactly the same way as in organic materials. Hence the radiocarbon method naturally suggests itself for the dating of mortars and plasters. The method is not suitable for gypsum or hydraulic mortars, since solidification and setting of these takes place with little or no carbon dioxide absorption.

Conflicting reports have been published on the reliability of the radiocarbon dating of mortar and plaster (Delibrias and Labeyrie 1965; Stuiver and Smith 1965; Baxter and Walton 1970). This is due to a number of limitations that have been encountered:

1. Use of calcareous sand would cause the incorporation of very "ancient" carbon in ill-defined proportions. This would give the mortar an apparent age different from, and much greater than, the true one.

2. Quicklime is produced from limestone from which carbon dioxide has been expelled by calcination. Inadequate calcination would leave some of the original limestone-carbon in the quicklime and would again lead to greater than true ages.

3. Mortar or plaster that have been subjected to damp over long periods might suffer from either the exchange of carbon with carbon from atmospheric

carbon dioxide or the depletion or deposition of carbonate from ground-waters.

Wood charcoal was used in the lime calcination process and is often present in mortars. If sufficient quantities of this wood charcoal for dating could be isolated from the mortar, its age should reflect that of the mortar in which it was found.

Dating of Iron

The basic forms of manufactured iron—bloomery iron, wrought iron, cast iron, and steel—are alloys of iron and carbon. The chemical form, concentration, and distribution of the carbon determine most of the properties of the alloy.

Carbon is added to iron at the time of manufacture. From earliest times up till about 200 years ago (when coal began to be used) charcoal was the universally used carbon additive in the smelting of iron. Smelting with charcoal requires the use of freshly cut wood. Since the carbon-14 activity indicates the date of death of the wood, it must be assumed that the date of manufacture of the iron will also be indicated by the same measurement. Table 17.5 shows the results of iron-dating samples analyzed to provide comparative historical and radiocarbon dates (Van der Merwe 1969). The consistency of the results shows that iron specimens can be accurately dated by the carbon-14 method. Analysis of this kind promises to provide extensive chronological information on the Iron Age.

TABLE 17.5   Radiocarbon Dates of Iron Samples

Sample and Location	Carbon-14 Age (Years B.P.)[a]
Cast iron, Saugus, Massachusetts, manufactured (1648-78 C.E.)	350 ± 60
Bloomery iron, Scotland, (83-87 C.E.)	1850 ± 80
Cast iron, Sian, China, Han Dynasty, (221 B.C.-220 C.E.)	2060 ± 80
Carburized steel, Yugoslavia (400-180 B.C.)	2130 ± 60
Cast iron, Honan, China, warring states period (480-221 B.C.)	2380 ± 80

[a]B.P.:before present.

Dating of Iron Slags

Few metallic iron specimens have been recovered at ancient iron smelting and manufacturing sites. Iron is easily corroded and sometimes only stains in the soil are found where iron artifacts were once buried. Slags, the refuse of smelting, on the other hand, consist mainly of silica, are almost indestructible, and are found in large quantities. Cinders are a mixture of slag and unburnt charcoal formed by bloomery furnaces. The charcoal can be extracted mechanically from cinders, thus providing an additional method of dating Iron Age sites. Results obtained by dating iron slags are listed in Table 17.6. They can be of use in establishing chronological sequences, as surveying devices at ancient smelting sites, and may help in the calculation of iron-production yields (Van der Merwe 1969).

TABLE 17.6   Radiocarbon Dates of Iron Slags

Sample and Location	Carbon-14 Age (Years B.P.)[a]
Slag from slag heap, Kgopolne, South Africa (1430 ± 60 C.E.)	520 ± 60
Slag found around iron bloomery furnace, Matsepe, South Africa (1870 ± 60 C.E.)	80 ± 60

[a]B.P.:before present.

The proper use of radiocarbon dating provides solutions to many archaeological problems, affording a chronology worldwide in scope. The accuracy of time measurement possible with carbon-14 is not as great as with dendrochronology or historical dating (see Fig. 17.5). This limitation must always be borne in mind if abuse of the technique is to be avoided.

Many of the so-called "invalid archaeological ages" determined with radiocarbon have been derived from samples of questionable origin, from improperly

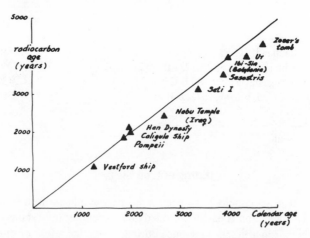

Fig. 17.5.   Known historic dates versus radiocarbon dates.

collected and stored samples, and from those contaminated by non-contemporaneous matter. Experience has shown that even in the face of controversy radiocarbon ages are generally reliable. But archaeologists should beware of the unqualified acceptance of radiocarbon ages derived from measurements on single archaeological specimens.

REFERENCES

Baxter M. S., and Walton A., *Nature,* **225,** 937 (1970).

Bennet, C. L., *Science,* **198,** 508 (1977).

Berger, R., *World Archaeol.,* **7**(2), 174 (1976).

Berger, R., Horney, A. G., and Libby, W., *Science,* **144,** 999 (1964).

Betancourt, P. P., Michael, H. N., and Weinsten, G. A., *Archaeometry,* **20,** 200 (1978).

Broecker, W. S., Kulp, J. L., and Tucek, C. S., *Science,* **124,** 154 (1956).

Delibrias G., and Labeyrie J., in R. M. Chatters, ed., *Proceedings of Sixth International Conference on Radiocarbon Dating,* CONF 650652, 1965, p. 344.

Hassan, A. A., and Ortner, D. J., *Archaeometry,* **19,** 131 (1977).

Hedges, R., *New Scientist,* 599 (1978).

Labeyrie, J., *La Recherche,* **7,** 1036 (1976).

Libby, W., *Science,* **120,** 733 (1954).

Libby, W., *Radiocarbon Dating,* Chicago U. P., 1955.

Longuin R., *Nature,* **230,** 241 (1971).

Nelson, D. E., Korteling, R. G., and Scott, W. R., *Science,* **198,** 507 (1977).

Protsch, R., Ph.D. dissertation, Columbia University, New York, 1973.

Pullman, A., in R. M. Chatters, ed., *Proceedings of Sixth International Conference on Radiocarbon Dating,* CONF 650652, 1965, p. 16.

Rubin, M., and Suess, H. E., *Science,* **121,** 481 (1955).

Sellstedt, H., Engstrand, L., and Gejval N. G., *Nature,* **212,** 573 (1966).

Stuiver M., and Smith C. S., CONF 650652, 1965, p. 338.

Suess, H. E., *J. Geophys. Res.,* **70,** 5937 (1965).

Van der Merwe, N. J., *The Carbon Dating of Iron,* Chicago U. P., 1969.

Ward, G. K., and Wilson, S. R., *Archaeometry,* **20,** 19 (1978).

Recommended Reading

Aitken, M. J., *Radiocarbon Dating, in Physics and Archaeology,* Interscience, London, 1961.

Barker, H., Radiocarbon Dating, *Nature,* **184,** 672 (1959).

Barker, H., Radiocarbon Dating, in E. Pyddoke, ed., *The Scientist and Archaeology,* Phoenix House, London, 1963.

Briggs, L. J., How Old is it? *Nat. Geog. Mag.* (August 1968).

Dyck, W., Recent Developments in Radiocarbon Dating, *Current Anthropol.*, **8,** 549 (1967).

Johnson, F., The Impact of Radiocarbon in Archaeology, *Science,* **155,** 165 (1967).

Kanren, M. D., Early History of Radiocarbon, *Science,* **140,** 584 (1963).

Libby, W. F., Radiocarbon Dating, *Endeavour,* **13,** 5 (1964).

Libby, W. F., Radiocarbon Dating, *Chemistry in Britain,* **5,** 548 (1969).

Smith H. S., Egypt and C-14 Dating, *Antiquity,* **38,** 32 (1964).

Books on Radiocarbon Dating

Chatters, R. M., ed., *Radiocarbon and Tritium Dating, Proceedings of the Sixth International Conference, Washington, D.C.,* CONF-650652, 1965.

Olsson, I. U., ed., *Radiocarbon Variations and Absolute Chronology, Proceedings of the Twelfth Nobel Symposium,* Uppsala University, Wiley, New York, 1970.

Radioactive Dating, Proceedings of a Symposium, Monaco, 1967, IAEC, Vienna, 1967.

Radiocarbon Dating, Proceedings of the Eight International Conference, **Royal Soc. of New Zealand, Wellington, 1972.**

The Impact of the Natural Sciences in Archaeology, A Symposium, London, 1969, **Oxford U. P., 1970.**

POTASSIUM-ARGON DATING

Potassium-argon dating is a method used to determine the age of rock-forming minerals in the time span 35.00-50.000.000 years. To be suitable for potassium-argon dating, a mineral must meet the following requirements: (1) contain potassium. (this need not be a major component of the mineral, but must be present in a concentration sufficient to permit accurate quantitative determination), (2) be resistant to weathering and dissolution by groundwaters, and (3) be capable of retaining argon within the framework of its crystal lattice.

The potassium-argon method of dating can be applied over a very wide time range. It is mainly a geological technique. In the archaeological context, potassium-argon age determinations are carried out on volcanic deposits related to the material considered. The method may yield valuable information on extinct species by providing dates for the geological beds in which their remains are found. The greatest use of potassium-argon dating to the archaeologist lies in its application to the dating of Pleistocene deposits, whose age is reckoned in hundreds of thousands of years. It has been invaluable in establishing a meaningful Cenozoic chronology and providing a time scale for the evolution of mankind, from the earliest known hominids onward.

Potassium consists of three isotopes: potassium-39, potassium-40, and potassium-41. The least abundant of these, potassium-40, (K-40) is radioactive. It decays to argon-40 (a stable, nonradioactive isotope) through the emission of a beta-particle. In any material containing the isotope potassium-40, therefore, the concentration of potassium will gradually be reduced, whereas the quantity of argon present will increase. The half-life of this process has been precisely determined: it is 1.25×10^9 years. The decay pattern of potassium-40 is shown in Fig. 18.1.

Potassium-40 is by far the most common of the natural radioactive elements. It is present in many minerals. The relative abundance of its isotopes is shown in Table 18.1. The decay of potassium-40 through the ages has led to an abundance of argon-40 in the atmosphere: about 1 percent of the atmosphere consists of argon-40, whereas all the other inert gases (helium, neon, etc.) together amount to 100 times less.

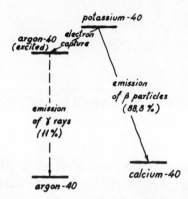

Fig. 18.1. Decay scheme of potassium-40. The two decay branches shown are: (1) beta or electron emission and (2) electron capture. The 0.2 percent unaccounted for in the model, decays by other modes.

TABLE 18.1   The Naturally Occuring Isotopes of Potassium and Argon

Isotope	Half-life	Relative Abundance (Percent)
Potassium-39	Stable	93.08
Potassium-40	1.25×10^9 years	0.00119
Potassium-41	Stable	6.91
Argon-36	stable	0.337
Argon-38	stable	0.063
Argon-40	stable	99.600

Argon-40, generated in a mineral containing potassium-40, is retained within the mineral's crystal lattice. Hence the age of a mineral can be determined by measuring the relative concentrations of potassium-40 and argon-40 in it. For the method to be reliable, the following postulates must be valid:

1. The radioactive decay of potassium-40 takes place at a constant rate, independent of the physical or chemical environment.
2. At time zero, ($t = 0$, the moment at which the measured time commences) there is no radiogenic argon in the mineral. That is, at time zero the mineral contains only argon with the same isotopic composition as atmospheric argon.
3. During the time measured, the potassium: argon ratio changes only as a result of the radioactive decay of potassium-40.
4. The period of formation of the mineral is much shorter than its total age.

METHODOLOGY OF POTASSIUM-ARGON DATING

The determination of potassium-argon age involves (1) sampling a mineral, (2) isolating the potassium-40 and argon-40, and (3) determining the relative amounts in which these elements are respectively present.

Sampling

The distribution of potassium and of radiogenic argon in a given mineral is not homogeneous throughout. This is because different grains of the same mineral are formed within the rock at different times. Ideally, therefore, the concentration of potassium and argon should be determined in the same sample. However, since the measurements of potassium and argon contents, respectively, have to be made on separate samples, it is important that every sample aliquot be truly representative.

The amount of sample required decreases with increasing potassium content. Because of the difficulty of obtaining a truly representative sample, it is not advisable to analyze samples too small or too large in weight; experience has shown that a range of 0.2-20 grams of mineral should be considered ideal. Rocks are usually composed of more than one mineral, which have to be separated before potassium and argon analyses are performed. Rock samples should, therefore, be larger in size than this range, and expert advice should be consulted before a choice of sample is made.

Potassium Determination

Many methods of potassium determination in minerals yield measurements of suitable precision for dating puposes. These include gravimetry, instrumental analysis, and physical methods. Simplicity and speed of performance are to be considered when evaluating an analytical method, but accuracy and precision are of overriding importance. Most potassium determinations are made nowadays using instrumental techniques, in particular flame photometry, in which the amount of potassium present in a sample is determined by measuring the intensity of the energy emitted by the potassium atoms when excited by a flame (see Chapter 3). Other instrumental methods used for potassium determination include isotope dilution (see following section on argon analysis) and optical and atomic absorption spectroscopy.

Argon Determination

The measurement of the amount of radiogenic argon in a mineral involves two successive steps: (1) extraction and purification and (2) measurement.

Argon is extracted from the sample by fusing the latter; on fusion, the gaseous element is released from the molten mineral. After extraction, reactive gases also released are next removed, whereas other inert gases may or may not be separated from radiogenic argon, depending on the analytical technique used.

Volumetric analysis, neutron activation, and stable isotope dilution have been used to determine the amount of radiogenic argon in minerals. At present, most laboratories concerned with potassium-argon dating use the isotope dilution technique, as it is the least susceptible to error. A quantity of argon of known isotopic composition is added to the sample being analyzed, and a mass spectrometer is then used to measure the isotopic composition of the resulting mixture. From this measurement the amount of argon-40 in the sample is calculated.

RESULTS

With a knowledge of the potassium-40 and argon-40 contents, the age of the mineral can be calculated using the relationship:

$$t = 1.709 \times 10^{10} \frac{\text{conc. Ar}_{\text{rad}}^{40}}{\text{conc. K}^{40}}$$

where t is age (in years); conc. Ar_{rad}^{40} is the concentration of radiogenic argon; and conc. K^{40} is the concentration of potassium-40. This relationship is a simplified form of a more complicated potassium-argon equation; it is perfectly adequate for the calculation of dates important to the archaeologist. The error arising from the use of this simplified equation (as distinct from experimental errors of measurement) is less than one percent at about 30 million years. The reliability of the age determined is expressed as a plus or minus figure, the estimation of which is explained elsewhere (see Chapter 17).

There is practically no upper limit to the applicablity of the potassium-argon dating method. The archaeologist is, however, concerned with relatively young minerals, which to be suitable for dating, should be rich in potassium. Experiments have shown that potassium-argon dating of rocks as little as 1,000-10,000 years old is feasible.

ERRORS

One of the basic assumptions in potassium-argon dating is that no radiogenic argon is present in the sample, other than that produced by the decay of potassium 40. The sample is also assumed to contain all the radiogenic argon generated within it. In other words, the sample must be a closed system. The largest errors in potassium-argon age determination, however, are related to the occurrence of either an excess or a deficiency of argon.

Loss of Argon

The effect of the loss of argon from minerals is a lowering of the potassium-argon age. Minerals can lose radiogenic argon from any (or all) of several causes. Generally, loss occurs by diffusion, whereby the gas diffuses out of the mineral into the surrounding environment by the random motion of molecules.

Geological factors can also cause argon loss of potassium-bearing minerals. Metamorphosis of rocks, which is accompanied by chemical and physical reconstitution, causes loss of argon. Weathering also leads to partial loss, whereas total mineral degradation will result in a total loss of argon.

Excess Argon

Extraneous argon (radiogenic argon additional to that generated by the decay of potassium-40) in a sample results in ages greater than true ones. Notwithstanding one of the basic assumptions of the potassium-argon method—that samples contain only radiogenic argon produced by the decay of potassium-40 in the mineral—a few anomalously old ages, attributed to extraneous argon, have been reported. Occlusion (incorporation) of argon-40 in minerals at the time of formation and diffusion of argon-40 into the mineral after formation have been suggested to explain this excess (Damon and Kulp 1958). To verify the existence of excess argon-40 in a sample, it is necessary to use independent methods for the dating of mineral, not always an easy task to perform.

ARGON-40-ARGON-39 DATING

The Potassium-Argon method, as described so far, is known as the "conventional method." Later studies lead to the development of the so called "*Argon-40-argon-39 method*" (Sigurgerson 1962; Merrihue 1965).

As with conventional potassium-argon dating, argon-40, derived from the decay of potassium-40, is measured to obtain the age of the sample. However, instead of measuring the potassium in part of the sample, as is conventionally done, the potassium content is determined in the same sample used for the argon determination and at the same time.

The determination of potassium is done by neutron activation: fast neutrons convert some of the potassium-39 in the sample to argon-39; the amount of argon-39 produced by irradiation is monitored and callibrated. To obtain the age of the sample, the argon is extracted and the ratio argon-40: argon-39 is measured in a mass spectrometer.

The advantages of this method are threefold: (1) it offers greater precission; (2) weathered samples—useless for dating by the conventional method—can be used to obtain accurate dates, and (3) detection is possible of samples that have either had excess argon in them or lost some of it since their formation (Curtis 1976).

SUITABILITY OF MATERIALS FOR DATING

Table 18.2 lists common minerals generally useful for potassium-argon dating. The data in the table shows that there is practically no upper limit to the range of applicability of the method. The archaeologist is, however, concerned with the lower limit. His interest lays with relatively young samples. To be suitable for dating, recently formed minerals should be rich in potassium.

TABLE 18.2   Minerals Useful for Potassium-Argon Dating

Mineral	Ages That Can Be Determined Using the Method (Years)
Anorthoclase	>50,000
Biotite	>200,000
Hornblende	>500,000
Plagioclase	>500,000
Sanidine	>10,000
Whole rock	>200,000

Fossil Material

It does not seem probable that meaningful dates of fossil material (bone, teeth, and shells) obtained by the potassium-argon method should be forthcoming; diffusion experiments have shown that radiogenic argon is easily lost from them. The few attempts to date fossils have yielded disappointing results: potassium-argon ages were significantly less than stratigraphic ages. (Lippolt and Gentner 1962, 1963).

DATING OF EARLY HOMINIDS

Potassium-argon dating has provided the opportunity for a comparative study of early man during a measurable span of time of several million years. The study has helped to establish beyond any doubt that man, as a toolmaker, is at least several million years old.

Java's Meganthropus

Fragments of an early hominid, the *Meganthropus*, were found near Modjokerto in Java, in 1952. Specimens of volcanic rock lying below a mandibular fragment of the *Meganthropus* are believed to be penecontemporaneous with it. These were dated by the potassium-argon method, and a date of 1.9 ± 0.4 million years was obtained. The date makes the *Meganthropus* comtemporaneous with African hominids of similar evolutionary development.

Zinjanthropus bosei

Olduvai Gorge in Tanzania, East Africa, has Pleistocene deposits rich in fossil and cultural remains. These are derived from a long sequence of stages in the early evolution of man. This sequence has been classified into five units, called Beds I-V, from the oldest to the most recent, respectively. In 1959 Dr. and Mrs. L. S. B. Leakey discovered in Bed I of Olduvai Gorge a fossil hominid skull of great antiquity—*Zinjanthropus boisei*—which was found in association with tools, flakes, and fossil and faunal remains. This was one of the most exciting archaeological discoveries of modern times. Two years later, the still older remains of a pre-*Zinjanthropus* hominid were found at a slightly lower level of Bed I.

Many of the deposits in the beds at Olduvai Gorge are volcanic in nature and thus suitable for potassium-argon dating. The results obtained illustrate the accuracy that can be achieved using this method, provided that cooperation between investigators from different disciplines is forthcoming.

Ages ranging from 1.57-1.89 million years were obtained for volcanic minerals found below the hominid remains. The ages of other minerals collected near the top of Bed I (above the hominid remains) ranged 1.02-1.38 million years (Leakey, Evernden, et al. 1961; Evernden and Curtis 1965). These results led to the conclusion that *Zinjanthropus* was about 1.75 million years old, far older than had previously been suspected!

A controversy over the true age of *Zinjanthropus* followed close on the publication of the dating results. Data obtained on samples from Bed I by other researchers conflicted with the results described. The controversy was only resolved after it was shown that some samples had been altered by weathering and were hence unsuitable for age determination. This example illustrates the need for careful selection of material for dating.

Independent confirmation of the potassium-argon dating of *Zinjanthropus* was obtained by fission track dating (see Chapter 19).

A consequence of the discoveries at Olduvai was the beginning of intensive field work in the search for still older ancestors of man. A large number of new hominid species were found and dated in the Lake Turkana area in Kenya and at the Omo Valley in Ethiopia. In both these places the hominid remains were found together with tools. At the Omo Valley potassium-argon dating gave an age of 1.9 million years. The early man from Lake Turkana was still more ancient: 2.6 million years old (Fitch and Miller 1970).

At the Lukeino Formation, in the North Kenya Rift Valley, a lower molar, clearly hominoid in structure was found in sediments overlain by a basalt formation. The latter was shown by potassium-argon dating to be 5.4 million years old. The sediment overlies mineral formations from which potassium-argon dates of 7.2, 6.8, and 6.7 million years were obtained. The hominid molar is thus "bracketed" by dated strata. It was probably deposited 6.5 million years ago (Pickford 1975).

These dates provide a complex and fascinating picture of hominid history. It seems reasonable to suspect however that hominids were in Africa and in South East Asia long before the dated specimens were entombed. Potassium argon dating will certainly help to provide a reliable view of early human development.

REFERENCES

Carr, D. R., and Kulp J. L., *Bull. Geol. Soc. Amer.*, **68**, 763 (1957).

Curtis, G. H., *World Archaeol.*, **7**(2), 198 (1976).

Damon, P. E., and Kulp, J. L., *Am. Mineralogist*, **43**, 433 (1958).

Evernden, J. F., and Curtis, G. H., *Current Anthropol.*, **6**, 343 (1965).

Fitch, F. J., and Miller J. A., *Nature*, **226**, 226 (1970).

Gentner, W., and Lippolt, H. J., in D. Brothwell and E. Higgs, eds., *Science in Archaeology*, Thames and Hudson, London, 1963, p. 72.

Leakey, L. S. B., Evernden, J. F., and Curtis, G. H., *Nature*, **191**, 478 (1961).

Lippolt, H. J., and Gentner, W., *Geochim. Cosmochim. Acta*, **26**, 1247 (1962).

Lippolt, H. J., and Gentner, W., in *Radioactive Dating*, International Atomic Energy Agency Publication, Vienna, 1963, p. 400.

Merrihue, C., *Trans. Am. Geophys. Union*, **46**, 125 (1965).

Pickford, M., *Nature*, **256**, 279 (1975).

Sigurgeirson, T., *Dating recent basalt by the potassium-argon method*, Report Physical Lab., Univ. of Iceland, 1962.

Recommended Reading

Dalrymple, G. B., and Lanphere, M. A., *Potassium-Argon Dating*, Freeman, San Francisco, 1969.

Schaeffer, O. A., and Zahringer, J., eds., *Potassium-Argon Dating*, Springer-Verlag, New York, 1966.

RADIATION-DAMAGE DATING

The changes effected in the physical and chemical properties of matter by the action of ionizing radiation are usually referred as *radiation damage*. These changes accumulate with time and can be used for the determination of age.

Radiation damage methods of dating include:

1. Thermoluminenscence (see chapter 5).
2. Fission-tracks.
3. Alpha-particle recoil tracks.

FISSION-TRACK DATING

Most naturally occurring minerals and many man-made materials contain minute quantities of uranium, a radioactive element. Uranium-238, the most common isotope of uranium, decays by alpha-particle emission, but one in every two million atoms of uranium-238 decays of spontaneous fission. In this case the atomic nucleus splits into two approximately equal parts, and a large amount of energy is released. In nonconducting solids, such as minerals and glasses, the fission fragments lose their energy by disrupting atoms in their path. Each fission fragment ploughs out a submicroscopic trail of damage (Wagner 1976). These trails have no visible or palpable effect on the material. However, if a freshly fractured and polished surface is viewed in an electron microscope (see Fig. 19.1), trails can be seen where uranium atoms have disintegrated near the surface. The trails are typically 30 angstrom units in diameter and 0.01 mm. long (Fleischer, Price, et al., 1965b).

The frequency of occurrence of spontaneous fission tracks increases with uranium concentration and with the age of the sample. Hence the age of a sample can be calculated by counting the spontaneous fission tracks and at the same time determining the uranium content. The practical difficulties involved in counting fission tracks with an electron microscope render this method of dating far too cumbersome to be useful. Fortunately, however, the tracks can be *"developed"* and rendered visible under an ordinary relatively simple chemical process.

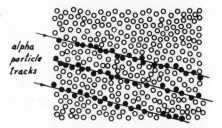

Fig. 19.1.   Radiation damage (schematic). Tracks caused by alpha particles in an insulating solid.

The energy imparted to the fission particles by the breakdown of the uranium atom is so great that not only is a hollow track left marking their direct path, but the entire crystal lattice in the vicinity of the fission track is dislocated and weakened (in the same way as a bullet passing through a sheet of safety glass, leaves a network of cracks radiating from a hole). A chemical reagent that disolves the material under examination will preferentially attack those regions that have been so weakened. By continued etching, the fission tracks can eventually be enlarged enough to be seen through an optical microscope. The geometry of etched tracks varies with the substance being examined, the chemical reagent used for etching, the temperature, and the duration of the etching. Etched tracks in obsidian are shown in Fig. 19.2.

STABILITY OF FISSION TRACKS

Fission tracks will fade and be erased in any material that is heated to a sufficiently high temperature. Some materials are more resistant to track erasure by heat than others. In some the temperature required for erasure is so low that over long periods of time tracks will fade at ambient temperatures, whereas in others tracks will be effectively permanent. Quartz, for example, will retain tracks at higher temperatures than mica. In some materials tracks will be retained even at temperatures as high as 400°C for a time equal to the age of the earth.

SUITABILITY OF SAMPLES FOR DATING

To be datable by the fission-track method, materials must satisfy certain requirements in regard to track permanence and track density. The age measured by the fission-track method is the time elapsed since the sample was at a temperature high enough to erase previously formed tracks. Materials suitable for dating are therefore those in which tracks do not fade until relatively high temperatures (say, 400°C) are reached. Materials less resistant to track fading may reveal information about heating episodes during their thermal history.

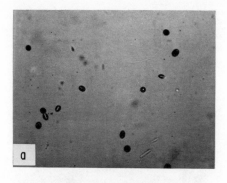

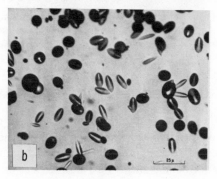

Fig. 19.2.   Fission-track dating. Finding natural tracks from the spontaneous fission of trace amounts of uranium-238 in materials such as the obsidian (a) has allowed the ages of minerals to be measured. Uranium content is measured by inducing further fission by neutron irradiation, as in (b).

The density of fission tracks is a function of the concentration of uranium and of the age of the sample. Age measurements are possible only if there is a track density sufficiently high to permit an accurate count to be made. In large, homogeneous samples a few tracks (5-10) per square centimeter may be adequate. In very small samples where the etched area is as small as 1/10,000 of a square centimeter, a density as high as 100,000 tracks per square centimeter may be necessary.

The typical concentration of uranium in minerals and glasses is about 1 ppm. This concentration yields a track density of 0.3 tracks per square centimeter per 1000 years of age. Since it is difficult to work with samples containing less than 5 tracks per square centimeter, "typical" specimens for dating are those of 10,000-100,000 years of age. In other words, the application of fission-track dating in archaeology is limited to samples that either have a high uranium content or are very old. Figure 19.3 illustrates the use of the fission-track method for mineral dating.

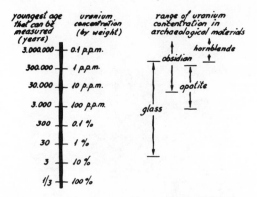

Fig. 19.3. Approximate limits of ages that can be determined by fission-track dating in material of different uranium content. (The oldest glass made by man dates back less than 5000 years. From the chart it can be seen that the only glass objects that can be dated by this method are those containing more than 100 ppm of uranium.)

METHODOLOGY OF FISSION-TRACK DATING

To find the age of a sample by the fission-track method, two values must be established: track density and uranium concentration. The first of these has already been discussed. Uranium concentration is determined by subjecting the sample to neutron bombardment in a nuclear reactor. Under irradiation, the number of uranium atoms undergoing fission is increased and fresh fission-tracks are formed. The concentration of uranium in the sample can be calculated from the difference in the fission-track density before and after irradiation. The total procedure, therefore, involves the following steps:

1. Sample preparation and spontaneous fission-track count.
2. Irradiation in a nucler reactor.
3. Counting of neutron-induced fission tracks.
4. Calculation of uranium content.

The preparation of the sample involves grinding the specimen to expose a fresh surface, which is then polished and finally etched by a suitable chemical reagent. A small chip of mineral or glass, a few millimeters in length along each edge can serve for fission-track age determination. Alternatively, it is possible to smooth and polish an inconspicuous surface of a larger object.

The etching reagent dissolves the damaged material adjoining the fission-tracks more rapidly than the remainder, and pits are formed where fission tracks intersect the surface. The density of spontaneous fission tracks is determined by counting the number of etch pits in a measured area of the sample surface, using an optical microscope. To count a statistically sufficient number to tracks, the procedure of polishing, etching, and counting is sometimes repeated a number of times on the same surface.

The specimen is then irradiated with thermal neutrons in a nuclear reactor or with a particle accelerator, and the thermal-neutron induced track density is determined in the same way as before. The fission-track age of the sample can now be calculated to a good approximtion using the relationship (Nishimura 1971):

$$t = 6.12 \times 10^{-8} \phi \frac{\rho_s}{\rho_i}$$

where t is the sample age in years; ϕ is the dose of thermal neutrons; ρ_s is the density of spontaneous fission tracks; and ρ_i is the density of induced fission tracks.

The age found by fission-track analysis is the time elapsed since the sample was last at a temperature sufficiently high to cause track fading. In most minerals and glasses the fission-track age indicates the time since solidification of the melt. In other cases, and especially in tools and artifacts, it may correspond to some moderate thermal event such as firing or burning. To measure the temperature above which fission-tracks fade out, a sample irradiated with thermal neutrons is polished, etched, and heated, in successive steps, at increasing temperatures until the etch pits are undetectable under a microscope.

ARCHAEOLOGICAL APPLICATIONS OF FISSION-TRACK DATING

Nearly all minerals and natural glasses reveal fission-tracks. The uranium content of man-made glasses, however, usually falls short of the levels required for many dating applications. The following are the situations in which fission-track dating is most likely to be of use in archaeology:

1. Tracing the provenance of minerals (e.g., obsidian cores and artifacts; see Chapter 6, on obsidian).
2. Determining the time of formation of natural glasses known to correspond with some activity of man.
3. Dating minerals known to have been reheated by some agent at a time that is of interest.
4. Dating of man-made glass that contains enough natural uranium. The uranium may have been added deliberately (e.g., as a colorant) or may have occurred naturally in the raw materials.
5. Verification of the authenticity of ancient man-made glass objects (Fleisher 1976).

The examples given in the following paragraphs illustrate instances in which the fission-track dating method has provided relevant information.

Dating of a Mesolithic Obsidian Knife

An obsidian knife blade was found at a cave in Elmenteita, East Africa. The knife originally struck from an obsidian core, was at some later date heated to a high temperature, thus being partially melted and acquiring the melted shape that can be seen in Fig. 19.4.

Fig. 19.4.   Knife made of obsidian in Mesolithic times. Length: 7.3 cm. The knife was heated in the distant past. The time of heating was dated by the fission-tracks technique.

Since high-temperature heating erases fission-tracks, those traces that are present must have been formed since the knife was heated. Examination of a surface of just over 5 cm² revealed only 17 fission tracks. After exposure of the knife to thermal neutrons and counting of the induced tracks it was possible to calculate that it had been heated 3,700 ± 900 years ago. The wide range of uncertainty, nearly 25 percent, is due to the low number of tracks counted. A count of 90 tracks for example, would have reduced the error to ±400 years; but the time required to do this would not have been justified (Fleischer, Price, et al. 1965a).

It was originally assumed that the knife had been burned in a mesolithic hearth. The fission-track date suggests, however, that the strong heating occurred during Neolithic occupation of the cave at Elmanteita; moreover, the date corresponds very closely to a carbon-14 date obtained for charcoal from a related Neolithic cremation site.

Dating of Ceramics

Spontaneous fission tracks were counted on the glaze of a potsherd from a bowl collected at a ceramic kiln in central Japan. The kiln is believed to have been used between the Kamakura (1190-1330 C.E.) and the Muromachi (1330-1570 C.E.)

periods. The date calculated from the counting was 1450 C.E., corresponding, therefore, to the middle of the Muromachi period.

Dating of Bed I at Olduvai Gorge

The tuff from Bed I at Olduvai Gorge includes a number of minerals: anorthoclase crystals, augite crystals, and porous glass. The uranium content of the anortho-clase was too low to be suitable for fission-track dating, and the cutting and polishing of augite proved too difficult. Attention, therefore, was concentrated in the glass (Fleischer, Price, et al. 1965c).

Two separate groups of glass fragments, both having a potassium-argon age of 1.76 million years, were used for the fission-track determination. The age was found to be 2.0 ± 0.5 million years; the 25-precent range of uncertainty resulted from the statistics of the track count. A second method of counting afforded an age of 2.03 ± 0.30 million years. Combination of the two results gives a final age for the Olduvai Gorge Bed I glass of 2.03 ± 0.28 million years, in agreement with the potassium-argon determination.

ALPHA RECOIL TRACKS

Some materials occasionally show an excess of fission-tracks, that is, a number of tracks over and above the number expected from the spontaneous fission of uranium-238. Using a special technique—phase-contrast microscopy—it was discovered that the discrepancy is due to the existence of a second type of track, very short and sometime invisible under normal illumination. Whereas fission tracks are produced by uranium-238, these short tracks are produced by the recoil of heavy nuclei that accompanies the emission of alpha-particles from uranium and thorium atoms (Huang and Walker 1967).

The most obvious use of such tracks is in dating samples in a way analogous to the fission-track method. There are some advantages to the alpha-recoil-track method: the alpha-recoil track density is higher than that of fission tracks, and hence the sensitivity of the method is high (Zimmerman 1971). Alpha-recoil tracks are also more resistant to thermal fading than are fission-tracks. The increase in sensitivity could be useful in the dating of teeth and bones, which cannot be done by the fission-track method because of the low uranium concentration in these materials. The resistance to thermal fading might be used to probe the thermal history of materials.

Thorium-232 and uranium-235 and 238 are elements whose nuclei are emiters of alpha particles. These are found in many substances that, in turn, may occur in natural or man-made material of archaeological importance; for example, these elements are found in mica, which often occurs in potsherds.

Investigation of alpha recoil tracks in potsherds from a wide geographical distribution in the United States revealed some interesting information. For example, there seems to be a direct and potentially predictable relationship between alpha recoil counts in a specimen and the time elapsed since the specimen was last subjected to fairly intense heating (Garrison, McGuinsey, et al. 1978). Further studies may make it possible to derive relevant data from many forms of fired earth, much of it previously considered undatable.

REFERENCES

Fitch, F. J., and Miller, J. A., *Nature*, **226**, 226 (1970).

Fleischer, R. L., *World Archaeol.*, **7**(2), 136 (1976).

Feischer, R. L., and Price, P. B., *Geochim. Cosmochim. Acta*, **28**, 1705 (1964a).

Feischer, R. L., and Price, P. B., *J. Geophys. Res.*, **69**, 331 (1964b).

Feischer, R. L., and Price, P. B., *Science*, **144**, 841 (1964c).

Feischer, R. L., Price, P. B., and Walker, R. M., *Science*, **149**, 383 (1965b).

Feischer, R. L., Price, P. B., and Walker, R. M., and Leakey, L., *Science*, **148**, 72 (1965c).

Feischer, R. L., Price, P. B., and Leakey, L.,*Nature*, **205**, 1138 (1965a).

Garrison, E. G., McGuinsey, C. R., and Zinke, O. H., *Archaeometry*, **20**, 39 (1978).

Huang, W. H., and Walker, R. M., *Science*, **155**, 1103 (1967).

Nishimura, S., *Nature*, **230**, 242 (1971).

Pickford, M., *Nature*, **256**, 279 (1975).

Price, P. B., and Walker, R. M., *J. Appl. Phys.*, **33**, 3407 (1962).

Silk, E. C. H., and Barnes, R. S., *Phil. Mag.*, **4**, 970 (1959).

Suzuki, M., *Zinruikagu Zassi (J. Anthropol. Soc. Nippon)*, **78**, 50 (1970).

Wagner, G. A., *Endeavour*, **35**, 2 (1976).

Watanabe, N., and Suzuki, M., *Nature*, **222**, 1057 (1969).

Zimmerman, D. W., *Science*, **174**, 819 (1971).

PART

V

ANALECTA

GEOCHEMICAL SURVEY OF ARCHAEOLOGICAL SITES

The continuing process by which human activity chemically modifies the composition of soil, in and around human settlements, is an interesting aspect of man's interaction with his environment that can be of considerable value to the archaeologist.

In areas of ancient human habitation the fertility of the soil is higher than that of the surroundings; dark deep soils define areas of intensive occupation with considerable precision. Latter-day Arab farmers at Ashdod in Israel or El Phosfat in Egypt, for example, used the soil that they excavated from ancient archaeological sites to fertilize their cultivated land. Farmers in the United States have noted that the soils of old Indian villages are more productive than adjacent soils (Thorp 1942).

The human occupation of land causes chemical changes in the soil. These changes are the result of the deposition of elements and compounds contained in refuse and excretions or produced by the decomposition of organic matter. The main sources of these added elements and compounds are human and animal excreta; in addition, they may include all sorts of waste materials and even the dead bodies of humans—usually buried—or of animals—usually eaten—but sometime allowed to decompose and occasionally buried.

The chemical elements that are of greatest importance in establishing the history of human habitation are phosphorus, nitrogen, and carbon. In desert, or agricultural land the concentration of phosphorus in the soil ranges 0.01-0.2 percent in the uppermost layers (say, the top 10 cm), whereas that of nitrogen ranges 0.1-1 percent. It has been calculated that a community of 100 people in an area of one hectare (10.000 m^2) will provide an annual increment to the soil of 125 kg of phosphorus and 850 kg of nitrogen (Cook and Heizer 1965). Relative to the amounts of these elements naturally present in the soil, these are considerable quantitites; they represent an annual increase of 0.7-7 percent of nitrogen and 0.5-1 percent of phosphorus, in the uppermost layer of soil.

Whereas nitrogen and carbon are added to the soil in relatively large quantities as a result of human habitation, they are also subject to depletion. Nitrogen is lost from the soil in a variety of ways and carbon, mainly through biooxidation. Not so phosphorus. Despite the fact that the natural concentration of phosphorus in the soil (and also the quantity added by human activity) is small, the loss of phosphorus is negligible in most soils. Phosphorus is said to be "fixed" in the soil; it is rendered insoluble by contact with the latter, even when applied to it in a very soluble form. Retention of phosphorus is highest in soils rich in calcium and iron, and in clays generally; but it is good even in coarse sands. Thus, of all the elements introduced into the soil through human occupation, phosphorus is the most accurate index to the extent, duration, and nature of human habitation.

PHOSPHORUS SURVEY

The technique of phosphorus survey is simple. Soil samples, usually of 10-20 cm depth, are collected at regular intervals according to a predetermined plan. After drying, the samples are analyzed for phosphorus, usually by a colorimetric or spectrophotometric method (see Chapter 4). The results are usually expressed as milligrams of phosphorus pentoxide (P_2O_5) per 100 grams of soil (Arrhenius 1931; Lorch 1940; Kiefman 1975).

PHOSPHORUS SURVEY OF ANCIENT SITES

The village of Stokkerupp, north of Copenhagen, Denmark, was deliberately destroyed in 1670. To ascertain the exact location of the ancient settlement, about 250 soil samples were taken from within an area of about 30 hectares and tested for phosphorus concentration (see Fig. 20.1). In the terrain surrounding the ancient site the test revealed concentrations of phosphorus varying from less than 5 to 50

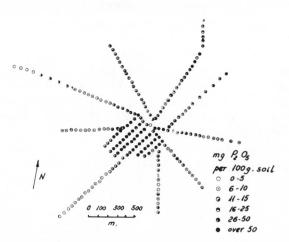

Fig. 20.1.   Geochemical survey in the village of Stokkerupp (north of Copenhagen). P_2O_5 analysis of about 250 soil samples made it possible to locate the ancient site.

mg of P_2O_5 per 100 g of soil. On the site itself, however, the readings averaged 100 mg and reached a maximum of 340 mg of P_2O_5 per 100 g of soil. The survey revealed that in the village center, 7 tons of readily soluble P_2O_5 per hectare had accumulated in the upper 50 cm of soil, making some 125 tons in all—about the same quantity of P_2O_5 as would be found in 500 tons of good-quality superphosphate fertilizer (Christensen 1940).

Samples from the site of abandoned medieval villages in Nottinghamshire, England, also showed a positive correlation between ancient habitation and the phosphorus content of the soil. One such site produced samples having more than 200 mg of readily soluble P_2O_5 per 100 g of soil, whereas the immediately adjacent land, under modern cultivation, showed a range of only 2-10 mg of P_2O_5 per 100 g of soil (Dauncey 1951).

Evidence for the Evolution of a Tell

A deep sounding through a tell at Sitagroi in northeastern Greece provided an opportunity to explain some geomorphological problems of archaeological interest. In the tell, whose age ranges 5400-2000 years B.C., layers of material up to 1.7 m thick were found between the concentration of floor levels. An explanation was sought for the formation of these impressive layers. To reveal variations in the intensity of occupation of the site, the total phosphate content of the soil was measured at different depths (Davidson 1973). Samples from layers between occupation levels were collected (Fig. 20.2 and Table 20.1), taking care to make

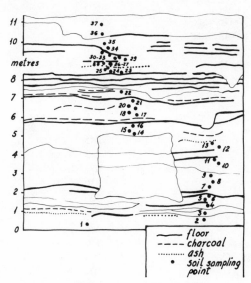

Fig. 20.2.   Stratigraphy of a deep sounding through a Tell at Sitagroi in northeastern Greece.

**TABLE 20.1   Phosphorus Contents of
Soil Samples from Tell at Sitagroi
(Northeastern Greece)**

Sample	Phosphorus Contents (mg of P_2O_5 per 100 g of Soil)
36	775
35	1078
34	1460
33	1026
32	1123
31	1150
30	998
29	860
28	1015
27	1070
26	980
25	943
24	943
23	870
22	1058
21	565
20	1085
19	705
18	600
17	695
16	1355
15	503
14	810
13	1200
12	1325
11	585
10	605
9	1205
8	845
7	845
6	573
5	930
4	700
3	1010
2	476
1	450
C_1	104
C_2	104
C_3	134

the number of sampling points proportional to the thickness and variability of the layers. Samples of soil near Sitagroi (C_1, C_2, and C_3 in Table 20.1) were also analyzed for comparison with the results obtained at the site itself.

The samples from the deep sounding contained much more phosphate than the local soil, indicating that the whole section was inhabited at some time. In the stratum containing samples 17-21 there were virtually no archaeological features, such as walls or floors; nevertheless, the phosphate content of these samples indicated that this section was not abandoned during the formation of the layers. The phosphate analyses confirmed that the tell evolved as a result of human occupation; man-induced processes, the most important probably being the decay of houses, have to be considered to account for the growth of the tell.

Other geochemical surveys of archaeological sites have been carried out. At Ventana Cave, Arizona (USA), the outline of a site was neatly plotted by using the results of soil phosphate analysis (Haury 1950). At an Adena site in the West Virginia (USA) samples from burial sites revealed the phosphorus level to be highest near, or at, the points of burial (Solecky 1951). A very limited number of samples was taken at Falerii Novi, Italy; the results obtained, though few in number, were enough to establish the limits of the ancient inhabited area (Lerici 1960).

PHOSPHORUS PROFILES

A settlement was localized in the course of phosphorus surveying at Brun, in the Stockholm region, Sweden. Figure 20.3 shows a map of the carefully surveyed area (Schnell 1960).

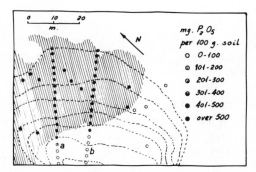

Fig. 20.3.   Geochemical survey at Brunn (near Stockholm). The highest concentration of P_2O_5 in the soil was found in the shadowed area.

A higher P_2O_5 content was found at the shaded area than in the surrounding terrain, as revealed by the numbers on the map, which express milligrams of P_2O_5 per 100 mg of soil. It is interesting to observe in the clear correlation that exists along the same profile between the P_2O_5 content and the number of sherds; the latter is an unequivocal indicator of the intensity of ancient human habitation.

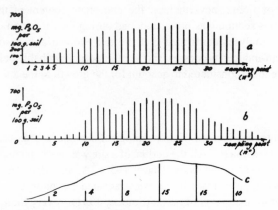

Fig. 20.4. Phosphorus profiles at Brunn (near Stockholm). (a) and (b) are plots of the variation of soil phosphorus content at points along lines (a) and (b), respectively, in Fig. 20.3. (c) shows the relationship between soil phosphorus content (continuous line) and the number of shards (verttical lines) found along profile (a).

Characteristic graphs of the type of Fig. 20.4 are well worth further research. It has been suggested, for example, that it is possible to distinguish between settlements of different economic character and that the graphs can even show changes in economy.

The number of samples and the distance between samples required for geochemical survey needs to be established separately for each case. The practicability of the method is limited by the relatively high cost of laboratory analysis. One solution to this difficulty might be to use a relatively crude method of field analysis with a large number of samples and then to select additional samples from important locations for more accurate analysis. A still better solution would be provided by the development of a sufficiently accurate technique of analysis *in situ*. This would make phosphorus surveys much more widely used than they are at present.

In addition to phosphorus, other elements have received attention as indicators of past human occupation. It has been shown, for example, that the vertical distribution of *calcium* and *magnesium* in previously inhabited soil can be distinguished, even after 2500 years, from that of uninhabited areas (Parsons 1962).

Copper and zinc have also received some attention, and it seems that there is some association between the concentration of these elements in the soil and habitation intensity (Sokoloff and Carter 1952).

The content of *phosphorus, nitrogen, potassium,* and *calcium* from soils of Alaskan archaeological sites were determined (Lutz 1951). In samples from the upper 15 cm of soil from an old village near the mouth of the Russian River, human occupancy had resulted in a 50-fold increase in the content of phosphorus. Increases in the content of the other elements, though of lesser magnitude, were also notable; nitrogen showed a sevenfold, potassium a fourfold, and calcium a 12-fold increase. In the soil of a Auke village northwest of Juneau there was an increase in the phosphorus content of about 175-fold. Nitrogen had increased about threefold, potassium sixfold, and the calcium content had doubled.

REFERENCES

Arrhenius, O., *Zeitschrift fur Pflanzenernahrund, Dungung und Bodenkunde,* Tiel B. 10 Jahrgang, 427 (1931).

Christensen, W., *Danmarks Geol. Unders. II,* No. 57 (1935).

Cook, S. F., and Heizer, R. F., 1965, *Studies on the Chemical Analysis of Archaeological Sites, University of California Publications in Archaeology,* Vol. 2, California U. P., Berkeley.

Dauncey, K. D. M., Adv. Sci., 33 (1951).

Davidson, D. A., *Archaeometry,* **15,** 143 (1973).

Haury, E. W., *The Stratigraphy and Archaeology of Ventana Cave, Arizona,* Arizona U. P., Tucson, 1950.

Kiefmann, H. M., *Informationen aus den Nachbarwissenschaften,* **6**(4), 1 (1975).

Lerici, C. M., *I Nuovi Metodi di Prospezione Archaeologica alla Scoperta delle Civilita Sepolte,* Milano, 1960.

Lorch, W., *Naturwissenschaften,* **28,** 633 (1940).

Lutz, H. J., *Am. J. Sci.,* **249,** 925 (1951).

Parsons, R. B., *J. Iowa Archaeol. Soc.,* **12,** 1 (1962).

Schnell, I., in Lerici, C. M., ed., *I Nuovi Metodi di Prospezione Archaeologica alla Scoperta delle Civilita Sepolte,* Milano, 1960, p. 121.

Solecki, R. S., *American Antiquity,* **16,** 254 (1951).

Sokoloff, V. P., and Carter, F. G., *Science*, **116,** 1 (1952).

Thorp, J., *Soil Sci. Soc. Am., Proc.,* **6,** 39 (1942).

Recommended Reading

Davidson, D. A., and Shackley, M. L., eds., *Geoarchaeology, Earth Sciences and the Past,* Westview, Boulder, Colorado, 1976.

PALEOCLIMATES AND PALEOTEMPERATURES

The constancy of climate was an accepted concept in the teaching of archaeology and anthropology until quite recently. At the beginning of this century, for example, it was widely believed that although there had been drastic climatic changes in the geological past, climates had been essentially constant for at least the last three millennia. Modern geological and geophysical studies have shown, however, that this is untrue and that there has, in fact, been a steady variation in both temperature and climate.

Instrumental measurements of temperatures were first carried out less than 300 years ago. The longest series of temperature measurements extant has been made in central England. Figure 21.1 shows the fluctuation of annual temperatures

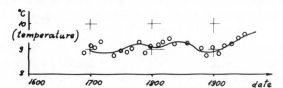

Fig. 21.1. Temperatures prevailing in central England over the last three centuries. (Manley 1958) The data were derived from overlapping records for several places. At none of these places were observations made throughout the entire period.

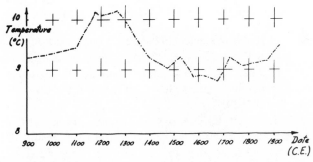

Fig. 21.2. Mean annual temperatures in central England over the last thousand years. These were calculated using mathematical equations connecting temperature with the relative frequency with which months of one or other extreme (mild or cold) are unambiguously recorded in available accounts.

recorded there since 1680. Using regression equations, temperatures prevailing in the last thousand years were calculated. The results (shown in Fig. 21.2) seem to verify climatic variations, especially the warmth of the Middle Ages between 1150 and 1300, when vineyards were cultivated in England for long periods and grain was grown in Iceland. The results also suggest that climate may have had more to do with the abandonment of many villages in England in the second half of the fourteenth century than did the Black Death (Lamb 1969). To the archaeologist, the knowledge of paleoclimates is important as it may contribute to the understanding of the evolutionary process and of the history of man's environment.

PALEOTEMPERATURES

Ancient temperatures can only be estimated from indirect measurements. Geological techniques have used inorganic as well as organic materials. Minerals and fossils, weathering processes, and pollen analysis have been used as indicators of ancient temperatures. Another means of studying paleotemperatures has been provided by the *isotopic geochemistry of oxygen*.

ISOTOPES AS TEMPERATURE INDICATORS

The physical and chemical properties of a chemical compound are dependent not only on the elements from which the compound is constituted, but also on the *isotopic composition* of these elements. The properties of two molecules of carbon dioxide of different isotopic composition are different. Those of $C^{12}O_2^{16}$, for example, are not identical to those of $C^{12}O_2^{18}$. The differences in the properties of the two molecules are so slight that when dealing with usual chemical properties, they may, for all practical purposes, be ignored. These small differences can, however, be utilized for the determination of the temperatures at which natural processes took place in the past.

STABLE ISOTOPES OF OXYGEN IN WATER

Oxygen, the most abundant element in the earth's crust, has three stable isotopes: oxygen-16, oxygen-17, and oxygen-18 (0-16, 0-17, and 0-18, respectively). Although the relative abundance of the isotopes varies with the source, oxygen-16 is always the most common variety. In air, for example, the ratio 0-16: 0-17: 0-18 is 99.759: 0.0374: 0.2039.

When water evaporates, the three oxygen isotopes do not go off at the same rate. The difference in mass between H_2O^{16}, H_2O^{17} and H_2O^{18} results in preferential evaporation of H_2O^{16} molecules from the surface of the water. The seas and oceans, in which this process has been going on for a long time, are thus richer in the heavier isotopes (see Fig. 21.3).

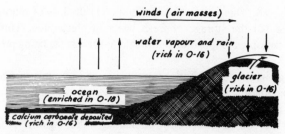

Fig. 21.3. Cycle of the oxygen isotopes in natural waters in a cold age. The lighter isotope (oxygen-16), preferentially evaporated from the ocean, is transported to land and eventually incorporated into glaciers. Residually greater concentrations of the heavier isotopes (oxygen-17 and oxygen-18) are left in ocean waters. This change is reflected in the isotopic composition of calcium carbonate deposits formed in the ocean.

Water-Carbonate Equilibrium

Atmospheric carbon dioxide dissolves in water. In marine waters the two components, water and carbon dioxide, are in chemical and isotopic equilibrium. The exchange reaction that takes place is:

$$\tfrac{1}{2}CO_2{}^{16} + H_2O^{18} \qquad \tfrac{1}{2}CO_2{}^{18} + H_2O^{16}$$

whereby carbon dioxide is enriched in oxygen-18. Hence materials such as calcium carbonate precipitated from marine water are also enriched in oxygen-18.

Oxygen Fractionation and Paleotemperatures

The sediments in vast areas of the ocean floor consist of carbonate deposits formed from the calcareous skeletons of aquatic animals (foraminifera and corals). These skeletons had their origin in carbon dioxide dissolved in the sea. The abundance of oxygen-18 in these sea-bottom carbonate deposits—caused by isotopic fractionation of oxygen—depends on two factors: (1) the isotopic composition of oxygen in marine waters when the deposits were formed and (2) the temperature of the marine waters at that time.

The degree of oxygen fractionation is very slight, but the two factors, isotopic composition and temperature, operate in the same direction. Calculations show that the ratio of the oxygen isotopes precipitated as carbonates at $0°$ is $0\text{-}18:0\text{-}16 = 1.026:500$, whereas at $25°C$ it is only $1.022:500$, if the ratio of the isotopes in water is $1:500$ (Urey 1948). That is, oxygen-18 is slightly more concentrated in the calcium carbonate than in the water; at lower temperatures the concentration is highest, diminishing with increase in temperature. During a glacial age, for example, the temperature of water diminishes so that the water becomes residually richer in oxygen-18. This is due to two factors: (1) the loss of marine water by evaporation and (2) the storage of fresh water in glacier ice. This change is reflected in the isotopic composition of the calcium carbonate being precipitated at that time.

The variations in oxygen-18 enrichment in marine carbonates are measurable. To determine paleotemperatures, therefore, it is necessary to compare the oxygen-18 enrichment of calcium carbonate of marine organism precipitated at a certain time, with the oxygen-18 enrichment of marine waters against an empirically determined temperature scale. In this way measurements of paleotemperatures, exact to within $\pm 1°C$, can be obtained. Sometimes, though with difficulty, measurements exact to within $\pm 0.5°C$ can be attained.

DETERMINATION OF PALEOTEMPERATURES

The technique necessary for the determination of paleotemperatures involves: (1) extraction of carbon-dioxide gas from carbonate samples (by acidification with a standard acid) and (2) examination of the carbon dioxide (by mass spectrometry) to determine the oxygen isotopic composition of the sample.

The difference in isotopic composition between the sample and standard carbon dioxide is then calcutated. The result is expressed in *per mil* (i.e., parts per thousand) and is designated as δ_{0-18}

$$\delta_{0-18}(\text{percent}) = \frac{R\,\text{sample} - R\,\text{standard}}{R\,\text{standard}} \times 1000$$

where *R sample* is the 0.18:0.16 ratio in the unknown sample, and *R* standard is the 0-18:0-16 ratio in the reference standard. The value $\delta = 10$, for example, means that the 0-18:0-16 ratio of the unknown sample is 10 per mil (i.e., 1 percent greater than the standard). The standard used is a material pure and homogeneous in 0-18:0-16 ratio. The paleotemperature can then be read off a previously callibrated scale.

Average values of δ_{0-18} for different temperatures are shown graphically in Fig. 21.4, called the *isotopic temperature scale* (Epstein, Buchsbaum, et al., 1953). The points in the curve were obtained by analyzing carbonate skeletons grown under known conditions. The relationship between the value of δ and the temperature is represented by the equation:

$$T = 16.5 - 4.3\,\delta$$

The measurement of paleotemperatures (see Fig 21.5) from isotope fractionation data has greatly broadened our knowledge of climatic trends over the last 100 million years and has made it possible to obtain detailed information on the glaciation cycles of the Ice Ages (Emiliani 1956, 1958).

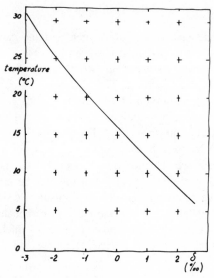

Fig. 21.4. Isotopic temperature scale: the relationship between ambient temperature and oxygen-18 content of contemporary calcium carbonate deposits; δ is the deviation (expressed in per mil) of the oxygen-18: oxygen-16 ratio of the sample from an accepted reference value.

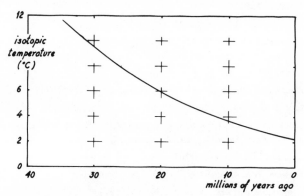

Fig. 21.5. Paleotemperatures. Oxygen isotope measurements show that the temperature of deep waters around Antarctica has been falling over a long period of time, indicating a gradual change from temperate to frigid climate in the south Polar region.

The possibility that isotope studies can give additional paleoclimatological data besides the determination of past temperatures has also been investigated. A study of past winds strength, based on oxygen isotope measurements, has led to the suggestion that it may have been the increase in storminess that led the Vikings, with the onset of a cold period, to abandon North Atlantic routes rather than seek a more southern route (Wilson and Hendy 1971).

Speloeothems

Measuring the isotope distribution of oxygen and the carbon-14 content of speleothems (concretions in caves), it has been possible to determine and date paleotemperatures in caves. As water percolates through soil it dissolves carbon dioxide, and carbonic acid is formed. The acid dissolves the limestone through which it pases, forming a solution of calcium bicarbonate. When this solution reaches a cavity, under certain conditions, calcium carbonate is precipitated, and stalagtites and stalagmites are formed. A difficulty in dating arises here, as the carbon laid down as calcium carbonate contains a mixture of ancient carbon (from the limestone carbonate, which contains no Carbon-14) and modern carbon (from plant roots and other organic materials); but this can be allowed for, and the Carbon-14 age can be estimated (Hendy and Wilson 1968). The variations in isotopic compostion of the oxygen in the calcium carbonate, from these cave concretions indicate alternating warm and cold periods, which are in good agreement with finds from other studies (Labeyne, Duplessy, et al. 1967).

REFERENCES

Emiliani, C., *Science, 123,* 924 (1956).

Emiliani, C., *Sci. Am., 58* (February 1958).

Emiliani, C., *Science, 154,* 851 (1966).

Epstein, S., Buchsbaum, R., Lowenstram, H. A., and Urey, H. C., *Bull. Geol. Soc. Am., 64,* 1315 (1953).

Flint, R. F., *Glacial and Quaternary Geology,* Wiley, New York, 1971.

Hendy, C., and Wilson, A. T., *Nature, 219,* 48 (1968).

Labeyne, J., Duplessy, J. C., Delibrias, G., and Letolle, R., in *Radioactive Dating and Methods of Low Level Counting,* IAEA, Vienna, 1967.

Lamb, H. H., *Nature, 223,* 1209 (1969).

Urey, H. C., *Science, 188,* 489 (1948).

Wilson, A. T., and Hendy, C. H., *Nature, 234,* 344 (1971).

AUTHENTICATION OF ANTIQUITIES

When an ancient object is separated from its archaeological context it looses much of its initial value to the scholar: location, position, and context convey a great deal of information about an object and also make its identification and authentication a much easier task. The authentication of isolated objects relies on the fact that antiquities carry within themselves evidence of the time and place of their manufacture. A positive approach to authentication should employ both artistic and scientific criteria, one complementing the other. Chemical and physical evidence should be used to reinforce stylistic and aesthetic considerations; scientific finding ought to be interpreted in the light of aesthetic standards. Close collaboration between the archaeologist and scientist is not merely desirable in this field—it is absolutely essential.

Antiquities and works of art have been appreciated and sought after by collectors since as far back as the heyday of ancient Rome. At most times the demand has exceeded supply, and forgers have set to work. The collector's insatiable appetite for works of earlier periods has promoted a lucrative commerce in copies, overzealous restorations, and imitations.

Sculptors working in Rome in the early centuries C.E. made replicas of greatly admired Greek works from preceding centuries, to satisfy the demands of wealthy collectors. One of the earliest references to forgeries of antiquities are the verses of the Roman fabulist Phaedrus, from the first century C.E.:

> Ut quidam artifices nostro faciunt saeculo,
> qui pretium operibus maius inveniunt, novo
> si marmori adscripserunt Praxitelem suo,
> trito Myronem argento.

> (Some artists of our times
> obtain a higher price for their works
> by inscribing on their new marble the name of Praxiteles
> or the name of Myron on worn out silver.)

The Middle Ages knew a brisk traffic in such items as false relics and fraudulent reproductions of precious stones. Stylistic and aesthetic criteria were unknown, however, and forgeries in the modern sense were unheard of. With the Renaissance came a new appreciation of antiquity and of its artistic values, which has remained with us ever since. From the end of the fifteenth century onward generations of forgers have toiled to supply the world with bogus antiquities.

Once the practice had been launched, the tradition of forging antiquities became continuous, although the objects of its attention have varied from time to time and from place to place. Michelangelo acquired fame early in life when he made a statue that he buried in the ground to give it an antique appearance. In 1496 the statue was sold in Rome as a classical sculpture (Condivi 1928)!!!

As the horizons of the archaeologist have widened and his techniques improved over the centuries, so have those of the forger. During the seventeenth century the art of ancient Egypt was rediscovered, and it stimulated a desire for the incredible and mysterious. Copies and variations of Egyptian sculpture, goldsmith's work, and bronze statues were produced in the seventeenth and eighteenth centuries. Roman mosaics looking still fresh after being lost and buried for many centuries, prompted numerous eighteenth-century artisans to compose "Roman" mosaics (Herbig 1933).

Antique gilded glass, consisting of gold leaf spread over even surfaces, was particularly in favor with early Christians. During the eighteenth century, when catacombs began to be excavated on a large scale, gilded glass became much sought after by collectors. As early as 1759 the Comte de Caylus, one of the founders of modern archaeology, wrote: "This technique [of gilded glass] has been reinvented in Rome a few years ago. I had the opportunity of examining some specimens which are very well done: one has utilised this device to dupe foreigners" (Caylus 1759).

The borderline between extensive restoration and outright forgery is difficult to draw. A common type of forgery consists, not of a newly made object, but of the remains of an old one, which serve as the foundation for a new work. The advantages of this method are obvious, since the basic materials have all the marks of age and authenticity. Occasionally, overzealous restorers create new styles by working derivatively from small fragments; ceramics and bronzes are often faked in this way, or supplemented by additional parts. Many of the generally accepted finds from Knossos owe much to the excessive zeal of restorers employed at the Knossos excavations in the early years of this century (Woolley 1962).

Increasing public interest in the artifacts of the ancient world has created a demand that ingenious forgers are only too ready to supply. Some forgers even proceed like clever businessmen, creating a demand for previously unheard objects. In the nineteenth century, an icon painter from Jerusalem invented the art of the Moabites, which for some years became a Prussian political issue (Clermont-Ganneau 1885; Yahuda 1944). Even the greatest archaeologists have their moments of unfortunate credulity. Sir Arthur Evans, the excavator of

Knossos, enthusiastically acquired the "Ring of Nestor," which chemical analysis later proved to be of modern origin.

AUTHENTICATION

Visual judgment, based on comparison and experience, usually suffices to detect all but the most painstaking forgeries. However, much has been done to free the examination of antiquities from exclusive reliance on aesthetic and stylistic standards. Many technical and scientific methods for authentication have been developed. The composition of materials, consistency and structure of decay products, techniques of manufacture, and internal structure of objects are all capable of providing objective criteria for the acceptance or rejection of archae-ological objects.

Evidence of authenticity may be either positive or negative. Positive evidence will indicate the approximate age of an object or define some physical or chemical property of the material from which it was made, thus helping to establish its authenticity. On the other hand, a demonstrated anachronism will constitute negative evidence, proving that an object could not have been made during the period assigned to it. Thus aluminum found in a purportedly ancient casting would clearly brand it as a recent forgery, since aluminum metal was first used only at the end of the nineteenth century.

Nonetheless, there is as yet no absolute scientific test by which to establish genuineness. The most reliable authentications embody a synthesis of artistic criteria, scientific evidence, and personal judgment based on experience. The following section describes some of the scientific and technical ways in which authenticity of ancient objects can be determined.

VISUAL EXAMINATION

Patina

The examination of ancient bronzes is usually begun with the patina, since it is widely accepted that patina is the most important evidence of authenticity. Change in the ancient surfaces of stone or ivory (less easy to define than in metals) often give these materials a characteristic antique look. This is often also referred to as *patina* and accepted as evidence of antiquity (Stross and Eisenloed 1965).

The importance of patina as an aid to authentication, however, has been overrated. Not enough consideration is given to the fact that a convincing patina can be fabricated with the aid of chemicals in a comparatively short time. It is possible to cause patina to be formed on the surface of a limestone object, for example, by first burying it in a wet salty soil impregnated with soluble iron salts and then exposing it to the atmosphere for a few weeks, and repeating the cycle several times. This patina could perhaps be differentiated from one formed slowly over many centuries because it is not very crystalline and not so strongly adherent to the stone. A false patina should, nevertheless, not be considered as *prima facie* evidence of forgery: false patinas are often applied to genuine antiquities on which remnants of the original patina may reemerge after cleaning. A false patina should, therefore, be removed before a piece is condemned out of hand.

Erosion

Negative evidence as to the spuriousness of surfaces is provided by erosion marks. The final polishing or finishing of a surface leaves distinctive marks, which are discernible under the microscope and often with the naked eye. It is amazingly difficult to remove traces left by the tools, instruments, and abrasive materials used in finishing surfaces, in such a manner that they cannot be detected under the microscope, even after a long period of use. Telltale traces of carving tools and polishing abrasives should be easy to establish as modern materials.

A modern forger trying to use the fastest method for completing his work would certainly use modern tools such as rasps, files, or cutting and polishing wheels, which leave parallel marks on the object abraded. Random or subparallel tool marks, on the other hand, indicate that a surface was produced with the help of a pointed tool of some type or by other primitive means.

Casting Techniques

Many ancient metal castings were made by the *cire-perdue* (lost wax) process, which uses one-piece molds. Modern fakes are often cast in halves; a casting fin or a line of filed solder material betrays them.

Cracks

Cracking is a frequent form of deterioration of ancient paintings. It is widely regarded as incontrovertible evidence of authenticity. Cracks in a paint layer occur as a result of chemical changes in the binding medium, which gradually loses its elasticity; it becomes stiff and unable to adapt itself to the distentions and contractions of the support. More severe cracks can be caused by short cycles of strong heat and high humidity than by centuries in an environment free from drastic changes.

Forgers cause stiffening of paint layers by heating the surface or by baking the finished object; they then expose it to low temperature. The stiff surface, unable to follow the rapid contraction of the support, breaks up and cracks. The fresh cracks are then filled with gray dust to enhance their antique appearance.

Differentiation of natural cracks from forged ones is not easy, especially since forgers know, at least as well as anybody else, the outline of the cracking pattern, so different for every support material. Crack patterns are sometimes made with a sharp or pointed instrument, but they are not sufficiently uniform to escape detection under the low-power microscope.

PHYSICAL EXAMINATION

X-Ray Radiography

X-Rays are absorbed in varying degrees by different materials. This makes them useful for examining internal structures without damaging the object being studied. The information obtained by radiography could otherwise only be obtained by damaging the specimen. Consequently, radiographs may help in the detection of forgeries or in the authentication of genuine objects. A small mummy found in the tomb of the Egyptian high-priestess Makare (of the twenty-first dynasty) was believed to be that of her daughter. An X-ray radiography revealed it to be a baboon (Harris and Weeks 1973). The radiograph of the central panel of a supposed fifteenth-century triptyc revealed open worm holes that had been stopped up and then covered with primer. Long machine-made nails were also clearly visible in the radiography. Anxious to employ old wood, the modern forger had painted on uneven, faulty pieces of worm-eaten wood that no fifteenth-century painter would have had reason to use (Kurz, 1967).

Occasionally, because of the thickness and density of the material being radiographed, the X-ray radiograph reveals nothing. In such cases, radioactive isotopes may be used to provide gamma rays of sufficient penetrating power to reveal the presence of hidden clues or spurious materials.

Ultraviolet Examination

Many materials fluoresce under ultraviolet light. Since no two materials produce the same fluorescence, ultraviolet examination may be used as a way of determining the heterogeneity of a surface. The technique is not suitable for detecting complete fakes but will show up patching and repaints by local differences in fluorescence. Faked patina on bronze can also be detected in this manner (Young 1970).

Marble objects from different sources, or that have undergone different types of weathering, may show differences in fluorescence, even though appearing similar in daylight. Thus it is possible to detect restoration in marble by observing contrasting areas under ultraviolet light.

The fluoresence of the surface of old and new ivory, respectively, is different; old ivory fluoresces yellow and mottled, and new ivory, bright purple (Rorimer 1931).

Outright fakes of Greek pottery are far less frequently encountered than skillful restorations, so painstakingly done that it is often a very delicate task to distinguish the new and ancient portions. Under ultraviolet light, the restored patches clearly shine out; the edges of the assembled fragments and the extent of their covering can be clearly seen. Parts that have been replaced by clay or plaster stand out; whereas the glaze of the pottery remains dark, plaster is brilliantly fluorescent.

CHEMICAL ANALYSIS

By the early nineteenth century chemical analyses were already being made of colors removed from old, or allegedly old, objects. However, accurate chemical analysis called for such large samples that the ancient objects being examined would be ruined as museum exhibits. Today all this has changed. Highly accurate and detailed analyses can often be carried out on samples of microscopic size. The rubbing of a freshly cleaned coin across the ground surface of a quartz plate or the prick of a hypodermic needle into the surface of a painting provide all that is needed for authentication. There are even a large number of nondestructive testing methods by which ancient objects may be analyzed for authenticity without removing any sample at all.

The usefulness of chemical analysis as an aid to authentication is directly related to the knowledge available regarding materials used by man at different periods and the impurities that they contain (de Wild 1929; Houtman and Turkstra 1964). Many materials in common use today were not known in the past. Pigments such as zinc white and Prussian blue were not known until the eighteenth century; others, such as cadmium yellow and synthetic ultramarine, only came into use in the nineteenth century. The detection of these materials in the pretended antiquity is pretty certain evidence of fraud.

Microchemical methods for the authentication of pigments were introduced more than 50 years ago and are still being used. More modern techniques, such as X-ray fluorescence and neutron-activation analysis, are used today for the identification of the components of pigments and call for extremely small samples.

Since man first began extracting metals from the earth, the techniques of refining have been steadily improved. For example, modern refining methods permit the economical removal of most of the small amounts of gold present in raw silver ores. Hence commercial silver refined during the last hundred years usually contains less than 0.1 percent of gold; silver produced very recently may contain even less than 0.01 percent. Clearly, analysis of the impurities present in metal objects may yield useful information for purposes of dating or authentication.

Sassanian Silver Art Objects

Hundreds of genuine Sassanian and Sassanian-style plates, pitchers, and bowls have been examined for technical and metallurgical characteristics. The data gathered on the gold content of the silver yield the following useful guidelines for authenticating similar objects, especially when technical considerations are added (Gordus and Gordus 1974): (1) gold is usually present in concentration higher than 0.6 percent, (2) copper (which makes up the balance of the alloy with silver and gold) does not usually exceed 6 percent of the total metal (Chapter 11), and (3) Sassanian plates or bowls are usually single walled.

These criteria, though very useful, are of course not absolute. However, if none at all of them are met, this is a strong warning that the object being examined may not be genuine.

Coin Forgeries

An illustration of a nondestructive analytical search for coin forgeries is provided by the following example. Legitimate moneyers and forgers alike often tried to save money by adulterating gold with less expensive metals, such as silver and copper. The forger's product would, as a rule, be very unlikely to contain the same proportions of gold, silver, and copper as the genuine coins. Thus the determination of the relative concentration of gold, silver, and copper in coins could help to separate forged pieces from genuine ones.

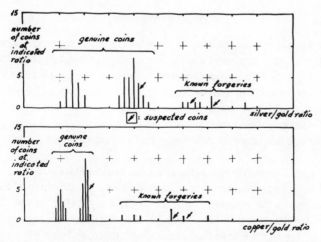

Fig. 22.1. Silver:gold and copper:gold ratios in a group of 50 coins that included genuine, forged, and suspect coins.

Neutron-activation analysis reveals the composition of the coins without more than scraping away a small area of the coin. The data collected from the analysis can be processed and the ratios silver:gold and copper:gold calculated. Such data, obtained from a collection of 50 coins, including a small group of definite forgeries and some suspected ones, is illustrated in Fig. 22.1 (Keisch 1972). Most of the coins—the genuine ones—fall into one of two well-defined groups, each with a definite silver:gold and copper:gold ratio. Coins known with certainty to be forgeries have distinctly higher ratios. Of three suspected forged coins, two have ratios that fall into the range of the known forgeries, but one turns out to be probably genuine.

DATING AND AUTHENTICATION

Radiocarbon Dating

Radiocarbon dating will usually provide a reliable answer when it is necessary to know the age of carbon-containing materials. In the case of artifacts, however, the age adduced by radiocarbon dating is not convincing proof of antiquity, since the method only yields information about the age of materials and not about the time when those materials were worked. Modern forgers have learned to use ancient wood as a starting materials for their work. It is not likely, however, that forgers would have taken the precaution of using antique wood before the method of radiocarbon dating was developed.

Thermoluminescence

Thermoluminescence measurements on a few milligrams of material scraped from fired clay objects, such as pottery and figurines, can readily establish whether one is dealing with genuine antiquities or merely with recent copies or fakes. Chemical tests of a terracotta statuette purporting to be of Etruscan origin (510-490 B.C.) showed that it could have been made in Italy but gave no hint as to its age. Thermoluminescence measurements indicated that the statue was homogeneous and not a pastiche but unquestionably a forgery of modern origin (Fleming 1966; Kohler 1967).

A group of small pottery objects (figurines, animals and facsimiles of bronzes), said to have been excavated near the village of Hui Hsien, province of Honan in China, appeared in the antiquities market in the early 1940s. Historically the objects were dated between 400 B.C. and 205 B.C. The figurines attracted a great deal of attention, reflected in the high prices they fetched, and copies were made in China; "genuine" and "fake" Hui Hsien figurines were soon being distinguished, and hence it is not surprising that a considerable amount of uncertainty as to authenticity and provenance surrounded the Hui Hsien objects. Chemical analysis and X-ray studies of the objects' clay offered no conclusive solution regarding authenticity (Young and Whitmore 1954). Investigation of the black surfaces and red pigments of the decoration could only be useful as guidelines (Benedetti-Pichler 1937; Plenderleith 1954).

Thermoluminescence results obtained from 22 Hui Hsien ceramic pieces (Table 22.1) showed that, without exception, they are of modern origin (Fleming and Sampson 1972).

TABLE 22.1 Ages of Hui Hsien Ceramic (Derived from Thermoluminescence Results)

Object	Maximum Age (Years)
Pair of horses	340
Shallow dish	30
Figurine	50
Figurine	295

The pieces studied are considered representative of the whole group of pottery objects. The question thus arises as to whether there are any genuine Hui Hsien objects.

Several examples, listed in Table 22.2, illustrate other applications of thermoluminescence in confirming the authenticity of established teracotta pieces (Fleming 1972).

TABLE 22.2 Thermoluminescence Dating of Well-known Terracottas

Description	Known Age (Years)	Thermoluminescence Age (Years)
Standing lady	465	475 ± 10
Pre-Columbian figure	1260	1.100 ± 300
Late Caeretean bust	1960	2.250 ± 100

TWO STUDY CASES OF AUTHENTICATION

The Piltdown Man

An illuminating example of a long-standing controversy settled by the use of scientific authentication is to be found in the case of the Piltdown man (Oakley and Weiner 1955; Millar 1972; van Eskweck 1972). This remarkable story began in

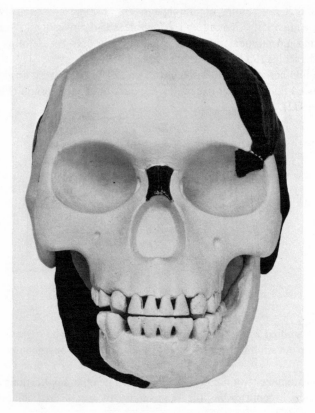

Fig. 22.2.   Piltdown man.

1912, when a human jawbone and fragments of braincase bones were found together with bones of various early mammals at a gravel dig at Piltdown in Sussex, England. The human bones were examined at the British Museum and it was concluded that they belonged to a 500,000 year old, hitherto unknown fossil ancestor of man which was named *Evanthropus dawsoni* or *Piltdown man* (Fig. 22.2).

There were, however, difficulties. Some authorities could not accept that both the braincase and the jawbone belonged to a single individual. The jawbone appeared to be more ape-like and the braincase, more human. The controversy over the Piltdown man occupied the scientific world for over 45 years. In 1949 a chemical method of dating fossils made the problem appear even more confusing; analysis of bones of the Piltdown remains revealed that neither the jawbone nor the cranial bones contained more than minute traces of fluorine, whereas fossils from the same gravel-bed contained a great deal (Oakley and Hoskins 1950). This find drastically reduced the greatest possible age of the Piltdown skull, from 500,000 to no more than 50,000. It also made an evolutionary absurdity of the Piltdown man, since its acceptance meant that in relatively recent times there existed in Piltdown a creature with no known ancestors and no known descendants.

Although the analyses of fluorine were accurate enough to prove that neither the braincase nor the jawbone were of the great antiquity previously supposed, they were not suficiently accurate to distinguish between a late fossil and a recent specimen. In 1953 a more refined analytical technique revealed that the fluorine in the braincase was just enough to account of its being ancient, but the jawbone and teeth were found to contain no more fluorine than modern bones (Fryd 1955). This being so, the organic protein content could be expected to be high. Not surprisingly, further chemical analysis confirmed that the jawbone and teeth contained the same amount of nitrogen as modern species. The braincase, on the other hand, contained much less. Used as a cross-check against the fluorine tests, this evidence was conclusive. Nevertheless, further tests were made to confirm that the jawbone was really modern. Electron microscopy revealed preserved fibers of organic tissues in a sample of jawbone, in contrast to the braincase, where all traces of such fibers had disappeared. So the Piltdown jawbone, originally supposed to date from before the Ice Age, was exposed as a bogus fossil of undoubted modern origin. It probably came from a partly grown female orangutan (a conclusion supported by independent anatomical investigations) deceitfully buried at Piltdown.

The dark brown skull bones were also fraudulently introduced into the Piltdown gravel. X-Ray analysis of the mineral structure of the human braincase revealed that the phosphate mineral component of the bones had partly changed to gypsum. But gypsum is not found in the Piltdown gravel. It has been proven, however, that when staining a partially fossilized bone with acid iron-sulfate solution, the bone is partially changed to gypsum. The Piltdown skull was stained in this way probably to match the gravel in which it was planted; hence the otherwise inexplicable presence of gypsum.

The human remains were accompanied by flint implements and animal fossils, all of which supported the early date attributed to the find. The flint implements were of a reddish color, matching that of the gravel, but spectrographic analysis revealed that they had been artificially stained with chromium and iron salts. Below the layer of stain there was found a white crust, whereas local gravel flints were brown throughout.

The animal fossils were also "planted" to suggest an early date. Radiometric analysis of bits of elephant teeth found at the site revealed that these were more than ten times as radioactive as genuine pre-Ice Age fossils found in deposits in East Anglia, England (Davidson and Bowie 1955). The red-brown color suggested that the animal fossils had also been artifically stained; and sure enough, chromium and iron were found on them.

Fig. 22.3.   The Metropolitan Museum's bronze horse.

An hippopotamus molar, previously supposed to be contemporary with the elephant's teeth, contained little fluorine and was also stained with chromium and iron salts. A dichromate solution was probably used by the forger in an attempt to assist the oxidation of the iron salt used to stain the specimens.

The age of the Piltdown man finds was finally determined by radiocarbn dating (Table 22.3). The braincase is probably less than 700 years old; the jawbone is even younger (de Vries and Oakley 1959).

**TABLE 22.3   Radiocarbon Dates of
Piltdown Remains**

Jawbone	500 ± 100 years
Braincase	620 ± 100 years

The results confirmed that the bones are of recent origin in geological terms. It is possible that the person who contrived the Piltdown hoax obtained the orangutan jawbone from a dealer in ethnographic materials. The Dyaks of Borneo are known to keep orangutan skulls as fetishes or trophies for many generations. One of these may have reached the forger via a dealer.

Thus a large body of scientific evidence established beyond any doubt that the Piltdown man was a bogus human ancestor. The whole group of bones and implements was shown to be of different chronological and geographical origin, fraudulently introduced into the Piltdown gravel. An interesting postscript to this extraordinary story is the disclosure that the Piltdown Man was simply a trap set by one eminent scientist to another (Collins 1978).

The Bronze Horse from the Metropolitan Museum

A bronze statuette of a horse 16 inches tall, owned by the Metropolitan Museum of New York (see Fig. 22.3), was considered for many years to be 2400 years old and one of the finest example of early Greek art. In 1967 a magnetic survey of the statuette indicted the presence of iron at certain points inside it. This observation prompted careful study that revealed a casting fin surrounding the sculpture. Additionally, X-ray radiography exposed an iron armature inside the statuette. As a consequence of this study, the horse was declared a modern forgery. The presence of a mold line and the use of iron wires to provide hollow spaces within the object were accepted as evidence that the casting was made by the "sand piece mold" process, a technique apparently not practiced until much later than the period of which the statuette was assigned.

Studies of casting techniques used by the Greeks were begun, and the authenticity of the horse was defended on the grounds that examples of genuine Greek bronzes containing iron armatures are known (Blumel 1969). Moreover, the composition and structure of the metal and the nature and extent of the corrosion were said to be consistent with the statuette being authentic.

The statuette had a well-solidified core made up of sand, clay, calcite, and mineral inclusions that would have been heated during manufacture. These materials are suitable for thermoluminescence dating and could provide evidence that might answer the question of the statuette's real age (Zimmerman, Yuhas, et al. 1974).

Previous radiographic work on the horse would, however, have given the core an artificial radiation dose that would have interfered with reliable authentication. Fortunately, a sizable sample of the core material had been removed before most of the radiographic work on the statuette was carried out. It was also established that during manufacture the core had been heated sufficiently to remove all stored geological thermoluminescence in the temperature range of interest. The thermoluminescence dating results proved that the horse was, in fact, made in antiquity.

Its age was finally established as lying somewhere within 2000-3500 years. This is sufficiently accurate to establish that the horse is not a modern forgery. The range of possible error is, however, too wide to establish with certainty whether the statuette is indeed a classical Greek sculpture of the fifth century B.C. or a later product of the Hellenistic or Roman Age.

REFERENCES

Benedetti-Pichler, A. A., *Ind. Eng. Chem., Anal. Sec.*, **9**, 142 (1937).

Blumel, C., *Archaeologischer Anzeiger*, **84**, 208 (1969).

Caylus, Comte de, *Recueil d' Antiquites*, **3**, 195 (1759).

Clermont-Ganneau, C., *Les Grandes Archaeologiques a Palestine*, 1885, p. 103.

Collins, J., *New York Times*, Nov. 7, 1978.

Condivi, A., *Michelangelo*, d'Ancona, Italy 1928, p. 59.

Davidson, C. F., and Bowie, S. H. U., *Bull. Br. Mus. Nat. Hist., Geol.*, **2**, 27 (1955).

de Vries L., and Oakley, K. P., *Nature*, **184**, 224 (1959).

De Wild, A. M., *The Scientific Examination of Picture*, 1929.

Engel, C. G., and Sharp, B. P., *Bull. Geol. Soc. Am.*, **69**, 478 (1958).

Fleming, S. J., *Archaeometry*, **9**, 170 (1966).

Fleming, S. J., *Naturwissenschaften*, **4**, 145 (1972).

Fleming, S. J., and Sampson, E. H., *Archaeometry*, **14**, 237 (1972).

Fryd, C. F. M.,*Bull. Br. Mus. Nat. Hist., Geol.*, **2**, 266 (1955).

Gordus, A. A., and Gordus, J. P., in C. W. Beck, Ed., *Archaeological Chemistry, a Symposium*, American Chemical Society, Washington, D. C., 1974.

Harris, J. E., and Weeks K. R., *X-Raying the Pharaohs*, Scribner, New York, 1973.

Herbig, R., *Romische Mitteilungen*, **48**, 312 (1933).

Houtman, J. P. W., and Turkstra, J., in *Radiochemical Methods of Analysis, Proceedings of a Symposium, Saltzburg*, Vol. 1, 85, IAEA, Vienna, 1964.

Keisch, B., *The Atomic Fingerprint*, p. 35, USAEC, Oak Ridge, 1972.

Keisch, B., and Miller, H. H., *Application of Nuclear Technology to Art Identification Problems*, Carnegie Mellon University, Pittsburg, Pa., 1972, p. 85.

Kohler, E. L., *Expedition*, **9**, (2), 16 (1967).

Kurz, O., *Fakes*, Dover, New York, 1967, p. 50.

Millar, R., *The Piltdown Men*, Gollancz, London, 1972.

Noble, J. V., *Metropolitan Museum of Art Papers*, New York, 1961.

Oakley, K. P., in D. Brothwell and E. Higgs, eds., *Science in Archaeology*, Thames and Hudson, London, 1963, p. 24.

Oakley, K. P., and Hoskins, C. R., *Nature*, **165**, 379 (1950).

Oakley, K. P., and Weiner, J. S., *Am. Sci.*, **43**, 573 (1955).

Plenderleith, H. J., *Far East. Ceramic Bull.*, **6**, (3), 1 (1954).

Rorimer, J. J., *Ultrviolet Rays and Their Use in the Examination of Works of Art*, Metroplitan Museum of Art, New York, 1931, p. 18.

Stross, F. H., and Eisenloed, W. J., *A Report on a Group of Limestone Carvings*, San Carlos, Calif., 1965.

van Eskweck, G., *Pleine Lumiere sur l'Imposture de Plitdown*, du Cedre, Paris, 1972.

Wooley, C. L., *As I Seem to Remember*, Allen and Unwin, London, 1962, p. 21.

Yahuda, A. S., *Jewish Quarterly Review*, **35**, 139 (1944).

Young, W. J., in S. Doeringer, D. G. Mitten, and A. Steinberg, eds., *Art and Technology, a Symposium*, MIT Press, Cambridge, Mass., 1970, p. 71.

Young, W. J., and Whitmore, F., *Far East.'Ceramic Bull.*, **6**, 8 (1954).

Recommended Reading

Fleming, S. J., *Authentication in Art; the Scientific Detection of Forgery*, The Institute of Physics, London, 1975.

Savage, G., *Forgeries, Fakes and Reproductions*, Barrie and Rockliff, London, 1963; republished, White Lion Publ. Ltd., London, 1976.

LITERATURE ON ARCHAEOLOGICAL CHEMISTRY

Any archaeologist might think of undertaking a research project in which chemistry, or other natural sciences, could play an important part. An essential preliminary to such an undertaking is to find out what is already known. Once the object of the research has been defined, the first things to be done are: (1) decide what information is wanted and (2) find out where it is available.

The sources of information about previous research are original published papers, review articles, collected abstracts, monographs, and reference books. The purpose of this chapter is to help the archaeologist unfamiliar with the literature of archaeological chemistry to find his way to the sources of information available to him.

REFERENCE AND GENERAL BOOKS

The first sources of information to consult are usually reference books. Archaeological chemistry is however in an early stage of development, and there are as yet no encyclopedic reference books in which all that is known about the subject can be found; there is nothing in the field that could be called "the bible of archaeological chemistry." There are a number of books that deal with certain chemical aspects of archaeology. Some are monographs, others reference books on a small scale, and still others, practical manuals. The more accepted are listed in the sections that follow.

General Works

Aitken, M. J., *Physics in Archaeology,* Oxford U. P., London, 1975.

Neuberger, A., *The Technical Arts and Sciences of the Ancients,* Methuen, London, 1969.

Singer, C., Holmyard, E. J., and Hall, A. R., eds., *A History of Technology,* Oxford U. P., London, 1954.

Tite, M. S., *Methods of Physical Examination in Archaeology,* Seminar, London, 1972.

There are two interesting books that explain in some detail, and in terms understandable to the layman, the present and past contributions of natural scientists to archaeology. These are:

Bieck, L., *Archaeology and the Microscope,* Butterworths, London, 1965.
Wilson, D., *The New Archaeology,* Knopf, New York, 1975.

Materials and Technology

Aitchison, L., *A History of Metals,* McDonald and Evans, London, 1960.
Bentley, K. W., The Natural Pigments, London, 1960.
Cornwall, I. W., *Soils for the Archaeologist,* Phoenix House, London, 1966.
Forbes, R. J., *Studies in Ancient Technology,* Brill, Leiden, 1965.
Gettens, R. J., and Stout, G. L., *Painting Materials,* Van Nostrand, New York, 1942.
Hodges, H., *Artifacts,* J. Baker, London, 1968.
Lucas, A., *Ancient Egyptian Materials and Industries,* E. Arnold, London, 1962.
Reed, R., *Ancient Skins, Parchment and Leathers,* Seminar, London, 1972.
Shepard, A., *Ceramics for the Archaeologist,* Washington, 1956.
Rosenfeld, A., *The Inorganic Raw Materials of Antiquity,* Weidenfeld and Nicholson, London, 1965.
Tylecotte, R. F., *Metallurgy in Archaeology,* E. Arnold, London, 1962.

Conservation

Organ, R. M., *Design for Scientific Conservation of Antiquities,* Butterworths, London, 1968.
Plenderleith, H. J., *The Conservation of Antiquities and Works of Art,* Oxford U. P., London, 1966.
Urbani, G., ed., *Problemi di Conservazione,* Compositori, Bologna.

DATING

Fleming S., *Dating in Archaeology,* Dent, London, 1977.

Authentication

Fleming S., *The Scientific Detection of Forgery,* The Inst. of Physics, London, 1975.

REVIEWS AND MONOGRAPHS

Reviews are surveys of the literature on a subject published during a certain period, whereas *monographs* are full discussions of a special, relatively restricted subject.

Both types of publication are very useful in that they normally contain references to the literature. The following is a classified list of reviews and monographs on subjects of archaeological-chemical interest.

Collected Reviews and Monographs

Brothwell, D. and Higgs, E., eds., *Science in Archaeology,* Thames and Hudson, London, 1969.

Michael, H. N., and Ralph, E. K., eds., *Dating Techniques for the Archaeologist,* MIT Press, Cambridge, Mass. 1971.

Pyddoke, E., ed., *The Scientist and Archaeology,* Phoenix House, London, 1963

Various authors, *The Conservation of Cultural Property,* UNESCO, Paris, 1968.

Materials

Caley, E. R., *Analysis of Ancient Glasses,* Corning Museum of Glass, Corning, 1962.

Caley, E. R., *The Analysis of Ancient Metals,* Pergamon, New York, 1964.

Winkler, E. M., *Stone: Properties, Durablity in Man's Environment,* Springer, Wien, 1973.

Decay and Conservation

Brown, B. F., Burnett, H. C., Chase, W. T., Goodway, M., Kruger, J., and Pourbaix, M., eds., *Corrosion and Metal Artifacts. A Dialogue between Conservators and Archaeologists and Corrosion Scientists,* N.B.S. Special Publication, No. 479, U.S. Dept. of Commerce, Washington, D. C., 1977.

Dating

Darlimple, G. B., and Lanphere, M. A., *Potasisum Argon Dating,* Freeman, San Francisco, 1969.

Libby, W. F., 1958, *Radiocarbon Dating,* Chicago U. P., 1958.

Van der Merwe, N. J., *The Carbon-14 Dating of Iron,* Chicago U. P., 1969.

Several general scientific journals often publish reviews of archaeological interest. Among these are: *Scientific American, Science, Nature,* and *American Scientist.* These publications take care to use language that is not overloaded with technical jargon and often provide useful leads to pertinent original literature.

PRIMARY SOURCES OF INFORMATION

Scientific periodicals, which publish original work, are the prime means through which scientific information is kept up-to-date and propagated. The material that appears in scientific periodicals differs from all other types of literature as it is relatively unorganized, sometimes turning up matters of interest in the strangest places.

The list of chemistry journals that occasionally publish works of interest to the archaeologist is too voluminous to be given here. A few periodicals specialize in archaeological chemistry and the application of physical sciences to archaeology:

Archaeometry, published twice yearly by the Research Laboratory for Archaeology and the History of Art, Oxford University, encourages its authors to write with the nonspecialist in mind. It publishes accounts of research, review articles, and discussions on the significance of results reported.

Pact is a new journal published by the European Study Group on Physical, Chemical and Mathematical Techniques Applied to Archaeology. The editors of the journal, which is sponsored by the Council of Europe, Strasbourg, France, undertook to publish at least one issue a year; the first was published in 1977.

Studies in Conservation, published quarterly by the International Institute for Conservtion of Historic and Artistic works, is dedicated mainly to technical problems of conservation. However, studies on materials and techniques of archaeological importance also see light under its cover.

Technical Studies in the Field of the Fine Arts is a new version of an earlier periodical. Between 1932 and 1942 the Fogg Museum at Harvard University published *Technical Studies,* a publication that offered original contributions and review articles by research workers concerned with the analysis and experimental investigation of archaeological material. Many of the articles have not been superseded to this day. The whole series has again been made available through a reprint issued by the Garland Publishing Company in New York.

Proceedings of Scientific Meetings

Until fairly recently, scientific periodicals were the only primary source of information about original work in archaeological chemistry. They are in recent years being increasingly supplemented by other basic literature forms. Thes include *reports, theses, commercial publications,* and *proceedings of scientific meetings.* Some of the more important proceedings that have been published in book form are:

Allibone, T. E., Weeler, M., Edwards, I. E. S., Hall, E. T., and Werner, A. E. A., eds., *The Impact of the Natural Sciences in Archaeology,* Oxford U.P., London, 1970.

Beck, C. W., ed., *Archaeological Chemistry Proceedings of a Symposium,* American Chemical Society, Washington D. C., 1974.

Berger, R., ed., *Scientific Methods in Medieval Archaeology,* California U. P., Berkeley, 1970.

Bishay, A., ed, *Recent Advances in Science and Technology of Materials,* Vol. 3, Plenum, New York, 1974.

Brill, R. H., ed., *Science and Archaeology,* MIT Pres, Cambridge, Mass., 1971.

Brown, B. F., Burnett, H. C., Chase, W. T., Goodway, M., Kruger, J., and Pourbaix L., eds., *Corrosion and Metal Artifacts,* N. B. S. Spec. Publ. 479, Washington, D. C., 1977.

Hall, E. J., and Melcalf, N., eds., *Chemical and Metallurgyical Investigations of Ancient Coinage,* Royal Numismatic Soc., London, 1972.

Heizer, R. F., and Cook, S. F., eds., *The Application of Quantitative Methods in Archaeology,* Quadrangle Books, Chicago, 1960.

Levey, M., ed., *Archaeological Chemistry,* Pennsylvania U. P., Philadelphia, 1967.

Olson, I. U., ed., *Radiocarban Variations and Absolute Chronology,* John Wiley and Sons, New York, 1970.

Thomson G., ed., *Proceedings, New York Conference On Conservation of Stone and Wooden Objects,* IIC, 1970.

Timmons S., ed., *Preservation and Conservation: Principles and Practices,* The Preservation Press and the Smithsonian Inst. Press, Washington, D.C., 1976.

Young, W. J., ed., *Application of Science in Examination of Works of Art,* Boston Museum of Fine Arts, Boston, Mass., 1965, 1973.

ABSTRACTS

It is impossible in practice, for a single person, to go through more than a small fraction of the newly published peridical literature that may contain something of interest for him.

Eventually, much of the information that first appears in periodicals finds its way into books. Between the publication of an article in scientific periodicals and the incorporation of its content into a book, up-to-date information can be obtained from "Abstracting journals." These are invaluable aids to locating information scattered in the vast amount of literature currently published. There are two abstracting journals that cover the scientific disciplines most likely to be germane to the archaeologists' work. These are: (1) *Art and Archaeology Technical Abstracts,* and (2) *Chemical Abstracts.*

Art and Archaeology Technical Abstracts is published biannually by the International Institute for Conservation. The publication abstracts current literature that deals with the technical examination, investigation, analysis, restoration, preservation, and technical documentation of objects having historical significance. Also abstracted is that portion of the technological literature that reports data on the physical and chemical properties of the substances that play important roles in the structure of historic objects as well as those used in the treatment, repair, and preservation of these objects.

Chemical Abstracts is designed for the retrieval of chemical information. It is published by the American Chemical Society in biannual volumes divided into 26 parts. One of its sections, *No 20; History, Education, and Documentation,* contains abstracts of publications relevant to archaeological chemistry.

BIBLIOGRAPHY

Bleck, R. D., *Bibliography of the Archaeological-Chemical Literature,* Museum fur Ur- and Furhgeschichte Thuringens, Weimar, East Germany, 1967-1968.

INDEX